Long-Term Ecological Research

LONG-TERM ECOLOGICAL RESEARCH
CHANGING THE NATURE OF SCIENTISTS

Edited by
Michael R. Willig
and
Lawrence R. Walker

Oxford University Press is a department of the University of Oxford. It furthers
the University's objective of excellence in research, scholarship, and education
by publishing worldwide. Oxford is a registered trade mark of Oxford University
Press in the UK and certain other countries.

Published in the United States of America by Oxford University Press
198 Madison Avenue, New York, NY 10016, United States of America.

© Oxford University Press 2016

All rights reserved. No part of this publication may be reproduced, stored in
a retrieval system, or transmitted, in any form or by any means, without the
prior permission in writing of Oxford University Press, or as expressly permitted
by law, by license, or under terms agreed with the appropriate reproduction
rights organization. Inquiries concerning reproduction outside the scope of the
above should be sent to the Rights Department, Oxford University Press, at the
address above.

You must not circulate this work in any other form
and you must impose this same condition on any acquirer.

Library of Congress Cataloging-in-Publication Data
Names: Willig, Michael R., editor. | Walker, Lawrence R., editor.
Title: Long-term ecological research : changing the nature of scientists /
edited by Michael R. Willig and Lawrence R. Walker.
Description: New York, NY : Oxford University Press, 2015. | Includes
bibliographical references and index.
Identifiers: LCCN 2015030157 | ISBN 9780199380213
Subjects: LCSH: Ecology—Research—United States. | Long-Term Ecological
Research Program.
Classification: LCC QH541.26 .L67 2015 | DDC 577.072073—dc23 LC record available at
http://lccn.loc.gov/2015030157

9 8 7 6 5 4 3 2 1
Printed by Sheridan Books, Inc., United States of America

We dedicate this volume to past, present, and future participants in the Long-Term Ecological Research program, as well as to those visionaries in the National Science Foundation and other governmental agencies who have contributed to the program's genesis and perpetuation.

Contents

xi   Preface, *Michael R. Willig and Lawrence R. Walker*
xv   List of Contributors

Part One Introduction and Overview

3   1. Changing the Nature of Scientists: Participating in the Long-Term Ecological Research Program, *Michael R. Willig and Lawrence R. Walker*

29   2. Sustaining Long-Term Research: Collaboration, Multidisciplinarity and Synthesis in the Long-Term Ecological Research Program, *Robert B. Waide*

43   3. Reflections on Long-Term Ecological Research from National Science Foundation Program Directors' Perspectives, *Henry L. Gholz, Roberta Marinelli, and Phillip R. Taylor*

Part Two H. J. Andrews Experimental Forest (AND) LTER Site

55   4. Streams and Dreams and Cross-site Studies, *Sherri L. Johnson*

63   5. Data, Data Everywhere!, *Susan G. Stafford*

73   6. Science, Citizenship, and Humanities in the Ancient Forest of H. J. Andrews, *Frederick J. Swanson*

Part Three Arctic (ARC) LTER Site

83   7. Bridging Community and Ecosystem Ecology at the Arctic Long-Term Ecological Research Site via Collaborations, *Laura Gough*

91   8. Long-Term Ecological Research in the Arctic: Where Science Never Sleeps, *John E. Hobbie*

99   9. Forty Arctic Summers, *Gaius R. Shaver*

Part Four Baltimore Ecosystem Study (BES) LTER Site

109   10. Of Fish and Platypus: If You Could Ask a Fish What It Feels Like to Swim, *J. Morgan Grove*

119   11. Long-Term Ecological Research on the Urban Frontier: Benefiting from Baltimore, *Steward T. A. Pickett*

Part Five Cedar Creek Ecosystem Science Reserve (CDR) LTER Site

129 12. Beneficiary of a Changed Paradigm: Perspectives of a "Next-Generation" Scientist, *Elizabeth T. Borer*

137 13. Listening to Nature and Letting Data Be "Trump", *David Tilman*

Part Six Central Arizona–Phoenix (CAP) LTER Site

147 14. The Socializing of an Ecosystem Ecologist: Interdisciplinarity from a Career Spent in the Long-Term Ecological Research Network, *Daniel L. Childers*

155 15. An Urban Ecological Journey, *Nancy B. Grimm*

Part Seven Coweeta (CWT) LTER Site

167 16. An Anthropologist Joins the Long-Term Ecological Research Network, *Ted L. Gragson*

Part Eight Florida Coastal Everglades (FCE) LTER Site

177 17. The Benefits of Long-Term Environmental Research, Friendships, and Boiled Peanuts, *Evelyn E. Gaiser*

185 18. Collaboration and Broadening Our Scope: Relevance of Long-Term Ecological Research to the Global Community, *Tiffany G. Troxler*

Part Nine Jornada Basin (JRN) LTER Site

197 19. A Dryland Ecologist's Mid-Career Retrospective on Long-Term Ecological Research and the Science–Management Interface, *Brandon Bestelmeyer*

205 20. Tales from a "Lifer" in the Long-Term Ecological Research Program, *Debra P. C. Peters*

Part Ten Konza Prairie (KNZ) LTER Site

215 21. A Forest to Prairie Transition as a Long-Term Ecological Research Scientist, *John Blair*

225 22. Growing Up with the Konza Prairie Long-Term Ecological Research Program, *Alan K. Knapp*

233 23. Born and Bred in the Long-Term Ecological Research Network: Perspectives on Network Science and Global Collaborations, *Melinda D. Smith*

Part Eleven Luquillo (LUQ) LTER Site

243 24. Confessions of a Fungal Systematist, *D. Jean Lodge*

251 25. A Glimpse of the Tropics Through Odum's Macroscope, *Ariel E. Lugo*

259 26. Taking the Long View: Growing Up in the Long-Term Ecological Research Program, *Whendee L. Silver*

Part Twelve Moorea Coral Reef (MCR) LTER Site

269 27. Kelp Forests, Coral Reefs, and the Long-Term Ecological Research Program: Synergies and Impacts on a Scientific Career, *Sally J. Holbrook*

277 28. The Long-Term Ecological Research Construct for Understanding Dynamics of Coral Reef Ecosystems and Its Influence on My Science, *Russell J. Schmitt*

Part Thirteen Niwot Ridge (NWT) LTER Site

289 29. Top of the World Collaborations: Lessons from above Treeline, *Katharine N. Suding*

Part Fourteen North Temperate Lakes (NTL) LTER Site

299 30. My Evolution as a Long-Term Ecological Research Scientist, *John J. Magnuson*

Part Fifteen Palmer Antarctica (PAL) LTER Site

309 31. Learning from a Frozen Ocean: The Changing Face of Antarctic Ocean Ecology, *Hugh W. Ducklow*

Part Sixteen Plum Island Ecosystem (PIE) LTER Site

319 32. Mysteries in the Marsh, *Anne Giblin*

325 33. Perspectives on a 30-Year Career of Salt Marsh Research, *James T. Morris*

Part Seventeen Santa Barbara Coastal (SBC) LTER Site

335 34. Evolution of an Information Manager, *Margaret O'Brien*

Part Eighteen Sevilleta (SEV) LTER Site

345 35. From the Long-Term Ecological Research Program to the National Science Foundation and Back: A Personal History of Long-Term Ecological Research Science and Management, *Scott L. Collins*

355 36. The Long-Term Ecological Research Stimulus: Research, Education, and Leadership Development at Individual and Community Levels, *James R. Gosz*

Part Nineteen Shortgrass Steppe (SGS) LTER Site

365 37. Long-Term Ecological Research and Lessons from Networked Lives, *John C. Moore*

Part Twenty Virginia Coastal Reserve (VCR) LTER Site

375 38. Networking: From the Long-Term Ecological Research Program to the National Ecological Observatory Network, *Bruce P. Hayden*

381 39. Sharing Information: Many Hands Make Light Work, *John H. Porter*

Part Twenty One Analysis and Synthesis

391   40. Coda: Some Reflections on the Long-Term Ecological Research Program, *William H. Schlesinger*

397   41. Scholarly Learning in an Ecological Setting: Applying the Knowledge, Attitudes, and Behaviors Framework to Perceived Outcomes from Participation in the Long-Term Ecological Research Program, *Mark A. Boyer and Scott W. Brown*

411   42. Exploring the Scientific and Beyond-Science Interactions of Long-Term Ecological Research Scientists, *Courtney G. Flint*

423   43. Long-Term Ecological Research Over the Long Term: A Historian's Perspective, *Christopher Hamlin*

433   44. Trade-offs of Participation in the Long-Term Ecological Research Program: Immediate and Long-Term Consequences, *Lawrence R. Walker and Michael R. Willig*

439   Index

Preface

Michael R. Willig and Lawrence R. Walker

We are ecologists who met in 1988 during the process of writing a proposal to fund a site in the Luquillo Mountains of Puerto Rico as part of the Long-Term Ecological Research (LTER) network, a National Science Foundation (NSF) program designed to support long-term, collaborative research that is integrated with education and outreach at 26 sites throughout the United States. Despite our different research interests at the time (Willig, community ecology of vertebrates; Walker, plant succession), we have collaborated with each other on various syntheses regarding disturbance ecology (e.g., Walker et al. 1996; Walker and Willig 1999; Willig and Walker 1999; Willig et al. 2012) and have shared many frustrating and rewarding moments in our more than 25-year history of interactions as part of the LTER program. In the process, we have become better colleagues and good friends, frequently searching for promising scientific reasons to continue productive collaborations. This book represents one such opportunity.

CONCEPTION, GESTATION, AND PARTURITION OF THE BOOK

We informally chatted with each other and our colleagues in the Luquillo LTER program about the ways in which we think that involvement in long-term ecological research changed us as scientists. We initially had contemplated editing a volume such as this one, but with a focus on scholars working in the Luquillo Mountains rather than those working throughout the entire LTER network of 26 sites. We decided that although the latter task was more challenging, the final product would have enhanced generality and broader appeal because it reached out to those involved in the LTER program at a cross-section of sites that represented different histories and management models.

Scientists are trained to write dry, didactic prose, but here we offered them the challenge of writing more personal and clearly retrospective essays, sometimes covering three or more decades of their lives. We recognized that there is a natural tendency to emphasize positive experiences and that an anonymous approach might have resulted in a more accurate balance of positive and negative responses. We also noted the difficulty that those with little or no research experience outside the LTER program had in evaluating its effect on their personal growth as scientists. Furthermore, we limited our participants to scientists and information managers who have been with the LTER program for at least one and a half funding cycles (9 years), leaving out relative newcomers and students. Despite these caveats, we believe that this collection of essays provides a fascinating glimpse into how the collaborative, long-term research conducted within the LTER program has affected participating scientists and altered their perspectives, attitudes, and practices. We believe that this collection of essays and the analyses based on them by scholars outside of the biophysical sciences will be of broad interest, not only to ecologists and environmental biologists, but also to other biologists, science administrators, educators, historians, social scientists, information managers, and network designers. In short, this book explores ways in which a pioneering experiment, the LTER program, has affected the nature of scientists who have been funded by it.

PERSONAL REFLECTIONS OF M. R. WILLIG

My experiences in the Luquillo Mountains began early in my career (summer of 1982) with support from a sequence of three research fellowships from the Department of Energy's Science Programs administered by Oak Ridge Associated Universities. At that time, my interests were rather narrowly focused on evolutionary ecology of vertebrates, but I had strong quantitative expertise and considerable tropical experience. The need or desire to collaborate with those working on taxa other than animals was not apparent to me at that time, and working with ecosystem scientists, geoscientists, atmospheric scientists, or social scientists was not a direction that I would have remotely considered to be a part of my future. Long-term research meant a 2- or 3-year project, and collaboration mostly involved those with similar interests. Participating in the working group that crafted our first and declined LTER proposal would change all of that. I found myself surrounded by those who I knew very poorly and with whom I did not share a vocabulary, much less a common view of the important ecological questions. In retrospect, I was rather myopic and certainly poorly prepared to engage in a career that would ultimately involve suites of multidisciplinary collaborators dedicated to exploring long-term ecological interactions. The LTER program allowed me to appreciate the diversity of interactions (abiotic and biotic) that play out over the long term to determine the structure and functionality of complex ecological systems. By analogy, I was finally watching long segments of a film to understand the plot and its dénouement—rather than focusing on one or a few actors during a few frames of the movie! This led to excitement as well as considerable challenges, both personal and scientific. My hope is that this book captures the excitement and challenges faced by others who may have encountered this same phenomenon from a variety of perspectives and experiences.

PERSONAL REFLECTIONS OF L. R. WALKER

I am enriched by the numerous collegial experiences with fellow ecologists at the LTER site in Puerto Rico (most notably as a member of a "disturbed plant group"), but I am also frustrated by the pace and pitfalls of collaborative research. Working on primary succession on landslides in Puerto Rico has left me somewhat out of the mainstream research focus of the site (secondary succession in disturbed tropical forests). I never did convince the active stream ecologists in the group that landslides were just downhill flows of carbon and nutrients, although I certainly had fun trying! Nevertheless, there have been plenty of opportunities for me to collaborate with other researchers studying the same site, particularly in areas of synthesis (e.g., of hurricane effects on the Luquillo Mountains), with wonderfully dedicated technicians and students who collected much of the long-term data on landslides, and with colleagues from Puerto Rico and elsewhere who were interested in placing Puerto Rican ecology in a global context.

ACKNOWLEDGMENTS

We are grateful to the many contributors to this book, who took time from their busy schedules to reflect on their personal journeys, to provide overviews of the LTER program and network, or to integrate the plethora of contributed perspectives into a synthetic overview. We also acknowledge the editors at Oxford University Press, who supported our efforts to produce a somewhat unusual but innovative book about the LTER program and the various reviewers who provided useful input into the content and exposition of the chapters. In addition, we gratefully acknowledge the skills and dedication of S. J. Presley, who assisted with many administrative aspects associated with the production of the book. Finally, we were both supported by grants DEB-0620910 and DEB-1239764 from NSF to the Luquillo LTER program. In addition, M. R. Willig was supported by NSF grant DEB-1354040 to the University of Connecticut, and L. R. Walker was supported by a sabbatical from the University of Nevada Las Vegas.

REFERENCES

Walker, L.R., Silver, W.L., Willig, M.R., and Zimmerman, J.K., eds. (1996). Special Issue: Long Term Responses of Caribbean Ecosystems to Disturbance. *Biotropica*, **28**, 414–614.

Walker, L.R., and Willig, M.R. (1999). An introduction to terrestrial disturbances. In L.R. Walker, ed. *Ecosystems of Disturbed Ground*, pp. 1–15. Elsevier Science, Amsterdam, Netherlands.

Willig, M.R., Bloch, C.P., Covich, A.P., Hall, C.A.S., Lodge, D.J., Lugo, A.E, Silver, W.L., Waide, R.B., Walker, L.R., and Zimmerman, J.K. (2012). Long-term research in the Luquillo Mountains: Synthesis and foundations for the future. In N. Brokaw, T.A. Crowl, A.E. Lugo, W.H. McDowell, F.N. Scatena, R.B. Waide, and M.R. Willig, eds. *A Caribbean Forest Tapestry: The Multidimensional Nature of Disturbance and Response*, pp. 361–441. Oxford University Press, New York.

Willig, M.R., and Walker, L.R. (1999). Disturbance in terrestrial ecosystems: Salient themes, synthesis, and future directions. In L.R. Walker, ed. *Ecosystems of Disturbed Ground*, pp. 747–767. Elsevier Science, Amsterdam, Netherlands.

List of Contributors

Brandon Bestelmeyer, US Department of Agriculture–Agricultural Research Station Jornada Basin and Jornada Experimental Range Long-Term Ecological Research, New Mexico State University, Las Cruces, New Mexico 88003, bbestelm@nmsu.edu

John Blair, Division of Biology, Kansas State University, Manhattan, Kansas 66502, jblair@ksu.edu

Elizabeth T. Borer, Department of Ecology, Evolution, and Behavior, University of Minnesota, St. Paul, Minnesota 55108, borer@umn.edu

Mark A. Boyer, Department of Geography, University of Connecticut, Storrs, Connecticut 06269, mark.boyer@uconn.edu

Scott W. Brown, Department of Educational Psychology, University of Connecticut, Storrs, Connecticut 06269, scott.brown@uconn.edu

Daniel L. Childers, School of Sustainability, Arizona State University, Tempe, Arizona 85287, dan.childers@asu.edu

Scott L. Collins, Department of Biology, University of New Mexico, Albuquerque, New Mexico 87131, scollins@sevilleta.unm.edu

Hugh W. Ducklow, Lamont-Doherty Earth Observatory, Columbia University, Palisades, New York 10964, hducklow@ldeo.columbia.edu

Courtney G. Flint, Department of Sociology, Social Work, and Anthropology, Utah State University, Logan, Utah 84322, courtney.flint@usu.edu

Evelyn E. Gaiser, Department of Biological Sciences and Southeast Environmental Research Center, Florida International University, Miami, Florida 33199, gaisere@fiu.edu

Henry L. Gholz, Division of Environmental Biology, National Science Foundation, Arlington, Virginia 22230, hgholz@nsf.gov

Anne Giblin, The Ecosystem Center, Marine Biological Laboratory, Woods Hole, Massachusetts 02543, agiblin@mbl.edu

Ted L. Gragson, Department of Anthropology, University of Georgia, Athens, Georgia 30609, tgragson@uga.edu

James R. Gosz, College of Natural Resources, University of Idaho, Moscow, Idaho 83844, jgosz@uidaho.edu

Laura Gough, Department of Biological Sciences, Towson University, Towson, Maryland 21252, lgough@townson.edu

Nancy B. Grimm, School of Life Sciences, Arizona State University, Tempe, Arizona 85287, nbgrimm@asu.edu

J. Morgan Grove, Baltimore Field Station and Baltimore Ecosystem Study, Northern Research Station, US Forest Service, Baltimore, Maryland 21228, jmgrove@gmail.com

Christopher Hamlin, Department of History, University of Notre Dame, Notre Dame, Indiana 46556, chamlin@nd.edu

Bruce P. Hayden, Department of Environmental Science, University of Virginia, Charlottesville, Virginia 22904, bph@virginia.edu

John E. Hobbie, Marine Biological Laboratory, Woods Hole, Massachusetts 02543, jhobbie@mbl.edu

Sally J. Holbrook, Department of Ecology, Evolution, and Marine Biology and the Marine Science Institute, University of California Santa Barbara, Santa Barbara, California 93106, holbrook@lifesci.ucsb.edu

Sherri L. Johnson, Andrews Forest Long-Term Ecological Research and US Forest Service Pacific Northwest Research Station, Corvallis, Oregon 97330, sherrijohnson@fs.fed.us

Alan K. Knapp, Graduate Degree Program in Ecology and Department of Biology, Colorado State University, Fort Collins, Colorado 80526, alan.knapp@colostate.edu

D. Jean Lodge, Center for Forest Mycology Research, US. Forest Service, Northern Research Station, Luquillo, Puerto Rico 00773, djlodge@caribe.net

Ariel E. Lugo, International Institute of Tropical Forestry, Río Piedras, Puerto Rico 00926, alugo@fs.fed.us

John J. Magnuson, Center for Limnology, University of Wisconsin–Madison, Madison, Wisconsin 53706, jjmagnus@wisc.edu

Roberta Marinelli, Wrigley Institute for Environmental Studies, University of Southern California, Los Angeles, California 90089, rmarinel@dornsife.usc.edu

John C. Moore, Department of Ecosystem Science and Sustainability and The Natural Resource Ecology Laboratory, Colorado State University, Fort Collins, Colorado 80524, john.moore@colostate.edu

James T. Morris, Belle Baruch Institute for Marine and Coastal Sciences, University of South Carolina, Columbia, South Carolina 29208, morris@biol.sc.edu

Margaret O'Brien, Marine Science Institute, University of California Santa Barbara, Santa Barbara, California 93111, margaret.obrien@ucsb.edu

Debra P. C. Peters, US Department of Agriculture–Agricultural Research Station Jornada Basin and Jornada Experimental Range Long-Term Ecological Research Program, New Mexico State University, Las Cruces, New Mexico 88003, and US Department of Agriculture, Office of the Chief Scientist, Washington, DC 20250, debpeter@nmsu.edu

Steward T. A. Pickett, Cary Institute of Ecosystem Studies, Millbrook, New York 12545, picketts@caryinstitute.org

John H. Porter, Department of Environmental Sciences, University of Virginia, Charlottesville, Virginia 22904, jporter@virginia.edu

William H. Schlesinger, Cary Institute of Ecosystem Studies, Millbrook, New York 12545, schlesingerw@caryinstitute.org

Russell J. Schmitt, Department of Ecology, Evolution, and Marine Biology and the Marine Science Institute, University of California Santa Barbara, Santa Barbara, California 93106, russ.schmitt@lifesci.ucsb.edu

Gaius R. Shaver, Marine Biological Laboratory, Woods Hole, Massachusetts 02543, gshaver@mbl.edu

Whendee L. Silver, Department of Environmental Science, Policy, and Management, University of California Berkeley, Berkeley, California 94720, wsilver@berkeley.edu

Melinda D. Smith, Graduate Degree Program in Ecology and Department of Biology, Colorado State University, Fort Collins, Colorado 80526, melinda.smith@colostate.edu

Susan G. Stafford, Department of Forest Resources, University of Minnesota–Twin Cities, St. Paul, Minnesota 55108, stafford@umn.edu

Katharine N. Suding, Institute of Arctic and Alpine Research, Department of Ecology and Evolutionary Biology, University of Colorado at Boulder, Boulder, Colorado 80309, suding@colorado.edu

Frederick J. Swanson, US Forest Service, Pacific Northwest Research Station, Corvallis, Oregon 97331, fred.swanson@oregonstate.edu

Phillip R. Taylor, Research Advancement and Federal Relations, University of Southern California, Los Angeles, California 90071, prtaylor@usc.edu

David Tilman, Department of Ecology, Evolution, and Behavior, University of Minnesota, St. Paul, Minnesota 55108, and Bren School of Environmental Science and Management, University of California Santa Barbara, Santa Barbara, California 93106, tilman@umn.edu

Tiffany G. Troxler, Department of Biological Sciences and Southeast Environmental Research Center, Florida International University, Miami, Florida 33199, troxlert@fiu.edu

Robert B. Waide, Department of Biology, University of New Mexico, Albuquerque, New Mexico 87131, rwaide@lternet.edu

Lawrence R. Walker, School of Life Sciences, University of Nevada Las Vegas, Las Vegas, Nevada 89154, walker@unlv.nevada.edu

Michael R. Willig, Center for Environmental Sciences & Engineering and Department of Ecology & Evolutionary Biology, University of Connecticut, Storrs, Connecticut 06269, michael.willig@uconn.edu

PART ONE
INTRODUCTION AND OVERVIEW

1 Changing the Nature of Scientists

Participating in the Long-Term Ecological Research Program

Michael R. Willig and Lawrence R. Walker

INTRODUCTION

From the outside looking in, scientists are often characterized as old men in white laboratory coats, working in splendid isolation, usually within the confines of rather sterile-looking laboratories. Of course, this image was never quite accurate for ecologists, who abandoned white laboratory coats for more field-appropriate boots and khaki pants, but who nonetheless typically worked alone or with the benefit of a faithful field assistant (Figure 1.1a). The late 1900s was a time of rapid change in the way in which ecological research was conducted, in part because of opportunities for support from governmental agencies. Especially critical in effecting these changes was grant support that would allow scientists to comprehensively investigate the intricate and complex ecological interactions between organisms and their environment from a long-term and site-based perspective. Such efforts often involved large and diverse groups of scientists representing multiple disciplinary perspectives and investigative approaches (Figure 1.1b).

The US Long-Term Ecological Research (LTER) program, with support from the National Science Foundation (NSF), was one of the first governmental programs to catalyze long-term, site-based, multidisciplinary, and collaborative research. The scientific research arising from such support has been broad and deep, resulting in thousands of publications. The research insights have been integrated into a number of synthetic books, each dedicated to long-term research at a particular site in the LTER program (Knapp et al. 1998; Bowman and Seastedt 2001; Greenland, Goodin, and Smith 2003; Schachak et al. 2005; Magnuson, Kratz, and Henson 2005; Foster and Aber 2006; Chapin et al. 2006; Havstad, Huenneke, and Schlesinger 2006; Redman and Foster 2008; Lauenroth and Burke 2008; Brokaw et al. 2012). In contrast, the effects of the LTER program's many innovations on the participating scientists have not been explored in a comprehensive or systematic fashion. This book provides a window into how scientists have changed as a consequence of participation in the LTER program.

The LTER network of sites, begun in 1980, effectively implemented the first effort by the NSF to systematically fund long-term, site-based environmental research. The LTER program has successfully facilitated studies by environmental scientists from multiple disciplinary backgrounds (based on observations and experiments) who evaluate the structure and functioning of ecological systems at decadal or longer time frames.

FIGURE 1.1 Ecological research over the past half century has changed from a model involving one or a few scientists (Figure 1.1a) investigating disciplinary questions over local or short temporal horizons, to large collaborative groups (Figure 1.1b) investigating interdisciplinary questions over multiple spatial and temporal scales. (Artwork courtesy of Janine Caira.)

By supporting such approaches to environmental research projects, the LTER program has encouraged groups of scientists to work collaboratively to analyze multiple aspects of ecological problems (e.g., not only nutrient cycles but their interactions with plants, animals, and microbes, as well as their responses to disturbance and consequences to human well-being).

The LTER program, via the LTER network office, also pioneered web-based, centralized information management systems. It facilitated long-term monitoring studies and the storage, manipulation, and sharing of data and metadata among investigators within and among LTER sites, as well as with non-network scientists, educators, and natural resource managers. Although not explicitly formed as a synoptic network of sites for quantifying the same phenomena in the same manner at multiple locations, the LTER program, via activities of the LTER network office, has engaged scientists to compare and contrast system dynamics among sites with regard to salient environmental gradients. Moreover, the LTER network was among the first scientific organizations to promote a socioecological examination of how human and natural systems are coupled. The success of the LTER network is in part reflected in the extent to which other countries throughout the world have adopted a similar organizational structure and scientific agenda to broaden and deepen predictive understanding of socioecological systems. (The International LTER network is a multinational collection of LTER sites in various countries on most continents.) Finally, the LTER program has supported an innovative and effective outreach program that integrates research and education via its Schoolyard LTER program and a children's book series that is dedicated to site-based environmental themes. These many innovations have transformed the scientific nature of environmental research and education, and in doing so potentially have also changed how contemporary environmental scientists think and act as researchers, educators, and communicators to the public, all within a collaborative and multidisciplinary culture.

ORGANIZATION

The subsequent chapters in this book are organized into three parts. The first part comprises two chapters that provide an overview of the LTER program and network from the perspectives of those involved in science management at the LTER network office (Waide, Chapter 2) or at NSF (Gholz, Marinelli, and Taylor, Chapter 3). The second part includes 36 introspective essays in which participants in the LTER program examine how they have changed as a consequence of involvement in the program. The third part contains five chapters that comment on, integrate, analyze, or synthesize information contained in the second part. One chapter provides the perspective of an eminent ecologist (Schlesinger, Chapter 40); three chapters provide assessments from the perspective of behavioral scientists (Boyer and Brown, Chapter 41), a sociologist (Flint, Chapter 42), and an historian (Hamlin, Chapter 43); and the final chapter provides a synthesis by the coeditors (Walker and Willig, Chapter 44) that identifies emergent themes and suggests lessons that have been learned, especially as they presage directions for future environmental research and education.

SCOPE, FOCUS, AND LIMITATIONS

This book provides a broad assessment of the effects of a more than 35-year social, cultural, and scientific "experiment" (i.e., the LTER program) on the attitudes of its participants as well as on how they engage in research, teaching, service, mentoring, and outreach. It explores what happens when ecologists receive long-term funding in a network of site-based groups located across a diversity of biomes and habitat types. We wanted to know if and how such an innovative funding arrangement affected the scientists themselves: did perspectives, attitudes, or practices change as a consequence of participation in this "experiment?" We targeted participants in the LTER program who generally had at least 9 years of experience as principal investigators, co-principal investigators, or senior personnel, so that responses would be based on comprehensive participation in the program. We sought to involve one to three participants from each of a number of sites throughout the network, and to broadly represent sex and age groups, as well as disciplinary backgrounds. In some cases, authors began their association with the LTER program when they were graduate students or postdoctoral fellows. Some authors played a critical role in the formation of the program or its administration at NSF. Given our selection criteria, responses from potential participants, and publication timeline, not all LTER sites could be represented by chapters.

This book is an historical exposition from the perspectives of the participants of the broader impacts of LTER involvement on their professional and personal development. Although these essays represent data for integration and synthesis, they are not the product of a controlled experimental design. We did not randomize the selection of participants from among those with long-term associations with the LTER program. We have not asked the same kinds of questions to environmental scientists outside of the LTER program, or to those whose experiences within the LTER program were ephemeral. Moreover, we are not experts in crafting survey questionnaires, and there were no internal controls among the questions to detect consistency of opinion. Nonetheless, we did attempt to phrase questions so that they would be "outcome neutral," and to encourage critical responses that include both positive and negative components.

The main body of the book comprises the responses of a broad selection of scientists, including several data managers or educators, to a series of probing questions (see Chapters 4 to 39). Initially, eight questions were presented for consideration by the authors in crafting their introspective essays.

1. How has your LTER experience changed you as a scientist? Consider if the LTER experience has made you more or less collaborative, multidisciplinary, theoretical, insightful, comparative, synthetic, or place-based.
2. How has your LTER experience changed you as a communicator of science? Has your teaching changed? How about your perspective or involvement in outreach activities, service, or administration?
3. How has your LTER experience changed your scientific tool kit? Have you expanded your skills (e.g., in molecular techniques, geographic information system [GIS], statistics) due to the LTER experience?
4. How has your LTER experience changed your appreciation of or contributions to applied science? Are you more or less involved in forestry, conservation, natural resource management, restoration, or similar activities due to your association with an LTER site?
5. How has your LTER experience changed your mentoring? Has the LTER changed (1) how you mentor students or junior faculty members or (2) your attitudes toward or involvement in mentoring of students, faculty members, or staff?
6. How has your LTER experience changed you as a person? Has the LTER experience altered how and where you travel, your friendship circles, and your approach to other cultures or ethnic groups?

7. How has your LTER experience changed your attitudes toward time and space and their role in ecological studies?
8. Are there any other notable changes in your life that you attribute to your association with an LTER program?

After receipt of the first draft and to facilitate evaluations of responses, we asked the authors to rearrange the content of their essay so that responses would address 10 themes: (1) approach to science, (2) attitudes toward time and space, (3) collaboration, (4) applied research, (5) communication, (6) mentoring, (7) skill set, (8) personal consequences, (9) challenges and recommendations, and (10) conclusion. In addition, we asked each essayist to address the costs and benefits of collaborating on multidisciplinary projects and of engaging in network-level activities in the collaboration section; to include activities that target teachers in the mentoring section; and to provide advice, admonitions, or recommendations for aspiring ecologists or students as they begin a career, especially as it relates to leveraging experiences in the LTER program to good effect.

PARTING COMMENTARY

Ecological scholars of the twenty-first century increasingly seek to understand complex environmental dynamics at large spatial and long temporal scales, often doing so in the context of multidisciplinary projects supported by "big data," cyber-infrastructure, and next-generation computing capabilities. The extent to which particular projects succeed in transforming scientific knowledge or understanding will be assessed in part based on scholarly publications and a historical record of paradigm shifts in the published literature. The influence of these scientific challenges and associated projects on the nature of scientists as researchers, educators, or communicators with the public is at the heart of NSF's broader impacts. Without concerted efforts to capture such changes in the nature of scientists during transformative periods, a ripe body of information is quickly lost, and assessments may be relegated to retrospective essays such as those in this volume. Clearly, these kinds of narratives provide rich fodder for historical and social science research, but at the same time, they are limited in the extent to which they can provide accurate portrayals of the transformation process itself. Our hope is that this book will provide a window into the kinds of transformations that have occurred as a result of NSF's experiment of changing the culture of scientific research via long-term collaborative funding. Equally important, we hope that this work will stimulate more formal scholarly investigation of the ways in which initiatives or programs such as the LTER program affect broader impacts and do so in a manner that is contemporaneous rather than solely retrospective.

SITE CONTEXT

In 2013 when much of this book was being written, the LTER program comprised 26 sites, mostly located in the continental United States, although there were two sites in Alaska, two sites in Antarctica, one site in French Polynesia, and one in Puerto Rico (Figure 1.2). The network encompasses a diversity of habitats including deserts, estuaries, lakes, oceans, coral reefs, prairies, forests, alpine and Arctic tundra, urban areas, and production agriculture. We provide a brief, alphabetically arranged overview of each of the 19 sites that are represented in one or more essays in this book. The commentary, mostly derived from Peters et al. (2013), gives a link to each site's official web page, identifies the scientific themes that characterize each research portfolio, and describes the ecology of each site.

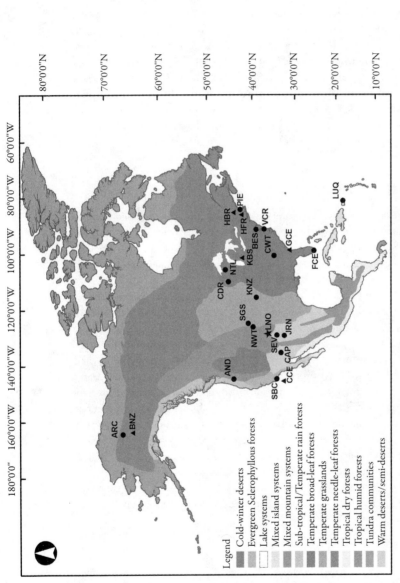

FIGURE 1.2 The Long-Term Ecological Research (LTER) network currently comprises 26 sites and a network office (LNO). Most of the sites (23) are disturbed across a diversity of biomes in North America (United States and Puerto Rico) and appear as circles (sites represented by essays) or triangles (sites not represented by essays) on the map. Two sites in Antarctica (McMurdo Dry Valleys site, MCM; Palmer Antarctica site, PAL) and one site in French Polynesia (Moorea Coral Reef site, MCR) are not represented in the graphic, although PAL and MCM are represented by essays in the book. Major biomes (Udvardy 1975) are represented by colors on the map; the LNO appears as a star. Three-letter acronyms identify particular LTER sites: AND, H. J. Andrews Experimental Forest site; ARC, Arctic site; BES, Baltimore Ecosystem Study site; BNZ, Bonanza Creek site; CEE, California Current Ecosystem site; CDR, Cedar Creek Ecosystem Science Reserve site; CP, Central Arizona–Phoenix site; CWT, Coweeta site; FCE, Florida Coastal Everglades site; GCE, Georgia Coastal Ecosystems site; HFR, Harvard Forest site; HBR, Hubbard Brook site; JRN, Jornada Basin site; KBS, Kellogg Biological Station site; KNZ, Konza Prairie site; LUQ, Luquillo site; MCM, McMurdo Dry Valleys site; MCR, Moorea Coral Reserve site; NWT, Niwot Ridge site; NTL, North Temperate Lakes site; PIE, Plum Island Ecosystem site; SBC, Santa Barbara Coastal site; SEV, Sevilleta site; SGS, Shortgrass Steppe site; and VCR, Virginia Coastal Reserve site. In

FIGURE 1.3 A tributary of McRae Creek within a stand of old-growth forest at the H.J. Andrews Experimental Forest Long-Term Ecological Research site. (Photo courtesy of Lina DeGregorio.)

H. J. Andrews Experimental Forest (AND)

This site (http://andrewsforest.oregonstate.edu/) was established as an Experimental Forest in 1948 and was part of the International Biological Program in the 1970s (Figure 1.3). In 1980, AND was funded as one of the original sites in the LTER program with a research focus on understanding how climate, natural disturbance, and land use, as controlled by forest governance, interact with biodiversity, hydrology, and carbon and nutrient dynamics. It is located in the western Cascade Range of Oregon in the 6,400-ha drainage basin of Lookout Creek, a tributary of the Blue and McKenzie Rivers. Elevation ranges from 410 to 1,630 m. The climate is wet in winter but warm and dry in summer. Precipitation occurs mainly as rain at lower elevations and as snow at upper elevations. The vegetation consists of coniferous forests that are intersected by streams. Forest age classes include 150- and 500-year-old stands developed after wildfire and younger stands established as Douglas fir (*Pseudotsuga menziesii*) plantations following clear-cutting in the 1950s through early 1980s.

Arctic (ARC)

The ARC site (http://ecosystems.mbl.edu/ARC/) was established in 1987 to understand and predict the effects of environmental change on the ecology of tundra, streams, and lakes (Figure 1.4). It is located at an elevation of 760 m in the northern foothills of the Brooks Range (Alaska), 254 km north of the Arctic Circle. Study sites occur within a 48-km^2 core watershed but are also dispersed across more than 1,000 km^2 of the tundra, including three other watersheds and a large burned area. The climate is typical of arctic regions, with a mean annual air temperature of about –7°C and low precipitation (200–400 mm; half falling as snow). The snow-free growing season extends from late May to mid-September, but freezing temperatures are possible at any time. The region is underlain by continuous permafrost that exerts a major influence on the distribution, structure, and function of ecosystems.

ARC represents a treeless tundra, with tussock tundra predominating but also including wet sedge tundra, drier heath tundra on ridge tops, and river-bottom willow shrub communities. The streams in the area constitute the headwaters of the Kuparuk River, and oligotrophic lakes of various ages are abundant.

FIGURE 1.4 Toolik Lake and the surrounding landscape at the Arctic Long-Term Ecological Research site. (Photo courtesy of James Laundre.)

FIGURE 1.5 West Franklin Street neighborhood and landscape at the Baltimore Ecosystem Study (BES) Long-Term Ecological Research site. (Photo courtesy of Steward Pickett, the BES LTER program.)

Baltimore Ecosystem Study (BES)

This site (http://www.beslter.org/) became part of the LTER program in 1997, with a mission to investigate metropolitan Baltimore as a socioecological system (Figure 1.5).

BES conducts research and educational activities in Baltimore City and surrounding counties. The project focuses on several watersheds to organize research from both spatial and functional perspectives. For example, the Gwynns Falls watershed encompasses 17,150 ha and drains into the Chesapeake Bay. The climate is humid temperate, with four distinct seasons, including cool winters (January average, 2.1°C) and hot, humid summers (July average, 27.1°C). Precipitation comes mostly as spring and summer rains; winter snowfall is sporadic, with total annual precipitation averaging 1,060 mm. The city is warmed by an urban heat island effect that lessens irregularly with distance from its center.

The watershed includes agricultural lands, recently suburbanized areas, established suburbs, and dense urban areas having residential, commercial, and open spaces. Vegetation of the watershed has changed from predominantly forest before European settlement to primarily herbaceous. There are no original stands of forest in the Baltimore area, although a reference second-growth forested watershed was established in a park in Baltimore County. Research on stream restoration centers on the Minebank Run catchment. A study of the ecological effects of residential neighborhood greening and restoration is being conducted in a 364-ha storm drain catchment in Baltimore City.

Cedar Creek Ecosystem Science Reserve (CDR)

The CDR site (http://www.cbs.umn.edu/cedarcreek) was established in 1949, and it became part of the LTER network in 1982 (Figure 1.6). Its mission is to improve understanding of the processes that govern the dynamics and functioning of ecosystems along the boundary between prairie and forests, using experimental manipulations of drought and additions of carbon dioxide and nitrogen. The site, located 50 km north of Minneapolis, Minnesota,

FIGURE 1.6 Aerial photograph of experimental manipulation plots and the surrounding landscape at the Cedar Creek Ecosystem Science Reserve Long-Term Ecological Research site. (Photo courtesy of David Tilman.)

encompasses an area of 22 km² at an elevation of 277 m. The climate is continental, with cold winters and hot summers. Precipitation (660 mm per year) is spread fairly evenly throughout the year. The mean July temperature is 22.2°C, and the mean January temperature is −10°C. Soils derive from a glacial outwash sandplain, and upland soils are nitrogen poor. Numerous nutrient addition experiments in old fields and native savanna document that nitrogen is the major limiting resource to plant growth. Much of the site comprises wetlands, including swamps, acid bogs, wet meadows, and marshes. Uplands include abandoned agricultural fields, savannas, and hardwood forests. Prescribed burns began in the oak savanna in 1964.

Central Arizona–Phoenix (CAP)

The CAP (http://caplter.asu.edu) site was established as part of the LTER network in 1997, with a mission to study human interactions with the environment in central Arizona and the Phoenix metropolitan area. Phoenix is now the fifth largest city in the United States (Figure 1.7).

The study area (6,400 km²) is at the confluence of the Salt and Gila Rivers (331 m). With less than 180 mm of annual rainfall, which falls in winter and during summer monsoon storms, CAP is situated in an arid landscape with a high rate of evaporation, and the urbanized areas comprise a growing urban heat island with increasing nighttime temperatures. The site is located in a basin that once supported lowland Sonoran Desert vegetation and riparian ecosystems, as well as agriculture. Two plants, blue palo verde (*Parkinsonia florida*) and saguaro cactus (*Carnegiea gigantea*), visually dominate the desert portions of the landscape and distinguish the Sonoran Desert from other North American deserts. Plant associations in the urban portions of the CAP area contain many non-native species typical of the American lawn. The chief constraint on plant growth is water availability, which varies with substrate conditions in the desert but is almost entirely controlled by people in the urban area. Water quality and quantity, air quality, drought, and urban heat island effects are studied to examine the feedbacks between human decisions and environmental characteristics. In this context, land use change is a major driver of ecological patterns at multiple scales.

FIGURE 1.7 The Phoenix metropolitan area (background), situated within the Sonoran Desert (foreground), represents the landscape of the Central Arizona–Phoenix (CAP) Long-Term Ecological Research (LTER) site. (Photo courtesy of Eyal Shochat and the CAP LTER program.)

FIGURE 1.8 Image depicting a typical landuse (foreground) and the broader landscape (background) in the Caler Creek Watershed that are typical of the Coweeta Long-Term Ecological Research site. (Photo courtesy of Ted Gragson.)

Coweeta (CWT)

The CWT (http://coweeta.ecology.uga.edu/) site was established by the US Forest Service as the Coweeta Hydrologic Laboratory in 1934, with an initial emphasis on the effect of forestry management practices on the hydrological cycle in small, experimental watersheds of the Coweeta Basin (Figure 1.8).

This basin is located in the Nantahala Mountain range of western North Carolina, and consists of a 1,626-ha bowl-shaped basin ranging in elevation from 670 to 1,592 m. The Coweeta LTER program, starting in 1980, initially focused on the basin but evolved from a site-based to a site- and region-based program by 1994 to better address the synergisms between substrate, society, and biology that underlie disturbance processes across the environmental gradients characteristic of southern Appalachia. Summer temperatures rarely exceed 30°C and decrease from low to high elevation. Nonetheless, summer temperatures on the higher peaks are more similar to those in central New England, 1,400 km to the north, than they are to the lower Piedmont, only 100 km to the east. Precipitation is abundant, averaging above 1,800 mm per year and generally increasing 5% per 100 m of elevation along an east–west axis. Importantly, local mountain effects create wet zones and rain shadows that can differ by 100% in precipitation. The dominant vegetation is temperate deciduous forest, although an intermixing of northern and southern taxa results in one of the most biologically diverse regions of North America. The region represents an ideal natural laboratory for examining extensive and transient human disturbances (e.g., fire and logging) through localized and permanent disturbances (e.g., agriculture and urbanization), contributing to the growing understanding of how human practices influence forest and stream ecosystems at multiple scales.

Florida Coastal Everglades (FCE)

This site (http://fcelter.fiu.edu/) was established in 2000, with a mission to understand how population- and ecosystem-level dynamics of the coastal Everglades are controlled by water source, water residence time, and local biotic processes (Figure 1.9). Located in southern Florida, FCE occurs where a rapidly growing human population (> 6 million) lives in close proximity to—and in surprising dependence on—the Florida Everglades. FCE is entirely within the boundaries of Everglades National Park, the third largest wilderness area (6,110 km²) in the continental United States and location of the largest mangrove ecosystem in the Western Hemisphere.

FIGURE 1.9 Aerial view of the coastal landscape that represents the Florida Coastal Everglades Long-Term Ecological Research site. (Photo courtesy of Stephen Davis.)

Although elevational variation in the Everglades is small, water flows from about 2 m at the northern boundary of the park to the estuaries. Average temperatures range from 12°C (winter lows) to 32°C (summer highs), with summer rains and total annual precipitation of 1,520 mm. The soils are deep, poorly drained peats. Wetlands include marshes, sloughs, and wet prairies, and uplands support tree islands, tropical hardwood hammocks, and pine forests. Because the coastal Everglades cover a large area that is, in effect, topographically flat, it is susceptible to dramatic changes in response to sea level rise. Hurricanes and storms are common, and add pulse disturbance to the slow press disturbances of rising sea level. The Everglades Restoration Project seeks to return the existing Everglades to a healthy and stable state so that it can continue to provide critical ecosystem services to human populations. FCE research focuses on population and ecosystem dynamics in the oligohaline regions of Taylor Slough and Shark River Slough, where freshwater and estuarine ecosystems meet to form ecotones.

Jornada Basin (JRN)

The JRN site (http://jornada-www.nmsu.edu/), located 37 km northeast of Las Cruces (New Mexico), became part of the LTER program in 1982, with a mission to understand the key factors and processes that control ecosystem dynamics and desertification in Chihuahuan Desert landscapes (Figure 1.10).

The study site includes the Jornada Experimental Range (78,000 ha), operated by the US Agricultural Research Service (ARS), and the Chihuahuan Desert Rangeland Research Center (22,000 ha), operated by New Mexico State University. Data have been collected since 1915, and vegetation records date to the mid-1800s. JRN (1,300–1,400 m) has mild winters (as low as 13°C) and hot summers (as high as 36°C). Annual precipitation averages 260 mm, with about 52% of rain occurring in summer. Extreme droughts are recurrent climatic phenomena with profound influence on the vegetation. The five major plant communities at JRN differ in degree of desertification and include upland and lowland, and a

FIGURE 1.10 Creosote shrubland is one of the five major plant communities of the Chihuahuan Desert that compose the Jornada Basin Long-Term Ecological Research site. (Photo courtesy of Debra Peters.)

series of desertified shrublands, including tarbush (*Flourensia cernua*) on lower piedmont slopes, creosotebush (*Larrea tridentata*) on upper piedmont slopes and bajadas, and honey mesquite (*Prosopis glandulosa*) on the sandy basin floor. Grazing by livestock was historically the predominant land use in the region, although urbanization has been increasing.

Konza Prairie (KNZ)

The KNZ site (http://www.konza.ksu.edu/) became part of the LTER program in 1980, with a goal of understanding how three key drivers (i.e., fire, grazing, climatic variability) affect ecological patterns and processes in tallgrass prairies (Figure 1.11). The Konza Prairie Biological Station (3,487 ha), at 330 m elevation in the Flint Hills of northeastern Kansas, is a C4-dominated grassland with a continental climate characterized by warm, wet summers and dry, cold winters. Steep slopes in the area are overlain by shallow limestone soils that are unsuitable for cultivation. These soils overlay as many as 10 distinct layers of alternating limestone and shale, contributing to the complex subsurface hydrology of the region. Because mean annual precipitation is sufficient to support woodland or savanna vegetation, periodic drought, fire, and grazing are important in maintaining the grassland. The vegetation is over 90% native tallgrass prairie dominated by perennial C4 grasses. Numerous subdominant grasses, forbs, and woody species contribute to high floristic diversity. Gallery forests grow along major stream courses. Long-term studies include a replicated watershed-level experiment, in place since 1977, which explicitly incorporates the major factors influencing mesic grasslands in a long-term experimental setting, as well as numerous plot-level experiments and stream weirs and groundwater sampling locations.

Luquillo (LUQ)

This site (http://luq.lternet.edu/) became part of the LTER program in 1988, with a mission to study the long-term effects of natural and human disturbances on tropical forests

FIGURE 1.11 Landscape image of the Konza Prairie Long-Term Ecological Research site, with a pair of bison in the foreground. (Photo courtesy of Edward Raynor.)

FIGURE 1.12 Landscape image of the forests of the Luquillo Mountains, including suburbanizing areas in the lowlands, that represent the Luquillo Long-Term Ecological Research site. (Photo courtesy of Jerry Baurer and Oxford University Press.)

and streams in Puerto Rico (Figure 1.12). Research in this area dates back over 100 years, with LUQ being one of the most intensively studied tropical forests in the world.

The site is located in the Luquillo Mountains (300–1080 m; 11,330 ha), which harbors the largest area of primary forest and the most pristine rivers in Puerto Rico. The climate is subtropical maritime, and moderated by trade winds that maintain relatively constant air temperatures. Rainfall is in excess of 100 mm each month, although periods of lower rainfall occur between February and April and periods of higher rainfall occur in September. Severe hurricanes occur on average every 60 years. Drought (< 100 mm/month) recurs on decadal scales. Dominant soils are deep, highly weathered and leached clays with low pH and base saturation. The vegetation is evergreen broadleaf subtropical forest, with 240 tree species that vary in abundance along the elevational gradient to form distinctive forest types based on species composition and structure. Natural disturbances include hurricanes, landslides, tree falls, floods, and droughts. Human disturbances include changes in land use and land cover change, climate change, and species introductions. Forest cover on the island was reduced to about 5% in 1950, but with industrialization and abandonment of agriculture, has recovered to about 42%.

Moorea Coral Reef (MCR)

This site (http://mcr.lternet.edu/) joined the LTER program in 2004, with a mission to understand the long-term consequences of disturbance and changing climate regimes on coral reef ecosystems (Figure 1.13). MCR comprises a complex of coral reefs and lagoons that surround the 60-km perimeter of Moorea, a small volcanic island in French Polynesia, about 20 km west of Tahiti in the South Pacific. Average temperatures range from 21° to 31°C, with little seasonal variation. Most of the 1,760 mm of rainfall per year occurs between November and February.

Major coral reef types are easily accessible at MCR. Reefs are dominated by massive *(Porites* spp.*)*, branching *(Pocillopora* spp., *Acropora* spp.*)*, and encrusting *(Montipora*

FIGURE 1.13 Staghorn coral represent dominant components of the underwater landscape that is typical of the Moorea Coastal Reserve Long-Term Ecological Research site. (Photo courtesy of Melissa Holbrook.)

spp.) coral species that are periodically disturbed by cyclones (e.g., 1982, 1991, 2010), outbreaks of crown-of-thorns sea stars (*Acanthaster planci*) that consume coral (e.g., 1991, 2008), and coral bleaching events (e.g., 1991, 1994, 2002, 2003). Like coral reefs worldwide, those in Moorea are highly vulnerable to ocean warming and acidification. To date, the reefs of Moorea show high resilience to disturbances, making them an outstanding model system for exploring long-term dynamics and ecosystem processes.

Niwot Ridge (NWT)

The NWT site (http://culter.colorado.edu/NWT/) became part of the LTER program in 1980, with a mission to understand the effects of climate change on the ecological and hydrological processes in high-elevation areas in the Colorado Front Range of the Rocky Mountains (Figure 1.14). Research began at the site in the 1940s with the return of World War II veterans who had extensive experience in cold-region logistics. NWT is approximately 35 km west of Boulder, Colorado, and elevation ranges from 2,895 to 3,814 m. The research area (34 km^2) is bounded on the west by the Continental Divide and comprises extensive alpine tundra and a variety of glacial landforms. Snowfall accounts for more than 80% of precipitation. The interactions among wind, snow, and high topographic relief result in a mosaic of moisture availability. The vegetation includes subalpine forests and meadows, and patches of krummholz in the abrupt transition between forest and tundra. The major research area is the Saddle, with its western half being a snow accumulation area (up to 10 m in some years) and its eastern half remaining free of snow for most of the winter. Changes in abundance and species composition of the native flora and fauna of these mountain ecosystems are potential bellwethers of global change. A suite of short- and long-term experiments

FIGURE 1.14 Image of scientists collecting climatic data under harsh conditions typical of the Niwot Ridge Long-Term Ecological Research site. (Photo courtesy of John Marr and the National Science Foundation.)

provide insight into how alpine tundra and lakes respond to changes in climate and nutrient loading.

North Temperate Lakes (NTL)

The NTL site (http://lter.limnology.wisc.edu/) became part of the LTER program in 1981, with a mission to understand the ecology of lakes in relation to atmospheric, geochemical, landscape, and human processes (Figure 1.15). NTL comprises two geographically distinct regions: the Northern Highlands Lake District (NHLD) and the Yahara Lake District (YLD). These districts lie in formerly glaciated terrain of northern and southern Wisconsin, respectively. Research focuses on seven lakes ranging from small (< 1 ha) dystrophic lakes to large (> 1,500 ha) oligotrophic lakes in the north, and four eutrophic lakes in the south. The regional climate is continental, with cold, snowy winters (often reaching −40°C) and warm summers (typically reaching 33°C). Lakes are the focal landforms of both regions, providing unique habitats, ecosystem services, and foci for human activity.

The NHLD, one of the most lake-rich regions of the world, is largely forested and sparsely settled. Outdoor recreation centered on the 7,600 lakes of the region is a mainstay of the economy. The YLD is an agricultural, but urbanizing, landscape with scattered remnants of presettlement ecosystems. Ecological research began in the YLD in the 1880s and in the NHLD in the 1920s. Long-term, whole-lake experiments provide insights about lakes responses to changes in land cover and climate.

Palmer Antarctica (PAL)

The PAL site (http://pal.lternet.edu/) joined the LTER program in 1990 as the first marine pelagic site in the LTER network (Figure 1.16). It is situated at Palmer Station on the south coast of Anvers Island on the western Antarctic Peninsula.

FIGURE 1.15 Aerial photograph of the aquatic and terrestrial components of the landscape that represents the North Temperate Lakes Long-Term Ecological Research site. (Photo courtesy of Carl Bowser and Silver Pixel Images.)

PAL encompasses a large region extending several hundred kilometers offshore, with several circumpolar pelagic habitats, including the continental shelf within the marginal ice zone covered seasonally by sea ice and the open ocean beyond the continental shelf break, as well as a nearshore zone influenced by glacial meltwater. Within the nearshore zone are small islands that have become deglaciated in the last few centuries. Seabirds, including penguins, petrels, and skuas, inhabit these islands along with mosses and

FIGURE 1.16 Sea ice and glacier-covered mountains in Marguerite Bay (Andelaide Island) represent typical land- and sea-scapes of the Palmer Antarctica Long-Term Ecological Research site. (Photo courtesy of Hugh Ducklow.)

two species of vascular plants. The nearshore waters abound in large marine mammals including seals, orcas, humpback whales, and minke whales. Palmer Station is occupied by humans year-round, but most scientific activity is concentrated in the austral spring and summer. Research focuses on understanding the dynamics of the Antarctic marine ecosystem as it is affected by interannual variations in sea ice, documenting and predicting ecosystem responses to rapid climate change.

Plum Island Ecosystem (PIE)

This site (http://ecosystems.mbl.edu/PIE/) joined the LTER program in 1998, with a mission to (1) develop a predictive understanding of long-term responses of watershed and estuarine ecosystems to changes in climate, land use, and sea level and to (2) apply this knowledge to the management and development of policy that aims to protect the natural resources of the coastal zone (Figure 1.17).

The coupled watersheds and estuary of Plum Island Sound are located near the Boston metropolitan region (Massachusetts). The Ipswich River (400 km^2), Parker River (160 km^2), and Rowley River (40 km^2) basins are areas of low-relief and expansive wetlands. Temperatures are highly seasonal, ranging from lows of 5°C in January to highs of 18°C in July. Temperatures are highly seasonal. Water temperature ranges from a low of 2°C in February to a high of 19°C in July, whereas air temperature ranges from an average winter minimum of −7°C to an average summer maximum of about 28°C. Rainfall (total, 900 mm) is evenly distributed throughout the year. The estuary includes salt marsh habitats, fresh marshes, intertidal flats, and open-water tidal creeks and bays. Marine species diversity is low, with half the number of fish species found in areas south of Cape Cod. The Plum Island Sound estuary supports productive commercial and recreational fisheries for soft-shell clam

FIGURE 1.17 Tidal creek and salt marsh habitats are typical of the landscape at the Plum Island Ecosystem Long-Term Ecological Research site. (Photo courtesy of David Johnson.)

and striped bass. Land-use composition in the watershed during 2001 was approximately 46% forest, 34% urban or suburban, 10% agriculture, and 10% wetland or water.

Santa Barbara Coastal (SBC)

The SBC site (http://sbc.lternet.edu/) became part of the LTER program in 2000 with a mission to understand the linkages among ecosystems at the land–ocean margin, particularly as demonstrated by giant kelp forests (Figure 1.18). The principal study site is the semiarid Santa Barbara coastal region, which includes steep watersheds, small estuaries, sandy beaches, the neritic and pelagic waters of the Santa Barbara Channel (5,850 km^2), and the habitats encompassed within it. SBC is characterized by a Mediterranean climate, with mild, moist winters and moderately warm, generally rainless, summers. Winter rainstorms provide the majority of freshwater input to rivers, streams, and the nearshore marine environment. One of the more notable habitats is shallow rocky reefs dominated by giant kelp (*Macrocystis pyrifera*) forests, which are among the most productive ecosystems in the world. The amount of nutrients and organic matter delivered to these forests from the surrounding ocean and adjacent land varies in response to short- and long-term changes in climate, ocean conditions, and human use. Variation in the supply of nutrients and organic matter interacts with physical disturbance and biological interactions to influence the abundance and species composition of the forest inhabitants and the ecological services that they provide.

Sevilleta (SEV)

The SEV site (http://sev.lternet.edu/), founded in 1988, studies the effects of variability, extremes, and directional change in climate in arid land ecosystems (Figure 1.19).

FIGURE 1.18 Giant kelp forest and fish compose typical seascapes of the Santa Barbara Coastal Long-Term Ecological Research site. (Photo courtesy of Ron McPeak and University of Santa Barbara Digital Collections [item 16822].)

Research is focused at the Sevilleta National Wildlife Refuge, an ungrazed refuge of about 300 km^2 located 80 km south of Albuquerque (New Mexico), where the transitions among shortgrass steppe, Chihuahuan Desert grassland and shrubland, juniper savanna, and pinon–juniper woodland facilitate the study of controls on community and ecosystem structure and function.

Precipitation is a dominant driver of ecosystem processes, with mean annual precipitation of about 250 mm, 60% of which occurs during the North American Monsoon (July–September). Multiyear droughts recur every 50 to 70 years. High light and low relative humidity favor soil drying, and biotic activity occurs in bursts following precipitation

FIGURE 1.19 The Blue Grama Meteorological Station (foreground) on the Sevilleta Long-Term Ecological Research site, looking east toward the Lost Pinos Mountain Front. (Photo courtesy of Doug Moore, the Sevilleta Photo Archives, and the National Science Foundation.)

events. Studies of pulse dynamics have addressed the effects on soils, nutrient cycling, vegetation structure, and species interactions among habitats and across biome transition zones, and have emphasized long-term manipulative experiments.

Shortgrass Steppe (SGS)

This site (http://www.sgslter.colostate.edu/) was established on a foundation of research data and infrastructure produced by the US Department of Agriculture (USDA) in 1939 and the International Biological Program (1968–1974), and was a member of the LTER network from 1982-2014. Research at the SGS site sought to understand how climate, natural disturbance, physiography, and human activities in the shortgrass steppe drive structure and processes within the ecosystem (Figure 1.20). Monitoring campaigns, manipulative experiments, and models were used to study population and community dynamics of plants and animals, conceptualize and measure nutrient cycling, and identify important drivers in the system. Core field studies were conducted on the USDA–ARS Central Plains Experimental Range in Nunn, Colorado and expanded to sites within the Pawnee National Grassland of the US Forest Service and Great Plains region.

SGS (1538 ml) encompasses 63 km² of low hills, broad valleys, and ephemeral streams. The climate is semiarid with a mean annual temperature of 8°C and 340 mm of precipitation per year. Approximately 70% of the precipitation falls during the growing season (April–September), with high variability in amount and location. The site is dominated by short grasses (64%), succulents (21%), and dwarf shrubs (8%). Blue grama *(Bouteloua gracilis)* predominates and contributes 60% to 80% of plant cover and net primary productivity. The disturbance regime includes a long history of grazing by large herbivores, periodic drought, and infrequent fires. Over time, intense selection by grazing and drought has created an ecosystem with low-standing vegetation and biological activity and organic matter concentrated below-ground. Currently, grazing by domestic livestock is the primary

FIGURE 1.20 Panoramic view of the landscape from the top of Owl's Peak in the Central Plains Experimental Range (the Front Range of the Rocky Mountains appears in the background) that represents typical habitat in the Shortgrass Steppe Long-Term Ecological Research site. (Photo courtesy of John Moore.)

use of native grassland, which occupies about 60% of the shortgrass steppe, but row-crop agriculture, energy development, and urbanization continue to increase as land uses.

Virginia Coast Reserve (VCR)

The VCR (http://www.vcrlter.virginia.edu/) became a part of the LTER program in 1987 with a mission to understand how long-term changes in climate (storms, temperature), sea level, and land use affect the dynamics and biotic structure of coastal barrier systems and the services they provide to humans. VCR extends over 110 km on the eastern shore of Virginia from the Maryland border to the mouth of the Chesapeake Bay, and is characteristic of coastal barrier ecosystems along much of the Atlantic and Gulf Coasts (Figure 1.21).

The coastal barrier landscape is heterogeneous and extremely dynamic, with barrier islands, shallow lagoons with extensive tidal flats, tidal marshes, and mainland watersheds. Transitions between ecosystems can be abrupt with threshold responses to external drivers. Some ecosystems show inherent bistability, which means that they have two alternative stable states. The islands of VCR are among the most dynamic in the United States; lateral accretion and erosion rates are as high as 40 m per year—the highest along the mid-Atlantic seaboard. The shallow seaward slope of the landscape (< 0.1%) makes it a particularly sensitive location for studying responses of intertidal marshes to sea-level rise, which, at 4 mm per year, is the highest along the Atlantic Coast. The maritime temperate climate has temperatures ranging from 1° to 31°C, and receives circa 1100 mm of precipitation that is distributed fairly evenly across the year. Long-term storm records indicate that storminess has increased since the late 1880s, with about 15 extratropical storms per year hitting the VCR. At the turn of the twentieth century, the barrier island and lagoon system supported one of the most prosperous farming and fishing communities in the country. Towns on the islands were abandoned after the Great Storm of 1933, and the scallop fishery collapsed with the loss of seagrass around the same time.

FIGURE 1.21 Aerial photograph of the coastal barrier landscape that represents the ecosystems of the Virginia Coast Long-Term Ecological Research site. (Photo courtesy of John Porter.)

Restoration during the last decade has resulted in more than 1800 ha of healthy seagrass meadows, which once again support scallops.

NETWORK DYNAMICS

The LTER network was relatively stable prior to 2014. At that time, it had been 23 years since a site was lost from the network, 10 years since a site had been added to the network, and 17 years since the Network Office had moved from one hosting institution to another. As a consequence of recent events, the site composition of the LTER network has become quite dynamic, as has the structure by which it is organized to facilitate cross-site comparison, integration, and synthesis. A number of events illustrate this dynamism. At the end of 2014, the Shortgrass Steppe LTER site was no longer funded by NSF. Similarly, the Sevilleta LTER site learned in 2015 that it would not be renewed for funding by NSF. In response to the reduced size of the network and opportunities for expansion in marine systems, NSF recently (2016) issued a call for proposals to establish three new sites, two with a focus on ocean or coastal ocean ecosystems, and one with a focus on arid or semi-arid ecosystems. In addition, NSF did not renew the LTER Network Office at the University of New Mexico (funding expired in 2015). Rather, NSF issued a new solicitation for an LTER National Communications Office, charged with (1) coordinating research, education, and outreach activities across the network; (2) facilitating network governance; (3) fostering communication among LTER sites and with the broader scientific community; (4) promoting the dissemination of information and resources among LTER sites and to additional stakeholder communities; (5) organizing meetings and workshops; and (6) advancing the LTER program both nationally and internationally. The National Center for Ecological Analysis and Synthesis at the University of California Santa Barbara, was selected in 2015 by NSF to lead the network in these endeavors.

ACKNOWLEDGMENTS

We sincerely thank J. N. Caira and B. T. Klingbeil for producing Figures 1.1 and 1.2, respectively. We are also grateful to S. J. Presley, A. R. Sjodin, and B. T. Klingbeil for helpful comments on earlier versions of the manuscript. The principal investigators (see websites listed in site descriptions) of the various LTER sites that are included in this book provided commentary that improved site descriptions and assisted in securing representative photos of the sites. Finally, we gratefully acknowledge support from the NSF LTER program (grants DEB-0620910 and DEB-1239764). In addition, M. R. Willig was supported by NSF grant DEB-1354040 to the University of Connecticut, and L. R. Walker was supported by a sabbatical from the University of Nevada Las Vegas.

REFERENCES

Bowman, W.D., and Seastedt, T.R., eds. (2001). *Structure and Function of an Alpine Ecosystem.* Oxford University Press, New York.

Brokaw, N., Crowl, T.A., Lugo, A.E., McDowell, W.H., Scatena, F.N., Waide, R.B., and Willig, M.R., eds. (2012). *A Caribbean Forest Tapestry: The Multidimensional Nature of Disturbance and Response.* Oxford University Press, New York.

Chapin, F.S., III, Oswood, M.W., van Cleve, K., Viereck, L.A., and Verbyla, D.L., eds. (2006). *Alaska's Changing Boreal Forest.* Oxford University Press, New York.

Foster, D., and Aber, J.D., eds. (2006). *Forests in Time.* Yale University Press, New Haven, Connecticut.

Greenland, D., Goodin, D.G., and Smith, R.C., eds. (2003). *Climate Variability and Ecosystem Response in Long-Term Ecological Research.* Oxford University Press, New York.

Havstad, K.M., Huenneke, L.F., and Schlesinger, W.H., eds. (2006). *Structure and Function of a Chihuahuan Desert Ecosystem.* Oxford University Press, New York.

Knapp, A.K., Briggs, J.M., Hartnett, D.C., and Collins, S.C., eds. (1998). *Grassland Dynamics.* Oxford University Press, New York.

Lauenroth, W.K., and Burke, I.C. (2008). *Ecology of the Shortgrass Steppe.* Oxford University Press, New York.

Magnuson, J.J., Kratz, T.K., and Benson, B.J., eds. (2005). *Long-Term Dynamics of Lakes in the Landscape.* Oxford University Press, New York.

Peters, D.P.C., Laney, C.M., Lugo, A.E., Collins, S.L., Driscoll, C.T., Groffman, P.M., Grove, J.M., Knapp, A.K., Kratz, T.K., Ohman, M.D., Waide, R.B., and Yao, J., eds. (2013). *Long-Term Trends in Ecological Systems: A Basis for Understanding Responses to Global Change.* U.S. Department of Agriculture, Agricultural Research Service, Technical Bulletin Number 1931.

Redman, C., and D.R. Foster (2008). *Agrarian Landscapes in Transition.* Oxford University Press, New York.

Schachak, M., Gosz, J.R., Pickett, S.T.A., and Perevolotsky, A., eds. (2005). *Biodiversity in Drylands.* Oxford University Press, New York.

Udvardy, M.D.F. (1975). *A Classification of the Biogeographical Provinces of the World.* International Union for Conservation of Nature (IUCN), Morges, Switzerland.

2 Sustaining Long-Term Research

Collaboration, Multidisciplinarity and Synthesis in the Long-Term Ecological Research Program

Robert B. Waide

IN A NUTSHELL

1. The Long-Term Ecological Research (LTER) program was created by the National Science Foundation (NSF) to encourage comparative study of ecological phenomena that occur over decades and centuries.
2. The vision, mission, and goals of the LTER network have evolved to address current societal needs: to understand the dynamics of key ecosystems, to interpret effects on ecosystem services of importance to humans, and to forecast the effects of future ecosystem scenarios.
3. Challenges inherent in sustaining a long-term research program include building effective plans for research, governance, and transitions among generations of scientists.
4. The LTER program has been extremely successful at meeting its original goals, but increased expectations for cross-site and network-level synthesis are not yet fully realized.
5. The initial absence of a shared conceptual model inhibited progress toward large-scale synthesis. Recent agreement on a conceptual framework that includes multiple disciplines has begun to address this issue.

INTRODUCTION

On February 12, 1980, I began my first real job with the Center for Energy and Environment Research (CEER) at the University of Puerto Rico. I was hired to work on an ongoing program funded by the Department of Energy that had begun 17 years earlier under the direction of Howard Odum. Along with my colleagues, Laurence Tilly and  Douglas Reagan of CEER, and Ariel Lugo of the US Forest Service, I was also to participate in a proposal that had just been submitted for a new NSF initiative called the Long-Term Ecological Research (LTER) Program. None of us knew very much about this program, and the request for proposals described it in five short paragraphs. Not only was it a new program, it was very different from other NSF programs. Proposals needed to involve groups of investigators working on core research topics, and the proposed work should be coordinated in some way with a network of other sites. Principal investigators were warned to be prepared to make long-term time commitments. Little did we know.

Our first proposal was unsuccessful, as was the second. By 1986, I had become head of the Terrestrial Ecology Division at CEER. In that capacity, I represented the University of Puerto Rico as co–principal investigator, along with Ariel Lugo, on our third attempt at becoming an LTER site. Once again, we were unsuccessful, but this time we were encouraged by Tom Callahan, the LTER program officer at NSF, to submit again the following year. Finally, in 1988, we were chosen as the tropical forest site for the LTER network. I continued as co–principal investigator of the Luquillo LTER project for 9 years before I moved to New Mexico in 1997 to become executive director of the LTER network office, a position I held until 2016. Essentially, my entire academic career has focused on the LTER program in one way or another.

Herein, I introduce and provide a historical perspective on the mission, goals, and nature of the LTER program and the LTER network, giving a critical assessment of the extent to which the mission and goals have been accomplished. I will also assess how the LTER program needs to develop further to fully attain its goals. In the process, I will comment on how the LTER program has created a social network based on a culture of collaboration, multidisciplinarity, and mutual trust. In keeping with the nature of this book, I am approaching this charge from a personal perspective. The history I relate will be short on dates and names. For those interested in such details, several publications describe the history of the LTER program as it has evolved (Callahan 1984; Callahan 1991; Hobbie et al. 2003; Magnuson et al. 2006; Gosz, Waide, and Magnuson 2010; Waide and Thomas 2012; Gholz, Marinelli, and Taylor, Chapter 3).

THE LTER PROGRAM

The NSF created the LTER program in 1980 to provide an option for ecologists to use in understanding the long-term dynamics of ecosystem processes. Before the existence of the LTER program, efforts to study key, long-term ecological processes depended on ecologists' abilities to leverage funding from short-term programs or support from mission-oriented agencies or private sources to generate long-term data. The creation of the LTER program aimed at explicitly addressing certain types of ecological processes. The NSF understood that the existence of long-term data would allow the development of generalized theories to address what were envisioned to be critical issues of the day: disturbances including fire, grazing, deforestation, CO_2 increases, and acid precipitation (Callahan 1984).

The rationale for the LTER program was summarized by Dick Marzolf, the first chair of the LTER coordinating committee, as a suite of six premises (Marzolf 1982):

1. There are ecological phenomena that occur on time scales of decades or centuries— periods of time that are not normally investigated with research support from NSF.
2. Many ecological experiments are performed without sufficient knowledge of the year-to-year variability in the system.
3. Long-term trends in natural ecosystems are not being systematically monitored.
4. A coordinated network of sites is not available so that comparative experiments can be facilitated.
5. Examples of natural ecosystems where research was occurring are being converted to other uses incompatible with that research.
6. Advances in ecological research often oversimplify, ignore, or treat as constant or insignificant phenomena at higher or lower levels of organization.

These six premises encapsulate the most important scientific reasons for long-term research identified during the planning for the LTER program that took place between 1976 and 1979. However, as Callahan (1991) pointed out, the societal context for long-term studies is diverse and dynamic, and thus it mandates an adaptive approach to link emerging research opportunities with societal interests. As Callahan predicted, the rationale for the LTER program evolved as the interests of society changed, although the underlying scientific approach remained much the same. Recent trends include an increased emphasis on directional trends in factors driving ecosystem processes, changes in disturbance frequency and intensity, changes in biodiversity, ecosystem services, international and multidisciplinary studies, education, and data sharing. A recent description of the LTER program reflects these changes: long-term data "provide a context to evaluate the pace of ecological change, to interpret its effects, and to forecast the range of future biological responses to change" (Gosz et al. 2010).

ORIGINS OF THE LTER PROGRAM

The design of the LTER program incorporated a broad range of community input acquired through three workshops sponsored by the NSF. These workshops engaged a diverse group of ecosystem scientists and drew on the experience of existing long-term studies in the United States (e.g., Hubbard Brook, Lake Washington, Experimental Ecology Reserves) and abroad (e.g., Windermere, Rothamsted, International Biological Program). Participants in these workshops outlined an ambitious research program that included long temporal and broad spatial scales, as well as individual and group investigations. In the initial call for proposals for the first cohort of LTER sites, NSF focused on a more limited scope of activities that emphasized collection of data at a network of sites representing major biotic regions of North America and a pilot approach to identifying scientific, technical, and managerial problems that might arise in the conduct of long-term comparative research. This cautious approach to implementing the recommendations of the workshops reflected a sense that this new program was an experiment, whose results were unpredictable.

The initial expectations for the LTER program were modest. In describing these expectations, Callahan (1984) felt compelled to add a caveat: "There is always risk in forecasting future directions and expected benefits from basic research. Perhaps the greatest risk is that, by formalizing expectations, one may constrain the projects to fit them." Despite this concern, Callahan went on to describe the minimum expected outputs from long-term studies: inventories of the physical and biological states of the chosen research sites, documentation of the rates of fundamental ecological processes, description of the effects of disturbance on key variables, high-quality data sets that can be used to compare sites, and addition of investigators attracted by the accumulated data base. In more general terms, Callahan believed that if the program were successful: "a broad and stable platform will have been achieved on which planning for future ecological research can be based." He further hoped that success "will lead to new and improved ecological theories that can parsimoniously encompass far larger scales of space and time than before" (Callahan 1984).

IMPLEMENTING THE LTER PROGRAM

Because the LTER program represented a different approach to ecological science, the first cohorts of sites had to improvise solutions to some of the challenges of long-term research. Coordination among scientists required each site to agree on a motivating conceptual model (Pickett 1991). This remains a significant challenge, even today.

Because common conceptual models required data sharing, sites were obliged to develop mechanisms to make data accessible and a set of common protocols to ensure that data were available in a timely manner. Over time, open access to data has become a policy of the LTER network, and sites and the network office have invested substantial effort to make data from the LTER program more accessible. Effective sharing of data required adoption of data standards, coordination of approaches, recruitment of experts in information management, and shifts in the attitudes of scientists, who were accustomed to personally controlling access to data. The development of long-term measurement programs and experiments required significant planning, as well as strong, consistent management that focused on cooperation within and among sites.

KEEPING AN LTER PROJECT RUNNING FOR 100 YEARS

The greatest challenge in managing an LTER project is to maintain focus on long-term objectives while addressing new opportunities revealed by increasing knowledge of the study ecosystem. Understanding long-term ecological processes involves repeated measurements taken over decades; this requires discipline, constant supervision, close collaboration, and careful budgeting of time and resources. Even modest rates of inflation can limit the number of long-term measurements or experiments that a site can support. This problem is exacerbated by the subtle expectation that sites should include new ideas in every renewal proposal. The first decadal review of the LTER program (Risser et al. 1993) highlighted this problem:

> The expectations of each site increase incrementally as the accumulation of required research increases, as more demands are made on the site from other investigators who want to use the data or work at the site, and as the results of the LTER program become more widely known and receive increased attention and demand. Over the duration of the program, funding has increased, but not at the rate of the increasing demands on the science in the LTER program.

The problems of increasing expectations defined in the review have continued until the present. The budgets of LTER sites have roughly doubled since 1993, but the recent decline in federal support for research makes it clear that this trend may not continue much longer. This presents a dilemma that has no easy solution.

As understanding increases, new questions multiply. These new questions often require approaches that are more sophisticated or expertise from disciplines that are not supported by the LTER program. Multidisciplinary approaches have become commonplace over time, and most LTER sites would consider inclusion of multiple disciplines essential to understand ecosystem dynamics. For example, many LTER sites have found that microbial ecology and the social sciences provide interesting avenues for expansion of core research. In particular, the decadal plan for LTER recognizes that the involvement of social scientists in the LTER program is pivotal because so many of the focal ecosystems are strongly affected by societal policies and practices. In general, the LTER network considers involvement of other disciplines as an important opportunity to increase understanding of the complete range of factors that lead to ecosystem change.

The scarcity of long-term funding opportunities in many disciplines creates a barrier to multidisciplinary research. Most LTER sites have addressed multidisciplinary opportunities by using resources from NSF awards for core measurements and experiments that provide a baseline for associated projects funded through other sources or conducted by other scientists. Hobbie (2003) compared the LTER program to a fleet of research vessels that attracts diverse research projects, provides economies of scale, and stimulates collaborative science. On average, LTER sites host projects that bring in 2.8 times the support provided directly by the LTER program. Diversity in funding increases the importance of strong leadership to coordinate projects and participants funded by multiple sources.

The longest series of observations and experimental studies requires multiple generations of scientists, and thus science in the LTER program requires a level of altruism that is not necessary in short-term studies. Scientists initiating long-term studies may no longer be active when the final and potentially most interesting results become available. Planning for the transition between generations of scientists is an important challenge for LTER sites. Many sites address this problem by recruiting replacements from within the project. Likewise, changes in leadership are inevitable in long-term studies. Of the original 11 sites, only one (Cedar Creek) has had the same leader for the duration of the project, and new leadership will take over this site in 2016. Six other sites still remain active from the first two cohorts, and these sites have had three to six different leaders in the 30 years that they have existed.

A key challenge in sustaining a long-term research program is the difficulty in maintaining enthusiasm for long-term goals through changes in leadership, turnover in research teams, probationary periods, changing interests, and unpredictable interruptions in the research program. Initially, each LTER site had to develop its own approaches to these problems, but over time, successful solutions (e.g., adoption of formal plans for transitions in leadership, learning from the experience of sites that have been on probation) spread through the research community in the LTER network. Well-planned transitions in leadership are now the norm in the LTER program, and successful transitions minimize disruptions to research continuity. Even when unexpected changes in leadership occur, recruitment of new leaders from existing sites can maintain stability. The fear of an unsuccessful proposal and subsequent probation is reduced by support from other sites that have successfully addressed probation. Communication among sites, particularly through face-to-face meetings, rekindles enthusiasm by transmitting exciting new research approaches through the network. Inclusive leadership and an open decision-making process are important in making researchers feel valued. Positive feedback from colleagues and especially from funding agencies is important in sustaining enthusiasm. Most important, scientists in the LTER program have built a strong and supportive social network that facilitates sharing ideas about research, education, data, and project management.

MAJOR ACCOMPLISHMENTS

The most significant results from long-term research require decades or longer. Likens (1983) suggested that 20 or more years of baseline data may be required before meaningful hypotheses about long-term ecosystem behavior can even be formulated. Bearing this stricture in mind, it may be premature to draw conclusions about the success of LTER sites chosen in the past two decades. However, considerable evidence to assess progress exists for most sites.

Each of the current LTER sites has exceeded initial expectations of the program (Hobbie 2003). Sites have assiduously collected long-term data describing states of study systems, measured fundamental ecosystem processes, described the effects of the most important kinds of disturbance, produced high-quality data, and attracted many new investigators. Evidence for this stems from the fact that of the 29 sites chosen for the LTER network, only 5 have failed to be renewed, and only 1 site has become inactive in the past 20 years. Twenty-five sites have consistently demonstrated to reviewers, panels, and program directors that they have met the standards set by the NSF. Of the sites that have been placed on probation, all but two have subsequently demonstrated sufficient progress and scientific rigor to be renewed.

Scientists in the LTER program produced 16,192 peer-reviewed publications between 1982 and 2012. Authors produced synthesis books based on their research at 11 LTER sites. Six other books comparing methods or results across LTER sites are part of the LTER Oxford University Press book series, and LTER scientists have written 220 other topical books based on research supported by the LTER program. More than 1,600 graduate students have received advanced degrees for their research at LTER sites. The LTER network has become a recognized source of long-term data, producing 19,567 data sets distributed through the LTER Data Portal as well as through DataOne and the Oak Ridge National Laboratory Distributed Active Archive Center. LTER education programs offer opportunities for thousands of K–12 students to engage with LTER sites and scientists directly through the Schoolyard LTER Program and via the Internet through the LTER Education Digital Library. For example, the Schoolyard LTER Program reached 1,355 K–12 teachers and 40,918 students during the 2009–2010 school year (not counting those reached via social media (a single Google+ Hangout from Palmer LTER reached more than 10,000 individuals). LTER sites have identified more than 122 key findings from NSF-supported activities; for example, two of NSF's "Nifty Fifty Discoveries" came from research at the Hubbard Brook and Sevilleta LTER sites.

The North Temperate Lakes site, a member of the first cohort of sites, measured change in a variety of metrics over the course of their project (Magnuson et al. 2006). The rate of publication increased over time, from 7 to 31 papers per year, as did the percentage of multiauthored papers and the percentage of authors from outside the University of Wisconsin. Over time, the number of lakes studied and the number of years of data used per publication increased substantially. Eighty-one graduate students received degrees in the first 20 years of the project, and the attitudes of graduate students became more positive for many of the characteristics of research supported by the LTER program (e.g., regional studies, analysis of long-term data, time series studies, multiple site studies, whole-system experiments, and a variety of other factors).

As these numbers indicate, the prominence of the LTER program is primarily the result of the accomplishments of its constituent sites. The initial call for proposals indicated that sites chosen for the program would represent major biotic regions of North America, without providing any overarching conceptual approach to guide site selection. The process of choosing sites competitively from a national pool made sure that proposals from strong groups of researchers would be successful. Because each LTER project creates its own site-specific focus, research teams can take advantage of diverse scientific skills and unique site characteristics (Hobbie 2003). Criteria for success in the original call emphasized leadership, institutional support, physical facilities, site integrity, baseline information, bibliographic and reference collections, data storage, and publication record, but they did not include geographic distribution or biotic representativeness. Priority was given to sites identified in Experimental Ecological Reserves: A Proposed National Network, but only 10 of the 29 LTER sites ever selected for the program are mentioned in that document. It was not until the third competition in 1986, after 11 sites had been selected, that some biotic regions were given priority in the call for proposals. Beginning in 1990, competitions were more focused on biotic regions (e.g., Antarctic, urban, coastal sites), but these priorities were guided as much by fiscal as by conceptual considerations. Given the manner by which sites have been selected and deselected, it is hardly surprising that site science continues to dominate the LTER program after 33 years of operation.

The LTER Network

The rationale for developing a network of long-term research sites, as opposed to a collection of sites with unrelated goals, appears in the reports from the working groups that helped to define the LTER program. The fundamental reason for comparative studies across a network of sites is to evaluate the degree of generality of results so as to better understand the principles underlying the behavior of ecosystems (Pickett 1991). The first call for LTER proposals lists the "collection of comparative data at a network of sites" as one of the primary goals of the program. The call for proposals further indicated that sites should conduct coordinated studies on a series of five core research topics. These topics are known now as the LTER core areas and include:

1. Pattern and control of primary production
2. Dynamics of populations of organisms selected to represent trophic structure
3. Pattern and control of organic matter accumulation in surface layers and sediments
4. Patterns of inorganic inputs and movements of nutrients through soils, groundwater and surface waters
5. Patterns and frequency of disturbances

Thus a clear expectation existed that sites in the LTER program would compare results.

Despite the clear language on comparative studies across sites, the concept of the LTER program as a network of interrelated projects took some principal investigators by surprise. Leaders of the first cohort of LTER sites were invited to meet at NSF in October 1980 and were reminded that close cooperation among sites was a fundamental part of the LTER program (Magnuson et al. 2006). Subsequently, leaders of the six existing sites met in December 1980 at Konza Prairie to prepare a proposal to NSF to develop the comparative methods that all sites would need to adopt. Initially, governance of the network was by a committee of the principal investigators, the Coordinating Committee, led by the first elected chair, Dick Marzolf of the Konza Prairie site. To facilitate communication across sites, NSF provided funds for a coordinating office that was the prototype for the current LTER network office.

In 1981, the first LTER scientific meeting was held at the Jornada site in Las Cruces, New Mexico, with enthusiastic participation by scientists from the initial five sites. However, the ultimatum received from NSF to operate as a network created tension (Magnuson et al. 2006), especially when some sites came prepared with informal proposals to support intersite work. The tension intensified when five new sites were added to the network in 1982. NSF felt compelled to revisit the issue at the 1986 Coordinating Committee meeting. John Brooks, director of the Division of Environmental Biology (DEB) at NSF, pointedly remarked that new opportunities for the LTER program lay in network science and that henceforth network science would constitute part of the evaluation of sites (Magnuson et al. 2006). Brooks further remarked that the program would be assessed in 1990 to see if unique results were being obtained. He tempered his comments by committing $60,000 a year to the Coordinating Committee to foster network research (Magnuson et al. 2006).

Early Development of the Network

The Coordinating Committee identified a range of comparative approaches to employ in cross-site studies. The development of common standards for the collection of data facilitated comparative studies of ecosystem structure and process. However, as the number of sites and the diversity of ecosystems increased, common standards became harder to define and surrogate measures or dimensionless metrics were used in some comparisons (Magnuson et al. 1991). Comparisons across LTER sites located along environmental gradients were possible for grassland and coastal wetland sites but were more difficult across other biomes. Both LTER and non-LTER sites were included in large-scale experiments.

Despite some notable successes, the number of comparative studies across sites increased slowly. Risser et al. noted this situation (1993):

> The power of the network approach of the LTER program rests in the ability to compare similar processes (e.g., primary production or decomposition of organic matter) under different ecological conditions. As a result, scientists in the LTER program should be able to understand how fundamental ecological processes operate at different rates and in different ways under different environmental conditions. Some intersite comparisons have been conducted, but the power of the network of coordinated research sites has not yet been fully realized.

By 2003, analysis of jointly authored publications indicated that the sites in the LTER program had become a robust, fully connected network of research sites (Johnson et al. 2010). However, the vision of a fully integrated, question-driven network able to address critical long-term research issues to enable defensible environmental decisions and policies had yet to be achieved.

Growth and Maturation of the LTER Network

The LTER network has increased in size and diversity since 1981, while it also has matured as an organization. The network now numbers 26 sites and counts more than 3,400 past and present scientists and students among its participants, an increase of 630% since 1994. The annual budget approaches a million dollars per site (Waide and Thomas 2012). Scientists in the LTER program represent an increasing number of disciplines beyond the field of ecology, and the LTER network collaborates with long-term programs in more than 40 countries. Partnerships with federal agencies such as the US Forest Service, the US Agricultural Research Service, and the US National Park Service, as well as nongovernmental entities such as The Nature Conservancy, are fundamental to the vitality of research at LTER sites. Collaborations with new research efforts in the National Ecological Observatory Network (NEON), the Long-Term Agroecosystem Research Network, the Critical Zone Observatories, and the National Phenological Network amplify the impact of research supported by the LTER program and extend its reach to new communities of academic and citizen scientists. These and other accomplishments suggest that the LTER network will continue to be an important national asset for many years to come.

There was a clear expectation that LTER sites would conduct comparative studies to assess the generality of results and thus to better understand the principles underlying the behavior of ecosystems. The LTER network has invested considerable thought and effort into the process of stimulating collaboration, and empirical analyses show that collaborations between scientists at different LTER sites, measured by joint publications, have resulted in a highly connected research network (Johnson et al. 2010). NSF has provided funds for All Scientists Meetings, competitions for cross-site awards, and support for working groups through the LTER network office, all of which have increased the pace of collaborative science within the LTER program and with non-LTER scientists. In 2013 alone, 18 working groups conducted cross-site comparisons with support from the LTER network office. Many more collaborative projects were carried-out with support from other NSF programs or agencies. Development of a robust and coordinated model for information management, the LTER Network Information System, has made access to data for comparative studies simpler. Improved tools for communication among scientists across the LTER network reduce the cost of discovering possible collaborators and initiating collaborations. Through these efforts, the LTER network has incorporated cross-site science as a core element in its research agenda.

Planning for the Future

As the LTER program has matured, it has transformed from a loose federation of sites with similar purpose to a well-integrated network of sites with common goals. During the last decade, the LTER network has engaged in a continual process of planning for the future. In "Integrative Science for Society and the Environment: A Plan for Science, Education, and Cyberinfrastructure in the U.S. Long-Term Ecological Research Network," the LTER network provided a strategic framework for the key elements of the LTER program. This framework provides new vision and mission statements, as well as a revised set of goals for the LTER network. The vision of the LTER network includes a society in which exemplary science contributes to the advancement of the health, productivity, and welfare of the global environment that in turn, advances the health, prosperity, welfare, and security of our nation. The mission of the LTER network is to provide the scientific community, policy-makers, and society with the knowledge and predictive understanding necessary to conserve, protect, and manage the nation's ecosystems, their biodiversity, and the services they provide.

The implementation of the mission of the LTER network (U.S. Long Term Ecological Research Network 2007) involves six interrelated focal areas:

1. *Understanding*: to understand a diverse array of ecosystems at multiple spatial and temporal scales
2. *Synthesis*: to create general knowledge through long-term interdisciplinary research, synthesis of information, and development of theory
3. *Outreach*: to reach out to the broader scientific community, natural resource managers, policy-makers, and the general public by providing decision-support information, recommendations, and the knowledge and capability to address complex environmental challenges
4. *Education*: to promote training, teaching, and learning about long-term ecological research and the Earth's ecosystems and to educate a new generation of scientists
5. *Information*: to inform participants in the LTER program, as well as in the broader scientific community by creating well-designed and well-documented databases
6. *Legacies*: to create a legacy of well-designed and documented long-term observations, experiments, and archives of samples and specimens for future generations

The Integrated Science for Society and the Environment document presents a synthetic research framework for the LTER network (Collins et al. 2011). This framework combines the dynamics of social systems with those of ecological systems in a conceptual model that anticipates the major drivers of linked socioecological systems over the next century. The LTER network concludes that the long-term study of socioecological systems is of critical importance for the future management of ecosystems and the preservation of the services they provide. The impact of human social systems will continue to increase, and even those ecosystems that at present show little obvious effect of humans will eventually have to include social systems as part of their research agenda.

CHALLENGES TO FULFILLING THE POTENTIAL OF THE LTER NETWORK

Since the inception of the LTER program, tension has existed between independently developed, site-focused science and collaborative research addressing broader questions (Magnuson et al. 2006). This tension arises in part from the original definition of the LTER program by NSF that provided significant flexibility for investigators submitting proposals to the program. As Risser et al. (1993) point out: "This intentional flexibility has led to some ambiguity in the relative importance of site-specific long-term studies, cross-site comparisons, and furthering ecosystem theory through in-depth studies." Limited resources force sites to prioritize between continuing long-term site science described in proposals and developing new cross-site comparisons. NSF began to address this problem in 1988 by establishing an LTER research coordinator position in DEB to facilitate cross-site research. In response to recommendations of the first decadal review of the LTER program, NSF conducted special competitions in 1994 and 1995 for cross-site comparisons, synthesis at LTER and non-LTER sites, and international research. However, this brief effort fell victim to budget constraints. The second decadal review of the LTER program addressed the issue of support for cross-site research and reported that: "the level and nature of this support are perceived by the LTER community as not being sufficient, consistent, or focused enough to meet expectations" (Harris and Krishtalka 2002). During the last decade, LTER cross-site and network-level studies have been supported by other programs at NSF, other agencies, or creative, bottom-up collaborations (e.g., the Nutrient Network). However, no mechanism yet exists for funding long-term, network-level studies or experiments beyond the core budgets of LTER sites.

Several false starts have led to wariness among investigators in the LTER program regarding the commitment of NSF to cross-site research. The recommendations of the first decadal review raised expectations, but few of these recommendations were implemented by the network or received support from NSF. Competitions for cross-site science were never institutionalized as part of the LTER program. The popular triennial All Scientists Meetings, an important tool in planning for network-level science, were curtailed from 1994 to 2000. A program to augment sites to expand the disciplinary, historical, and geographic focus of the LTER program was eliminated after two sites were funded. Those two sites (North Temperate Lakes and Coweeta) made significant progress in expanding socioeconomic aspects of their research as well as expanding the geographic breadth of their work (Magnuson et al. 2006). The second decadal review urged the development of a strategic plan to better define the future goals of the LTER network, but the extensive investment in time and resources to develop that plan have not yet borne fruit. More recently, annual supplements for education, information management, and new technology have been constrained by budgetary uncertainty. Finally, NSF has recently signaled the need for a more strategic approach to cross-site synthesis by LTER sites.

Beyond the issue of funding, the ability of the LTER network to address grand challenge questions was limited from the outset by the absence of a shared conceptual model. The importance of such motivating conceptual models for long-term studies was well known during the formative years of LTER program. In a discussion on best practices for developing long-term studies, Pickett (1991) noted: "In considering the establishment of a network of LTS [long-term sites], it would seem even more critical to have a conceptual model." Although adherence to the five core areas facilitated comparisons, the absence of a focus for those comparisons limited the kinds of questions that could be addressed. Although long-term studies of penguins in Antarctica and warblers in eastern deciduous forest have produced important insights into the biology of those organisms and the ecosystems that they occupy, there has been little effort to find an intersection between those studies that contributes to a long-term understanding of the dynamics of trophic structure. A common conceptual model could help sites choose research approaches that would more easily lend themselves to subsequent comparison.

The absence of an overarching conceptual model had additional effects on the capability of the LTER network to address regional- or continental-scale questions. The piecemeal development of the network did not allow for strategic selection of sites to fill in gaps along environmental gradients or in underrepresented biomes. Although many non-LTER research sites that could plug these gaps exist, few have the resources to conduct long-term studies. Without a conceptual model, there can be no conclusion about even the number of sites needed for an effective network. For example, the existence of two forested LTER sites in the Northeast has generated a better understanding of regional ecosystem dynamics, and the existence of three desert sites has done the same in the Southwest. Sites with no nearby counterpart encounter more difficulty in scaling up their results to larger geographic areas.

The early expectations for comparative studies aimed at assessing the generality of results have broadened over time to include the expectation that the LTER network should also confront "the challenges of designing and operating a sustainable biosphere" (Risser et al. 1993) by conducting policy-relevant research at regional and national scales. This recommendation from the first decadal review also involved doubling the number of sites, expanding the levels of biological organization under study, adding the disciplines of physical and social science, developing new technologies, contributing to the nation's educational effort, and increasing the LTER program's budget by an order of magnitude. Although the funding recommendation was quickly dismissed, elements of the other recommendations persisted as reasonable expectations of the LTER network. Thus the second decadal review recommended that the LTER network should "become a research collaboratory, namely, a seamless, integrated continuum from site-specific to cross-site to network- and systems-level ecological research" (Harris and Krishtalka 2002). These are exciting images, but they cannot be achieved given the level of resources currently available to the LTER network.

During the first 20 years of its existence, the LTER network successfully met the initial goals set out by the NSF. Many indicators confirm that site science is strong and that comparative studies across sites are increasingly important. During the past decade, however, the LTER network has established a new set of goals through its strategic planning process. These new goals incorporate the enthusiasm for the LTER program that was expressed in decadal reviews but are tempered by the realities of funding long-term studies. These new goals have already stimulated increased network-level science, and they provide a realistic target to guide activities as LTER program continues into its fourth decade. Since its inception, the LTER program has been caught in an "either-or" dilemma. Either it remains faithful to the long-term site studies that have been initiated, or it refocuses its efforts on larger-scale questions. Some sites have addressed this challenge by leveraging multiple sources of funding to address local or regional policy issues (e.g., Harvard Forest). However, the resolution of this dilemma remains the ongoing challenge to fulfilling the potential of the LTER network. Until this problem is resolved, the ultimate goal of addressing large-scale, policy-relevant questions using the power of networked science will remain elusive.

ACKNOWLEDGMENTS

Support was provided by NSF award DEB-0832652 and by the Department of Biology at the University of New Mexico. J. Brunt, M. Servilla, and J. Vande Castle provided data on numbers of publications, data sets, personnel, and working groups as well as comments on the manuscript. S. Collins reviewed the manuscript and made a number of useful suggestions for improvements. L. Walker and M. Willig offered encouragement and strengthened the manuscript through their editorial comments. All statements and conclusions are the sole responsibility of the author.

REFERENCES

Callahan, J. (1984). Long-term ecological research. *BioScience,* **34,** 363–367.

Callahan, J. (1991). Long-term ecological research in the United States: A federal perspective. In P. Risser, ed. *Long-term Ecological Research: An International Perspective,* pp. 9–22. SCOPE, New York.

Collins, S.L., Carpenter, S., Swinton, S.M., Ornstein, D., Childers, D.L., Gragson, T.L., Grimm, N.B. Grove, J.M., Harlan, S.L., Kaye, J.P., Knapp, A.K., Kofinas, G. P., Magnuson, J.J., McDowell, W.H., Melack, J.M., Ogden, L.A., Robertson, G.P., Smith, M.D., and Whitmer, A.C. (2011). An integrated conceptual framework for long-term social-ecological research. *Frontiers in Ecology and the Environment,* **9,** 351–357.

Gosz, J.R., Waide, R.B., and Magnuson, J.J. (2010). Twenty-eight years of the US-LTER program: Experience, results, and research questions. In F. Müller, C. Baessler, H. Schubert, and S. Klotz, eds. *Long-term Ecological Research: Between Theory and Application,* pp. 59–74. Springer, Dordrect/New York.

Harris, F., and Krishtalka, L. (2002). *Long-Term Ecological Research Program Twenty-Year Review: A Report to the National Science Foundation.* Unpublished report.

Hobbie, J.E. (2003). Scientific accomplishments of the Long Term Ecological Research Program: An introduction. *BioScience,* **53,** 57–67.

Johnson, J.C., Christian, R.R., Brunt, J.W., Hickman, C.R., and Waide, R.B. (2010). Evolution of collaboration within the U.S. Long-Term Ecological Research Network. *BioScience,* **60,** 931–940.

Likens, G.E. (1983). Address of the past president: Grand Forks, North Dakota; August 1983: A priority for ecological research. *Bulletin of the Ecological Society of America,* **64,** 234–243.

Magnuson, J.J., Benson, B.J., Kratz, T.K., Armstrong, D.E., Bowser, C.J., Colby, A.C.C., Meinke, C.W., Montz, P.K., and Webster, K.E. (2006). Origin, operation, evolution, and challenges. In J.J. Magnuson, T.K Kratz., and B.J. Benson, eds. *Long-Term Dynamics of Lakes in the Landscape,* pp. 280–322. Oxford University Press, New York.

Magnuson, J.J., Kratz, T.K., Frost, T.M., Bowser, T.J., Benson, B.J., and Nero, R. (1991). Expanding the temporal and spatial scales of ecological research and comparison of divergent ecosystems: Roles for LTER in the United States. In P.G. Risser, ed. *Long-term Ecological Research: An International Perspective,* pp. 45–70. Chichester [England]: Published on behalf of the Scientific Committee on Problems of the Environment (SCOPE) of the International Council of Scientific Unions (ICSU) by Wiley.

Marzolf, G.R. (1982). Long-term ecological research. In J.C. Halfpenny, K. Ingraham, and J. Hardesty, eds. *Long-term Ecological Research in the United States: A Network of Research Sites,* pp. i. Long-Term Ecological Research (LTER) Network, Albuquerque, New Mexico.

Pickett, S. (1991). Long-term studies: Past experience and recommendations for the future. In P. G. Risser, ed. *Long-term Ecological Research: An International Perspective,* pp. 71–88. Chichester [England]: Published on behalf of the Scientific Committee on Problems of the Environment (SCOPE) of the International Council of Scientific Unions (ICSU) by Wiley.

Risser, P.G, Lubchenco, J., Christensen, N.L., Dillon, P.J., Gomez, L.D., Jacob, D.J., Johnson, P.L., Matson, P., Moran, N.A., and Rosswall, T. (1993). Ten-year review of the

National Science Foundation's Long-Term Ecological Research (LTER) Program. NSF, Directorate for Biological Scences, Arlington, Virginia. Unpublished report.

U.S. Long Term Ecological Research Network. (2007). The decadal plan for LTER: Integrative science for society and the environment. LTER Network Office Publication Series No. 24. Albuquerque, New Mexico.

Waide, R.B., and Thomas, M.O. (2012). Long Term Ecological Research Network. In R.A. Meyers, ed. *Encyclopedia of Sustainability Science and Technology*. Springer Science+Business Media, New York. (doi:10.1007/978-1-4419-0851-3)

3 Reflections on Long-Term Ecological Research from National Science Foundation Program Directors' Perspectives

Henry L. Gholz, Roberta Marinelli, and Phillip R. Taylor

IN A NUTSHELL

1. Evolution of the Long-Term Ecological Research (LTER) program has required highly motivated leadership in both the National Science Foundation (NSF) and the science community. It has also benefited from inspired leaders in other agencies.
2. Core research areas enable comparative study across highly diverse field sites.
3. The LTER program promotes integrative ecological research and is an important model for other environmental research programs.
4. New observational capabilities and emerging networks will change the operating environment for the LTER program in unforeseen ways.

INTRODUCTION

The conceptualization and implementation of the LTER program that began in the mid-1970s have depended on the dedicated guidance and input from a large number of individuals within NSF management, within other agencies (particularly the US Forest Service), and in the science and education communities that they serve. The authors served as NSF program directors for the LTER program, respectively, for 10, 8, and 14 years between 1997 and 2011, in the Biological Sciences Directorate (BIO), Office of Polar Programs (OPP), and the Geosciences Directorate (GEO). From that context, we offer our perspectives on this remarkable program.

Several central issues have dominated the development of the LTER program since its inception in 1980. These issues are the designation of core thematic research areas, the establishment of new sites and the expansion of NSF program involvement, the evolution of comparative and synthetic science across multiple LTER sites, the dynamics of top-down (NSF-driven) and bottom-up (principal investigator–driven) efforts that have coalesced to produce the present-day network, and the development of new environmental observing capabilities that should enhance the future scientific impact of the LTER program.

CORE THEMATIC AREAS FOR RESEARCH

The specification and emphasis on five core research areas (Waide, Chapter 2) as elements of the LTER program, which served as part of the initial rationale for the formation of the program, have varied over time and with changes in program management at NSF. Our consensus is that core research themes provide a major vehicle for integrative research, both comparative and synthetic, and additionally, serve as a strong guide for programmatic review. We recognize that dissenting views, held by some principal investigators, NSF program directors, and external program reviewers, are that the core areas are either too vague and not useful or too confining given the increasing diversity of projects in the program and the need to investigate other critical drivers of ecosystem processes. The relatively limited base budget of all LTER projects has forced many compromises in site-specific approaches to core measurements, as well as collateral data. At the same time, limited budgets provide incentives to seek additional support and develop new collaborations for expanding the overall scientific portfolio and value of each site. It is a double-edged sword.

From the NSF biophysical science perspective, core thematic areas have figured prominently in the review of LTER projects, particularly in the renewal process. Attention to core themes allows programs to identify potentially problematic areas within projects and recommend actions that align projects with goals of the LTER program. However, introduction of the social sciences into the mainstream of the LTER program as the two urban sites have matured has raised questions about the relevance of the original core areas. Social scientists were particularly outspoken during the community's strategic planning process from 2004 to 2008, suggesting that the core areas were no longer suitable or sufficient to effectively guide overall science in the LTER program that, in their view, increasingly required explicit acknowledgment and integration of the human dimension.

Given the increasingly accepted view that humans are essential parts of ecosystems, a tremendous amount of intellectual capital was invested by the community supported by the LTER program between 2003 and 2008 in formulating an encompassing interdisciplinary and transdisciplinary agenda for the full engagement and incorporation of social science research into the LTER program's existing ecological and environmental framework. Developed with NSF funding and logistical support from the LTER network office, the community derived an approach, called Integrated Science for Society and the Environment (ISSE) that was focused on how changes in climate, biogeochemical cycles, and biotic structure affect ecosystem services and dynamics with explicit feedbacks to human behavior. ISSE was an argument for substantial new NSF investments in this transdisciplinary research, both within the agency and external to the LTER program. It recognized the need for a major change in NSF's overall environmental science portfolio.

NSF's reaction to the possibility of a whole-scale revision or abandonment of the traditional LTER core research areas was strong and negative. The existing thematic basis for the LTER program is considered by NSF to be essential for the assessment and evaluation of progress within and across projects. Equally important, some projects (e.g., the two in the Antarctic) had very little opportunity to pursue a meaningful, site-specific social science agenda, or even regional scale agenda. As a result, an overlay of new themes has not been embraced at NSF.

Nevertheless, NSF continued to encourage the development of meaningful social sciences research within the LTER program throughout the 2000s via supplements provided by the Social, Behavioral, and Economic Sciences Directorate (SBE) to sites that proposed novel explorations of human–natural system connections. However, beyond cofunding the two urban LTER projects, a more sustained commitment of significant core support from SBE has not materialized, and the other participating programs were understandably unwilling to divert their core LTER funding more explicitly toward the human dimension. Current language in NSF's LTER solicitation states that projects "may elect to include social science research if it helps to advance or to understand key, conceptually motivated ecological questions." It is doubtful, however, that greater integration of social and natural sciences within the LTER program will occur to any significant degree in the absence of new committed resources from interested funding programs in NSF, or enhanced partnerships with other funding agencies.

In large part, because of the associated debate and discussion across the LTER network during the strategic planning, the core areas are now probably as centrally important as ever to overall program management, assessment, and support (see NSF 13-588). Nevertheless, there persists a strong interest in collaborative social science research across the network. A continuing intellectual, as well as practical, challenge to both the community and NSF is in articulating how and where new social science research should become a more central and embedded component of the LTER program.

ESTABLISHMENT OF NEW SITES

The addition of new sites has been of singular importance to the network, diversifying its ecological domain, expanding the intellectual breadth of participants, and attracting investment by programs outside of BIO. In this respect, the LTER program has served to coalesce entities within and outside of NSF, and it has greatly increased the impact of the program on NSF itself, collaborating agencies, and the ecological research community.

For its first 15 years, the LTER program was entirely supported by funds from BIO; new projects were added incrementally through open competitions. This limited the physical scope of new projects to primarily natural, continental (plus Puerto Rico) and largely terrestrial sites, with single agricultural, large river, wetland, and lake projects added over several competitions.

NSF resistance to targeted competitions was overcome when the involvement of NSF units outside of BIO justified targeting. Two Antarctic sites were the first to be added in this mode, with support coming entirely from outside of BIO. As with many aspects of the LTER program, this resulted from internal discussions by program directors and senior staff within NSF, combined with external community enthusiasm for adding new ecological systems that offered unique opportunities for both site-based, long-term studies and comparative ecological work. The primarily marine-focused Palmer LTER project (1990) and the terrestrial- and freshwater-focused McMurdo Dry Valleys LTER project (1993) were selected in two separate competitions focused on Antarctic ecosystems. From their initiation until 2012, these were supported largely from polar programs funds through Antarctic Organisms and Ecosystems (formerly Polar Biology and Medicine) along with the science and support allocations of the Antarctic Logistics section. In 2012, the Antarctic LTER projects were moved to the Antarctic Integrated Systems Science Program.

Increasing interest in research around processes at interfaces between coastal lands and water also gave rise to the NSF Land-Margin Ecosystem Research (LMER) program in the early 1990s, cooperatively sponsored by BIO and GEO. The LMER program was not designed as long term, but considerable cross-site activities between LMER and LTER projects developed, driven in part by investigators who were themselves simultaneously participating in LMER and LTER projects. The national and international popularity of projects in the LMER program, and the associated development of the International Geosphere-Biosphere Programme's Land–Ocean Interactions in the Coastal Zone (LOICZ) project, led some in the US community to recommend that all LMER projects be converted to LTER projects. Once again, NSF threw its support to expanding the LTER program into coastal margins. However, NSF also emphasized an open competitive process and peer review, rather than an automatic conversion of LMER to LTER projects. As a result, in 1997, GEO and BIO created a new partnership that led to four LTER projects emphasizing the land–water interface. These included one former LMER project (Plum Island Ecosystems in 1998), one fully revised LMER project (Georgia Coastal Ecosystems in 2000) and two entirely new projects (Santa Barbara Coastal and Florida Coastal Everglades in 2000). These four projects greatly enhanced comparative and integrative watershed-to-estuary work in the coastal zone, along with the existing Virginia Coastal Reserve LTER project.

In 1996, two original LTER projects (Coweeta and North Temperate Lakes) were expanded to be regional in scope so that more human-dominated landscapes and issues could be considered. This led to an open competition for new urban sites and establishment of the Baltimore and Phoenix LTER projects in 1998, facilitated by a collaboration where SBE and the Education and Human Resources Directorate partnered with BIO for support.

Marine ecologists within GEO and the community were also attracted to the idea of augmenting the LTER network with long-term research on fully marine sites (coastal, open-ocean, and deep-sea). Some of the interest arose because of substantial prior experience of marine scientists within the LTER program (primarily the Palmer site). However, there were also existing ecological and biogeochemical time-series research programs supported by NSF, the National Oceanic and Atmospheric Administration, and others delivering important findings on long-term change in ocean ecosystems. These included the California Cooperative Oceanic Fisheries Investigations, as well as time-series developed during the first decade of the US Global Change Research Program, associated with Hawaii, Bermuda, and Georges Bank. Long-term observations were clearly critical for understanding climate change impacts on marine ecosystem processes. Following the recommendations of an NSF-initiated community workshop at the University of Wisconsin hosted by the LTER network, GEO agreed to expand the marine focus of the LTER program through an open competition beginning with the coastal zone. This resulted in two new LTER projects established in 2004 (Moorea Coral Reef; California Current Ecosystems). No further movement on open-ocean projects was realized, despite some strong interests in the community and at NSF.

Enhancements and additions to the LTER network continue to be discussed and debated, and new funding partnerships are explored. A competition for exploratory urban research projects was held in 2009 as a collaboration between NSF (BIO and SBE each contributed 25%) and the US Forest Service (USFS) research program (contributing 50%). Urban Long-Term Research Areas–Exploratory Research (ULTRA-Ex) provided seed funding to 23 new urban initiatives in anticipation of a new call for jointly funded urban LTER projects. This call has not materialized for a variety of reasons, funding being the most significant. The USFS has been a strong partner in the LTER program, historically providing critical investments in the long-term observational infrastructure for five of the six forested sites (all established in the early years), as well as the urban project in Baltimore. These investments have enabled many insights to watershed biogeochemistry through collaborations between USFS and university researchers. Unfortunately, the funding for USFS research has greatly eroded, with the result that the LTER program faces a major challenge ensuring the continuity of core data sets based on USFS investments of the past.

By the turn of the millennium, the LTER program had a well-deserved reputation as a network that balanced experimental and observational approaches to produce groundbreaking research on long-term dynamics of ecosystem structure and function. The LTER program had grown in its first 20 years to include support from three principle core research programs and two associated facilities programs, across three NSF directorates. Together, those programs continued to grow the network. LTER partnerships have been an important vehicle for collaboration among program directors that oversee much of NSF's ecological and biogeochemical portfolios, and they have been important in the development of other cross-directorate environmental programs at NSF.

The result of open and targeted competitions for LTER grants is a network of sites (or projects, sensu stricto) that spans a very large spectrum of biomes and land uses. Nonetheless, the network is not synoptic, and certainly not balanced in representation or replication within particular biomes. The biophysical range of LTER sites has been an obvious strength of the network, enabling a unique scope of scientific contributions. Clearly, there will never be resources to simply add replication at the site level. However, greater balance in spatial coverage will be increasingly provided by emerging in situ environmental observing networks, such as the National Ecological Observatory Network (NEON), Ocean Observatories Initiative, and the Arctic Observing Network, as well as by the increasing capabilities of global remote sensing systems.

NETWORKING

Despite the entreaties of NSF leadership (beginning early from John Brooks as Division Director of what was then the Division of Biotic Systems and Resources, and Tom Callahan as the first NSF program director for the LTER program), the scientific investigators of LTER projects historically focused heavily on site-based, local science. For its first 20 years, the LTER program existed primarily as a loose federation of separate projects organized around core areas, with the foundational notion that some changes in ecosystems can only be detected via long-term measurements. During this time, research that was explicitly cross-site in nature was funded through separate competitive grants from various core programs at NSF.

Successes with this cross-site approach were manifest, for example, in the Long-Term Intersite Decomposition Experiment (LIDET). LIDET was funded by grants from the Ecosystem Studies Program, augmented by in-kind local institutional commitments and working group awards from the National Center for Ecological Analysis and Synthesis. LIDET was only marginally funded by the NSF LTER program. LIDET existed formally from 1990 to 2002 and involved a decade of coordinated field studies at all of the 17 terrestrial LTER sites that existed at its beginning, along with another 11 non-LTER sites purposefully included to increase biome coverage. In response to significant demand, BIO also ran a series of three open, "cross-site" competitions between 1994 and 2000 to stimulate research across the LTER network. The competitions were explicitly open to participation of non-LTER scientists and to research at non-LTER locations; funding was provided at the discretion of core BIO programs (i.e., programs other than LTER). Unfortunately, because proposal loads on core programs increased dramatically in the early 2000s, voluntary contributions from non-LTER programs soon became unsustainable.

For many sites, the existing LTER framework of observations and experiments has attracted considerable external (non-NSF) research funding that has leveraged the network itself. Such funds have been used to augment site-specific research activities of individual projects as well as to enable some cross-site comparisons and synthesis. It is very difficult to quantify the total leveraging power of the LTER program, but it is substantial. For example, one survey conducted by the LTER network office in 2001 estimated that additional financial support, 2.9 times the core NSF funding (averaged per site, and from all sources), is generated by LTER projects. In this respect, outside agencies and institutions have contributed substantially to broadening the financial support and the functionality of the network.

The commitment to networking was an element within the formal LTER review process for at least two decades, explicitly addressed in both mid-term reviews and renewal proposals. Evidence for this commitment has taken many forms over time, from researcher participation in cross-site research and syntheses, to contribution of time and effort from sites to LTER-wide governance. A central theme has emerged for networking across and beyond the LTER network, namely that project data be fully compatible with and available through the LTER Network Information System (NIS).

GOVERNANCE AND THE LTER PROGRAM

A significant step in moving the LTER program toward a more functionally integrated network was the establishment of the LTER network office (LNO) in 1993 as a mechanism to focus and centralize many network activities. Although this was a big step, establishing the LNO in the absence of a formal governance structure meant that little decision-making authority was ceded (the first LTER bylaws and data standards were not enacted until 2003; Waide and Thomas 2013.)

Complications from a loose governance structure have affected the evolution of LTER information management. The LNO was charged with developing an NIS, yet all of the individual sites have their own information managers, who are responsible to their principal investigators and researchers. These local information managers have, over time, developed sophisticated, but often highly site-specific, data systems on their own. With the goal of moving the NIS to completion, and thereby achieving a much greater centralization of LTER information management, programs across NSF pooled support for an award in 2009 to the LNO as part of the American Recovery and Reinvestment Act. The impacts of this investment on the integration of LTER information management have still to be realized, but it is clear that a truly centralized LTER NIS remains culturally distant to many participants, and that this has hindered its development. Current plans are for LNO to be reconceived via an open competition in 2017. In this context, the evolution of the LTER NIS is likely far from over.

NEON AND LTER

In the late 1990s, the concept of a NEON came to life. Conversations about the nature of NEON varied over the ensuing decade, but it was always envisioned as an NSF Major Research Equipment and Facilities Construction (MRE/FC) project, centrally managed to distribute infrastructure and equipment for making long-term observations of environmental and ecological parameters and processes in a synoptic manner at the continental scale. From the beginning, there was recognition that some overlap of NEON with LTER sites could be beneficial and was quite likely to occur. A number of prominent early NEON supporters were key participants in the LTER program. In fact, early NEON designs could be interpreted as an effort to capitalize on the LTER sites through an investment in standardized equipment, as called for in the first decadal external review report of the LTER program (Risser et al. 1993). Subsequent discussions of the LTER program and NEON at NSF often were mutually exclusive in an attempt to avoid confusion between the two. In retrospect, this unfortunately contributed to a sense of incompatibility and competition of missions, rather than optimizing possible synergies.

The formulation of NEON continues to develop, as does its relationship with the LTER program (and other programs or networks). Within NSF, NEON began and remains a BIO program, with no mandate to include coastal, marine, or polar systems. Approximately one third of the proposed NEON domains contain preexisting LTER sites. But beyond this physical coincidence, the synergies between NEON and the LTER program remain to be seen.

Recent federal budget constraints are likely to drive closer cooperation and collaboration between these two long-term research efforts. At present, NSF funding is flat, whereas NEON construction is still a work in progress. Moreover, there are associated uncertainties in future requirements for NEON operations and maintenance. It seems likely, however, that NEON and LTER together (along with many other field sites, projects, and networks) will be better able to answer the increasingly complex scientific questions each is being challenged to address.

At this time (mid-2014), BIO faces little prospect of obtaining new budgetary support for NEON-enabled research. It is more likely that the core research programs most related to NEON questions and data products (e.g., ecosystem studies, population and community ecology) will be tasked to support research using the valuable infrastructure assets of NEON. It is also likely that much of this research will be driven by the results and future directions of science in the LTER program. As one of the most organized ecological research networks in the world, the LTER program will provide a powerful interface between the community of science and NEON.

EFFECTS OF THE LTER PROGRAM ON NSF?

The LTER program has a remarkable history within the agency. Most of NSF's funding emphasizes fresh perspectives and novel approaches to foundational science, with typical research awards of 2 to 5 years, and sometimes the possibility of renewals for continuing innovative progress. Only rarely does NSF make longer-term research funding commitments, albeit pending schedules and rigorous review. Given that perspective, the most profound impact of the LTER program within NSF (and perhaps beyond), aside from new knowledge created, may well be its continued existence after 35 years and its indefinite future horizon. The LTER program is a highly visible and influential example of how NSF *can* commit and provide sustained financial support to a long-term, focused research program (as distinct from the long-term funding of facilities and infrastructure, such as Antarctic stations, ships, telescopes, and aircraft). The LTER program stands as a hallmark example, in the face of persistent criticism to the contrary from the science community, that both individual projects and thematic research programs can achieve long-term support by way of NSF's commitment, when combined with excellence in results and continued development of the ideas as assessed by periodic merit review.

In similar fashion, LTER awards were an early demonstration that NSF could make individual award commitments of significant duration. The 6-year award periods of LTER grants (initial and renewal awards) are the longest in the foundation. Fifteen years ago, LTER awards were used by NSF leadership as examples of the agency's sincere ability and willingness to enhance support for longer and larger research awards at a time when it was strongly arguing for higher budgets. The agency encouraged its units across all areas of science to assertively work to lengthen the duration of awards for compelling, well-designed, and well-argued projects for which longer awards made good sense. In parallel, NSF worked to persuade the broad US science community to embrace longer and larger awards both within proposals and through their review input. The LTER program had already successfully made this case.

The LTER program at NSF crystallized over time through the ground-up efforts of individual program directors, many on loan as "rotators" from the community, who worked to make connections across the foundation. They were generally aware of the research value and nature of the LTER approach and openly discussed interests in soliciting and supporting new projects within the purview of their ecological portfolios (e.g., Antarctica, oceans, coastlines and coastal margins, urban). At the same time, like-minded leaders in the various research communities engaged to assess and express community interest.

Although connections to the LTER program were not based on formal directorate-level interactions, they had considerable bearing on higher-level NSF discussions about other interdisciplinary research opportunities, especially regarding long-term investments in observation and experimentation. OPP's and GEO's partnership with the BIO-led LTER program remains unique and has led to increased interactions among the terrestrial, polar, and ocean ecological research communities.

As a specific example of its impact on NSF, the LTER program model has been translated into a growing interest of the earth sciences community in longer-term, interdisciplinary environmental systems research pertaining to the surface veneer of Earth —the so-called "critical zone." NSF's Critical Zone Observatories were created in 2007 by the Earth Science Division, with the LTER model very much in mind. Similarities in approach and purpose are very evident in these two distinct NSF-networked programs: a diversity of system types, common measurements, overlapping research agendas that are both site-based and more generally translatable by design, plans for highly integrated data and information systems, and a powerful example and influence that has led to international sets of aligned projects. More recently, BIO issued a new solicitation in 2010 for MacroSystems Biology, a special program designed in anticipation of NEON that explicitly calls for new regional-to-continental scale science that takes advantage of existing research platforms and data availability. The LTER program is explicitly recognized as an important potential partner.

It is arguable whether the LTER community that developed ISSE truly anticipated the attention that coalesced in 2010 within the broad portfolio of science, engineering, and education for sustainability (SEES). What we are confident to assert, having all participated in discussions and negotiations for the multiple components of SEES, is that the LTER and ISSE efforts kept humans, ecosystems, and environments at the forefront of many SEES research thrusts developed by NSF. The engagement of program directors in the earlier LTER community discussions about ISSE, as well as the considerable internal debates about how ISSE might be implemented, placed program directors in excellent positions to demonstrate the broad and compelling science questions that the research communities, including scientists in the LTER network, were proposing to address in partnership with other disciplines. NSF's many LTER program directors (rotators and permanent) were engaged in the cross-directorate, high-level negotiations and planning for new initiatives and subelements of SEES, and they served on working groups to develop new programs and solicitations (e.g., Sustainability Research Networks, Ocean Acidification, Coupled Natural and Human Systems, Coastal and Polar SEES, Dimensions of Biodiversity, Earth Systems Modeling, MacroSystems Biology). Many expected integration between the social science and environmental sciences.

From 1980 until 2012, when a new solicitation for LTER award renewals was issued, the administrative management of LTER across NSF was largely informal. The lack of a formal administrative structure allowed the cognizant program directors considerable latitude in management practices, reporting, public relations, negotiating with other partners (e.g., those in the international office), and establishing review criteria. For example, there was considerable flexibility (some might say too much) in setting standards for, and reviewing, information management, which led to considerable angst on the part of both project principal investigators and LTER information managers. Subsequent dialogue with NSF program directors led to a 2004 proposal from the community for the first guidelines for review and assessment of LTER information management programs. These guidelines were openly discussed and debated by NSF and LTER leaders, and resulted in a document, "Review Criteria for LTER Information Management Systems (version 1.0, 12 April 2005)" that was adopted in 2005 by NSF. For the first time, NSF was able to provide guidance in advance to all reviewers and panelists of LTER projects and the program as a whole.

Today, BIO maintains LTER as a formal core program, with a budget that is largely separate from related core programs, clearly highlighting its status within the directorate. In contrast, GEO manages LTER through separate grants from existing core programs (i.e., Antarctic Integrated System Science, Biological Oceanography), giving its investments a less distinct profile. Although these contrasting bureaucratic models can be debated endlessly, LTER proudly approaches the middle of its fourth decade.

ACKNOWLEDGMENTS

This manuscript is based on work done by H. L. Gholz while serving at the NSF. The views expressed in this paper do not necessarily reflect those of the NSF or the US government.

REFERENCES

Risser, P.G., Lubchenco, J., Christensen, N.L., Dillon, P.J., Gomez, L.D., Jacob, D.J., Johnson, P.L., Matson, P., Moran, N.A., and Rosswall, T. (1993). Ten-year review of the National Science Foundation's Long-Term Ecological Research (LTER) Program. NSF, Directorate for Biological Scences, Arlington, Virginia. Unpublished report.

Waide, R.B., and Thomas, M.O. (2013). Long-Term Ecological Research Network. In R.A. Meyers, ed. *Encyclopedia of Sustainability Science and Technology*, pp. 233–266. Springer-Verlag, Berlin.

PART TWO
H. J. ANDREWS EXPERIMENTAL FOREST (AND) LTER SITE

4 Streams and Dreams and Cross-site Studies

Sherri L. Johnson

IN A NUTSHELL

1. The influence of the Long-Term Ecological Research (LTER) program on my science has been to broaden my scope through exposure to long-term research and to encourage me to explore major questions across biomes.
2. Communication and outreach with natural resource managers and policy makers has given me insight into translation of science and shaped my research.
3. Through my experiences in the LTER program, I began collaborations with stream ecologists and biogeochemists across sites, which expanded into a high-profile research project that spanned several decades.
4. I encourage scientists to work at LTER sites because they are supportive science communities with a wealth of information to share.

PERSONAL OVERVIEW

Currently, I am a co-principal investigator at the H. J. Andrews Experimental Forest LTER project (AND) in Oregon and have been involved with LTER sites most of my professional life. In 1990, I began graduate research on freshwater shrimp responses to a hurricane at the Luquillo LTER site (LUQ) with Alan Covich, my PhD advisor at the University of Oklahoma. My involvement with LTER research expanded during my postdoctoral fellowship. Through the LTER All Scientists Meetings, I met Julia Jones and other researchers from AND. With their encouragement, I received a National Science Foundation (NSF) Postdoctoral Fellowship Grant in 1996 to examine stream temperature dynamics at AND. After several years at Oregon State University, I was hired by the US Forest Service (USFS) Pacific Northwest Research Station in 2001 as a USFS scientist for AND and became a co-principal investigator in 2002. I have had the benefit of being mentored for multiple years by Fred Swanson and have gradually assumed lead USFS responsibilities for AND.

As a stream ecologist, I have studied basic questions and applied issues involving water quality, water quantity, and stream food webs, primarily in forested streams. My research at the LUQ site has examined responses of fresh water shrimp to disturbances and their role in ecosystem dynamics. At AND, my research exploring patterns and controls of stream temperature began as a theoretical landscape-scale question and expanded to examination of temperature responses to flow paths, calculations of heat budgets, and policy implications of forest management (Johnson and Jones 2000; Johnson 2004). The cross-site study of the fate, uptake, and transformation of ammonium and nitrate isotopes in streams has greatly expanded my understanding of stream processes (Ashkenas et al. 2004; Mulholland et al. 2008; Sobota et al. 2012). In ongoing research, I am currently working with numerous collaborators to study effects of changing climate and forest harvest on streams (Argerich et al. 2013).

APPROACH TO SCIENCE

My involvement with the LTER program has influenced how I do research, and I so appreciate the science communities that develop at each LTER site. When I first began working at the LUQ site, I was amazed by the researchers and variety of projects that I learned about while at the field station. Perhaps it was because Hurricane Hugo had occurred just a few months previously and we were all there studying its effects, so it was a time of discovery. It was exciting to be with a community of researchers from a broad range of universities and agencies, sharing meals, laboratories, and working at an active field station with scientists who were exceptional in their respective fields.

LTER sites are places where so much has been learned already and so much more can be studied. Working at sites where major ideas have been developed gives me a sense of humility about the scientific work I am doing. Researchers who work on long-term projects produce data that provide a legacy of information to new researchers at a site. This background information increases understanding of processes and dynamics that occur over decades—far beyond what a single investigator can do. My research has benefited from previous scientists' efforts to collect data conscientiously and archive that data so it can be used by others.

Collecting long-term data can be viewed as a type of service because of delayed gratification in being able to address a key question. I have learned how collecting long-term data needs to be different than collecting data for a regular short-term grant or project. Without the details about the research process, quality controls, and other factors archived in metadata, future researchers would not be able to evaluate and use the data.

I believe that being involved in overseeing long-term research is a significant contribution to future generations of scientists.

ATTITUDES TOWARD TIME AND SPACE

LTER science is sometimes criticized because it focuses on long-term, place-based research. Long-term research is not usually the most glamorous topic that a scientist can pursue. Progress and accomplishments can be slow, and the day-to-day aspects can be tedious. But long-term research is crucial for understanding trends over time. With continued study of a place, new insights can occur (Dodds et al. 2012) and lead to increased understanding or new questions about biophysical dynamics and processes. Short-term research is not necessarily at odds with long-term research; both are needed and can be synergistic in addressing different aspects of a common question.

Place-based research can be ideal for the study of processes and dynamics. As we learn about site-specific processes, we and others can build on those findings to further explore related processes. Knowing a lot about selected places is informative, and the foundational information can be examined relative to a broader range of sites to explore variability and commonality.

COLLABORATION

My research at LTER sites has frequently expanded into collaborative projects, both interdisciplinary and cross-site collaborations. Some might say that working at an LTER site encourages collaboration. Or alternatively, maybe there is self-selection of researchers, in that those drawn to the LTER program might be more predisposed to collaborative science. Perhaps focusing on long-term projects requires openness to collaborations because that provides the best prospect for continuity of data over generations?

Interdisciplinary collaborations within AND have taught me a lot. The collaborations might be an outgrowth of more frequent interactions, because at each LTER site there are shared field studies. Proximity counts and can lead to new interactions. The frequent local LTER group meetings also encourage us to learn more about each other's research and to be curious about common questions. The interdisciplinary collaborations could also be an outgrowth of the longer term nature of the LTER program, which allows time to ask broader questions. Ecological questions often require interdisciplinary perspectives and research at LTER sites benefits from having scientists from multiple disciplines working together.

The LTER stream ecology group has been an amazing example of positive and productive cross-site collaborations spanning several decades. Beginning in early 1990 at the LTER All Scientists Meetings, LTER stream ecologists began to identify ourselves as a group interested in cross-site collaborations for addressing relevant ecological questions. Several prominent leaders in stream ecology were very active at their respective LTER sites. They were keen on implementing cross-site comparisons and developed proposals that were funded by the NSF, and the cross-site Lotic Intersite Nitrogen Experiments (LINX) began.

I learned much about collaboration from the leaders of the LINX project. The leaders set a tone of respectful collaboration, inclusiveness, and mentoring that has been a model for training several generations of stream ecologists. Unlike many academic situations, where competitive interactions are the norm, the premise of our LINX stream group was true collaboration. Open sharing of ideas was encouraged, and communication by all participants, ranging from senior professors to beginning graduate students, was respected. As a young scientist, I felt assured that any individual contributions would be recognized and that there was no risk of being scooped by sharing ideas or data with others in the group. It was the first collaborative group that I have been part of where authorship ground rules were discussed early, put in writing, and revisited regularly. We have copied this idea for several subsequent collaborations, where we established a publication agreement to address how data would be used and authorship would be shared by members of the group as well as how data-sharing with the broader scientific community would be accomplished (Argerich et al. 2013). Collaborations can be challenging, and having been involved in several projects where the group was not proactive about defining authorship or data sharing, I regretted that we did not agree on a plan in writing at the beginning. Over time, I have realized that the LINX collaborative framework was singular. Not many groups, within the LTER program or beyond, have been as effective in creating long-lasting research affiliations that are as positive and productive.

It is important to not assume that research within the LTER program is synonymous with collaboration or that collaborations are the best way to make scientific progress. There are costs and benefits of collaborations and multidisciplinary research. Meetings and discussions can take a lot of time and are not always productive—there are frequent false starts and dead ends. Budgets for collaborative research projects can be very expensive, and researchers often have to compromise about details because the budgets will not support everyone's ideal design. Collaborations make scientists dependent on others, and synchronizing priorities can be challenging. Individual research is important in many settings. Even with these challenges, collaboration has been important to me because it is a great way to be constantly learning and to be addressing questions beyond what I could do independently. LTER sites often encourage exploration of research topics within sites and across sites by groups that might not have connected otherwise.

APPLIED RESEARCH

As a research scientist with the USFS, I am very involved in applied research. A large part of my job is designing and carrying out research as well as interpreting the science to inform forest management. I work with natural resource managers frequently, both to share information and to collaborate in planning new projects to address current questions. My focus is on forest–stream interactions and how forest management in riparian areas may be affecting stream ecosystems, aquatic food webs, and water quality. This topic can be controversial, and I have learned a lot about environmental policy and have seen lots of examples where science does not get incorporated into policy. But I also am involved with applied research where the adaptive management loop is completed and science findings are being used to revise and update forest practices in riparian areas.

Because our LTER site is based at the H. J. Andrews Experimental Forest and within a national forest, we are able to take advantage of long-term, small-watershed experiments of forest management that started in the 1950s. Six of the LTER sites are located in USFS experimental forests, and these experimental forests were active research sites for several decades before they became NSF-funded LTER sites. Natural resource managers are our major stakeholders for this science because of its relevance to management of national and state forests. Several managers have told me that they specifically applied for their current jobs because they wanted to be part of this unique research–management partnership. The synergy between basic and applied research coming from these long-term studies has been very productive and informative for both scientists and managers.

COMMUNICATION

The majority of my communication and outreach efforts are focused on informal education and outreach to adult learners because I am a scientist who works for the federal government and does not teach at a university. Therefore, the LTER program has not really had an impact on my teaching and communication. However, I am involved in field courses (Figure 4.1) and field tours with a variety of universities and occasionally coteach a field class at AND through Oregon State University. I also participate in the LTER Schoolyard Program at the site to work with teachers during field courses.

MENTORING

I feel lucky to be involved in research at AND and LUQ sites and to have the opportunity to be mentored by researchers I might not have had access to otherwise. This type of exposure to leading ecologists and "big science" can be a boost to the careers of young researchers. It can happen in many places in addition to within the LTER program; for example, many scientific societies make mentoring a priority. But the LTER program has been key for my career.

The mentoring I have received has shaped how I, in turn, mentor younger scientists. I work informally and formally with young faculty, postdoctoral fellows, technicians, and graduate students to encourage them to participate in our group's various research projects. Because I was a small business owner for many years before I entered graduate school, I find I sometimes have a different perspective than traditional academics and less of a hierarchical view of science and supervisory relationships. I try to be approachable and look for opportunities to discuss new and old ideas with younger scientists, discuss their professional development, and encourage them to take as much

FIGURE 4.1 A busy field tour in Watershed 1 at the H. J. Andrews Experimental Forest Long-Term Ecological Research site. (Photo courtesy of Sherri Johnson.)

responsibility in their areas as they are willing to. When reviewing articles and proposals, I am fairly exacting and try to politely push authors to be as clear as possible.

SKILL SET

My time in the LTER program has taught me a lot about working with long-term data. For long-term, place-based data to be most useful and for the data to be comparable to studies in the future, many details are necessary. But those that are most useful are the ones where the data and methods have been safely archived and clearly annotated. Working with and learning from other researchers who collect long-term data has increased my willingness to share data, in part because of the emphasis on data-sharing within the LTER program and in part due to NSF guidelines for making data publically available. I also have learned the importance of details in archiving my data. For example, while working with long-time series of temperature data from AND, we researchers realized that there were substantial changes in precision and accuracy with different types of sensors over time. Whether these details were archived affected how confident we were in the data and greatly influenced our analyses and interpretations of trends in climate. This reinforces my belief that scientists, field technicians, and information managers need to be very attentive to details to create long-term data records that are consistent over time.

Increasingly, researchers use data that they have not collected. When they are deciding what data from others is worth using, how do they judge what is good? LTER sites have been key in training researchers how to archive and document their own data. This background is often useful in knowing what details they should be looking for when using others' data. I have seen multiple examples of incorrect interpretations using data from the AND site because researchers did not take the time to read through the detailed metadata or pay attention to the quality assurance codes. Also, I have learned to be increasingly cautious about using publically available data that are not well documented when doing meta-analyses.

PERSONAL CONSEQUENCES

My experiences in the LTER program have introduced me to new opportunities and new collaborators and therefore broadened my world. I am the type of person who likes working with others, and my role at AND has been a good fit for me. But my role at AND also involves a lot of service—overseeing long-term studies, helping manage site dynamics, encouraging new research, finding funding, and doing outreach. Some colleagues have suggested that this amount of service would damage my scientific career and potential for advancement. However, I think that there have been more benefits than detriments to my career because of the LTER Program. If AND were not an LTER site, I would not have been exposed to so many ideas and experiences; a broader range of disciplines and perspectives; and new colleagues, scientists, and educators.

CHALLENGES AND RECOMMENDATIONS

Finding funding for long-term studies is increasingly a major challenge for me and others. Much of the data at LTER sites that has been collected over the longest periods have been obtained in partnership with federal agencies. The historical data were collected and thankfully archived due to the foresight of early researchers. More recently, though, the federal agencies are not able to provide such consistent funding for long-term data collection. Federal budgets are shrinking, and budget horizons are months instead of years, which leads to lots of extra work to keep consistent measurements coming from long-term sites. The budget cuts for maintenance of long-term studies are occurring at the very time that questions about changing climate are high profile and continuous long-term data are so limited.

Recruiting and training the next generation of LTER researchers are important. Site-based research can sometimes be perceived as exclusive by those who are not involved. Existing researchers at LTER sites have an ongoing task to be open to new researchers and to share information about how to begin research at each site or through the LTER program. To those not affiliated with a LTER site, it is not obvious how research projects are proposed and approved at a site or how LTER research funding is allocated. LTER funding at AND is primarily dedicated to collecting and maintaining the core data, which are publically available and used as base data in many projects; funding for new research or researchers is very limited. Because researchers devote many hours to LTER research for very little financial benefit, new LTER scientists tend to be those who are excited about opportunities for collaborations and place-based research and are creative in leveraging research funding from other grants or projects.

CONCLUSION

What my career path might have been if I had not been involved with two LTER sites is difficult to know—an alternate reality would be necessary. I do know that being involved with the LTER program has benefited me personally and professionally. By comparing my experiences with fellow graduate students who followed more traditional academic paths or with colleagues who are affiliated with USFS experimental forests that are not LTER sites, I realize I have had more opportunities and exposure to new ideas, situations, and collaborations because of my long involvement in the LTER program.

My recommendations for young ecologists are: (1) make sure you have lots of interesting questions to study—questions that are so exciting that you cannot stop thinking about them; (2) explore sites that have long-term data—it can be so helpful to be able to build on what others have learned at a place; and (3) if you are someone who likes working with others, find fun and interesting colleagues and pursue collaborative research as part of your research portfolio.

ACKNOWLEDGMENTS

I am grateful for being part of the LTER programs at H. J. Andrews Experimental Forest (funded by NSF DEB-0823380 and other awards, US Forest Service Pacific Northwest Research Station, and Oregon State University) and Luquillo Experimental Forest (International Institute for Tropical Forestry, University of Puerto Rico, DEB-9411973 and other awards), as well as for interactions with thoughtful colleagues and mentors at both sites.

REFERENCES

Argerich, A., Johnson, S.L., Sebestyen, S.D., Rhoades, C.C., Greathouse, E., Knoepp, J.D., Adams, M.B., Likens, G.E., Campbell, J.L., McDowell, W.H., Scatena, F.N., and Ice, G.G. (2013). Trends in stream nitrogen concentrations for forested reference catchments across the USA. *Environmental Research Letters*, **8**, 014039. (doi:10.1088/1748-9326/8/1/014039)

Ashkenas, L.R., Johnson, S.L., Gregory, S.V., Tank, J.L., and Wollheim, W.M. (2004). A stable isotope tracer study of nitrogen uptake and transformation in an old-growth forest stream. *Ecology,* **85**, 1725–1739.

Dodds, W.K., Robinson, C.T., Gaiser, E.E., Hansen, G.J.A., Powell, H., Smith, J.M., Morse, N.B., Johnson, S.L., Gregory, S.V., Bell, T., Kratz, T.K., and McDowell, W.H. (2012). Surprises and insights from long-term aquatic datasets and experiments. *BioScience,* **62**, 709–721.

Johnson, S.L. (2004). Factors influencing stream temperatures in small streams: Substrate effects and a shading experiment. *Canadian Journal of Fisheries and Aquatic Sciences,* **61**, 913–923.

Johnson, S.L., and Jones, J.A. (2000). Stream temperature response to forest harvest and debris flows in western Cascades, Oregon. *Canadian Journal of Fisheries and Aquatic Sciences,* **57**, 30–39.

Mulholland, P.J., Helton, A.M., Poole, G.C., Hall, R.O., Jr., Hamilton, S.K., Peterson, B.J., Tank, J.L., Ashkenas, L.R., Cooper, L.W., Dahm, C.N., Dodds, W.K., Findlay, S., Gregory, S.V., Grimm, N.B., Johnson, S.L., McDowell, W.H., Meyer, J.L., Valett, H.M., Webster, J.R., Arango, C., Beaulieu, J.J., Bernot, M.J., Burgin, A.J., Crenshaw, C., Johnson, L., Niederlehner, B.R., O'Brien, J.M., Potter, J.D., Sheibley, R.W., Sobota, D.J., and Thomas, S.M. (2008). Stream denitrification across biomes and its response to anthropogenic nitrate loading. *Nature,* **452**, 202–205.

Sobota, D.J., Johnson, S.L., Gregory, S.V., and Ashkenas. L.R. (2012). A stable isotope tracer study of the influences of adjacent land use and riparian condition on fates of nitrate in streams. *Ecosystems,* **15**, 1–17.

5 Data, Data Everywhere!

Susan G. Stafford

IN A NUTSHELL

1. My association with the H. J. Andrews Experimental Forest (AND) Long-Term Ecological Research (LTER) program defined my research and shaped my professional character. Creating the data and information system for the AND program became the focus of my research, and ensuring that protocols and methodology were established for effectively stewarding long-term data and metadata across the LTER network became my mission. Designing the system in such a way that it could continue to evolve in tandem with the growth and evolution of the LTER program was an enjoyable challenge!
2. Recognizing the need to ensure that data were initially collected, managed, and coordinated with research objectives to produce statistically sound results from future analyses became the cornerstone of my research methods courses and workshops.
3. Research at the AND site was conceptually positioned at the interface of basic and applied research. Communicating research results across a wide spectrum of

stakeholders, including state, national, international, and nongovernmental organizations, was commonplace. Watching the leadership of the AND program cultivate, nurture, and build these enduring partnerships provided invaluable lessons for effective communication with diverse and numerous constituencies. Learning how to work toward successful conflict resolution by tackling issues directly to find common ground has served me well throughout my career.

4. Direct participation, early in my career, in the collaborative and openly inclusive atmosphere of the AND program honed my skills for working with multidisciplinary and interdisciplinary teams. I embrace this approach in my research, teaching, and management style to this day.

5. I credit my early leadership opportunities afforded by the LTER program as contributing significantly to my overall professional success. I encourage all young researchers, from student to junior faculty, to seek out LTER sites that are nearest to them intellectually or geographically and build their network of professional colleagues early in their careers. They will not be disappointed!

PERSONAL OVERVIEW

One of the most significant aspects of my professional career was being associated with the AND site of the LTER program. The AND site has literally defined my research focus and shaped my professional identity. The timing of my arrival as a biometrician and consulting statistician in the Forest Science Department at Oregon State University (OSU; the administrative home of AND) in 1979 allowed me to fill a critical need of the newly formed site that was launched in 1980.

The AND site followed on the heels of the International Biological Program (IBP), successfully positioning itself as a place for studying forested ecosystems—particularly coniferous forests. The IBP era fostered great enthusiasm for doing collaborative research at the H. J. Andrews Experimental Forest, but left deep scars resulting from problems in the way data had been "managed." Post IBP, many data records were either inaccessible or very nearly so. This frustration catalyzed the effort to ensure that data and metadata from LTER sites would not suffer the same fate; and it did not. Quite to the contrary, the AND site became one of the flagship LTER sites for data and information management (IM).

APPROACH TO SCIENCE

The focus of my work was to ensure that all data collected were catalogued and managed in an organized and coordinated fashion such that statistically sound results could be drawn from future analyses. Protocols were established for effectively stewarding long-term data and metadata (Stafford et al. 1984). The Quantitative Sciences Group (QSG), which I created, led this effort, and the Forest Science Data Bank (FSDB) (Stafford et al. 1984; Stafford 1998) housed these long-term records.

While at OSU, we instituted a structured approach that integrated IM into the full research continuum (Stafford, Brunt, and Michener 1994; Stafford et al. 1986). To maximize efficiency, all research projects began by meeting with a statistician to identify and define research objectives; determine formatting requirements for FSDB data archival and storage; document and edit data on collection; and finally, analyze, interpret and synthesize the results. Meeting with a statistician during the final synthesis stage provided the added value of ensuring that each project had the statistical rigor necessary to withstand the potential political criticism that often accompanied the prominent nature of work at the AND site in the late 1990s (Geier 2007).

This was an exciting time in Oregon! In addition to establishing the QSG and FSDB, we secured additional National Science Foundation (NSF) funding for an "integrated science workbench for ecosystem research." We designed the system in such a way that it could continue to evolve in tandem with the growth and evolution of the LTER program (Stafford, Spycher, and Klopsch 1988). FSDB catalogs and protocols became prototypes for IM systems of future LTER sites. Eventually, most of the original Andrews IBP data were incorporated into the FSDB, making more than five decades of data and metadata easily accessible to other researchers.

One of the additional benefits of working with researchers, graduate students, and postdoctoral fellows in the LTER program, as well as those from other OSU departments, was my development of a personal interest in working with multidisciplinary and interdisciplinary teams and bringing this thoughtful and insightful approach to my own research. My experience in the LTER program did more than merely influence my ideas about how environmental research should be done—it established the gold standard! At the AND site, collaboration and inclusiveness were canonical principles that I naively thought were applicable to all LTER sites. I was mistaken. In the mid-1980s, while on an NSF site visit to another LTER site, I innocently asked a young researcher how resources were allocated and whether his requests were factored into final budget decisions. After a long pause, he answered, "Well, it is hard to participate in budget decisions when meetings are announced via e-mail late Friday evening (long after everyone has left) for a meeting very early Monday morning before most of us come in." (Recall that this was long before the era of cell phones with instant messaging and texting!)

The culture at the AND site tended to attract other like-minded researchers with an a priori tendency to trust others already in the group (Grier 2007). Over time, I came to realize that the collaborative culture at the AND site was an anomaly compared to that at most of the other LTER sites.

After directing the QSG and FSDB for nearly two decades, I left OSU to take a position as Department Head of Forest Sciences at Colorado State University (CSU) in 1998, home of the former Shortgrass Steppe (SGS) LTER program. In 2002, I assumed the Deanship of the College of Natural Resources at the University of Minnesota, home of the Cedar Creek (CDR) LTER site. At the SGS and CDR sites, I did not experience or observe the collaboration or inclusiveness that was so characteristic of the AND site.

At early IM (Information Management) meetings, I often mused that we needed a sociologist-in-residence to watch how well, or how poorly, we were doing, and how well sites were addressing the human dimensions of the ecological questions under study. The early LTER sites did not include the human dimension. It soon became abundantly clear, however, that the integration of social sciences within the LTER network was an excellent idea. Initially, there was an intranetwork augmentation competition to add an additional $1 million to the base budget of an existing LTER site. The North Temperate Lakes (NTL) site in Wisconsin and Coweeta (CWT) site in Georgia were the "winners." The second approach came a few years later, with an open competition for two new, truly urban sites. The Central Arizona–Phoenix (CAP) and Baltimore Ecosystem Studies (BES) sites were selected and joined the network. To me, these foreshadowed the Coupled Natural Human Systems (CNH) program at NSF (tridirectorate-funded program shared among the NSF Directorates of Biological Sciences; Geosciences; and Social, Behavioral, and Economics).

Anecdotally, after CAP and BES sites joined the LTER network, I served on an NSF Committee of Visitors (COV) for the Division of Environmental Biology—the NSF research division responsible for the LTER program. In the COV sample of proposals that we reviewed, it happened that both types of awards were included—one to an "augmented" LTER site and one to an urban LTER site. In the former, the human dimension was layered on top of the existing research at the LTER site; in the latter, human dimensions were integrated from site inception and embedded into the project's underlying fabric. I remember the comments of the lone sociologist serving on the COV panel, who said the integrated approach taken at the urban site was the better approach. That observation stuck with me. When developing subsequent interdisciplinary teams—whether for research or teaching—I always made sure that *all* disciplinary domains were included from the onset (e.g., social scientists as well as ecologists) to ensure socioecological factors were considered.

ATTITUDES TOWARD TIME AND SPACE

The underlying vision for the LTER network was based on the recognition that many ecological phenomena occur on a temporal scale too great to be studied within a single standard grant (i.e., 3–5 years) and could in fact, exceed the span of an individual's career of three to five decades. The LTER program was designed to study these kinds of phenomena. Taking a longer-term perspective added a necessary complexity to the research. Consequently, I appreciated from an early age that taking a more comprehensive and holistic approach, although seemingly more difficult logistically, often yielded the most insightful results.

COLLABORATION

During my early career at the AND site, I learned the value of working collaboratively. The AND site was managed uniquely but effectively. Monthly meetings were unlike anything I had ever experienced, with never a hint of *Robert's Rules of Order*! Characterized by friendly, free-flowing discussion over a wide range of timely topics followed by open debate that led to consensus-driven decision-making, active participation was encouraged and guests always invited. Although usually the youngest by age and experience, as well being the only female researcher, I never felt excluded. Quite to the contrary, I had a voice on a level playing field and felt empowered! On reflection, this mirrored the inclusive management style I adopted for my own, treating all with respect and professional honesty. As QSG director, I recruited talented professionals including US Forest Service (USFS) partners. As the QSG grew to meet challenging quantitative and technology needs, our work ethic (based on collaboration, fairness, and inclusiveness) never wavered.

It is obvious that my experiences in the LTER program influenced my penchant for working collaboratively. Another aspect, however, should be mentioned and that is that one can be perceived as "too collaborative." I fell into this category. As an administrator, I would routinely reach out to others to build partnerships and coalitions with faculty and deans across campus. Despite several successes, my overly collaborative style was perceived by some as a lack of true leadership. To them, I was not demonstrating what I refer to as "alpha behavior." This conundrum highlights the conflict between: (1) a truly collaborative style, that is reaching out to colleagues with similar interests, goals, and aspirations, and (2) the traditional approach that is all too common in academia (in my opinion) of needing to (or even being encouraged to) establish a "singular" recognition for one's self or their academic unit.

APPLIED RESEARCH

Research at the AND site was conceptually positioned at the interface between basic and applied research. Interactions with various stakeholder groups including state, federal, international, and non-governmental organizations were commonplace and part of everyday site business. Our monthly meetings also included reports from the Blue River District Ranger, USFS.

Further demonstration of the strong partnership with USFS came when the Willamette National Forest embraced the goals of ecosystem management. Watching leadership at the AND site cultivate, nurture, and build these enduring partnerships provided invaluable lessons for working effectively across a broad array of stakeholders and external constituents. My office was housed in the Pacific Northwest Research Station, and we worked in a National Research Forest. This provided me with a firsthand view of "researchers" working with "managers." Although some of these interactions initially were contentious, over time I saw both sides come to respect, listen to, and work with each other effectively. This was an early life lesson in not shying away from seemingly intractable differences, but rather working to find common ground and build working relationships, an invaluable lesson as I continued in my career.

COMMUNICATION

The AND site was a "tempest" in a teapot with respect to communication of science in the 1980s and 1990s. During my tenure, controversy swirled around ecosystem management (e.g., protecting spotted owl habitat vs. cutting old-growth timber). Local papers routinely reported on the public debate with "environment" pitted against "economy." Local communities across Oregon with a timber-based economy were often angered by restrictions and cutbacks in logging because of the impact it was having on people's livelihoods as well as the ripple effect on other sectors of the local economy. Because several of my husband's family made their living from the timber industry, these issues hit close to home. I saw complex and controversial topics addressed firsthand in public forums. My colleagues at the AND site served as excellent role models by showing less concern for political backlash than for providing public statements supported by sound science. When science stays only in the laboratory, communicated among colleagues, it falls far short of reaching its potential audience (Figure 5.1).

The concept of the LTER Schoolyard Program was ingenious: to use the uniqueness of the LTER network to connect long-term scientific research with science education. Recognizing how integral education was to the mission of the LTER program created a masterful learning opportunity. I consider the Schoolyard Program a true jewel in the crown of the LTER network. For a modest initial investment of $15,000 per site, there is great impact and potential for leverage. The Schoolyard students are some of the strongest and most compelling ambassadors for the LTER program. At CSU, I worked indirectly with several undergraduate students on their projects and I made sure that during various NSF site visits that their work was showcased prominently. Often, reviewers commented that that was one of the highlights of their visits.

MENTORING

My experiences in the LTER program directly influenced the way that I assess junior faculty for tenure and promotion—not penalizing them per se for high numbers of coauthored papers where they were not lead author. I look for their intellectual contribution

FIGURE 5.1 Susan Stafford (H. J. Andrews Forest Long-Term Ecological Research [LTER] site) and Steward T. A. Pickett (Baltimore Ecosystem Study LTER site) meeting with Congressional Representative of Pennsylvania, Chaka Fattah, during the Biological and Ecological Sciences Coalition Congressional Visits Day in Washington, DC. (Photo courtesy of Julie Palakovich and American Institute of Biological Sciences.)

to the research rather than merely their position in the author string. As department head and dean, I learned to "read between the lines" to discover an author's true contribution to the work.

As dean, I used what I learned to invest in the professional development of my department heads and unit leaders, with an emphasis on encouraging their leadership. To me, mentoring encompasses a wide range of skills and activities: from active listing to modeling the behavior you want to see to providing opportunities for growth and advancement. I should mention that I had the added perspective of often being the first woman in the position I occupied, which provided the opportunity and responsibility to be a role model for other women students (undergraduate as well as graduate), postdoctoral fellows, junior faculty, and other professionals. The value of mentoring was instilled in me from the very beginning of my career. I found the AND site to be a very welcoming professional environment. It became clear that personal relations of trust and respect were critical components of retention in the AND community. Looking back, I thrived in that environment.

As cofounding chair of the IM committee with Bill Michener from the former North Inlet LTER site in South Carolina, it was critical to create a sense of community for all information managers, especially because they were dispersed at sites with different levels of expertise and support. I viewed it as "paying it forward" for the mentoring that I received. With NSF support, Michener and I institutionalized the annual information managers' meetings that persist to this day, and that are used as a model to create cohesion and singleness of purpose within a diverse cadre of researchers who value collegiality and professionalism. In the early days of the IM committee, we discussed "doing in-reach" for our newest information managers. In retrospect, we were providing "mentoring" by developing orientation and training sessions for them as their sites joined the network.

SKILL SET

As one of the pioneers in IM, the "LTER experience" was my scientific tool kit. I spoke about it before it was popular to talk about data and IM. At a Coordination Committee (CC) meeting (now the Science Council) at Harvard Forest, I gave a presentation (as chair of the Data Management Committee) to the assembled researchers and suggested that data management be recognized as a core area of concentration at each site (to be added to the others, including primary productivity and disturbance). The room went silent! At lunch, Jerry Franklin (then CC chair), draped his arm around my shoulders and in a fatherly way and said, "Susan, I know you think data management is important, but really now—a core area?" It was considered blasphemous then, but look where IM is today and the significance it plays within the LTER network and in all NSF proposals.

PERSONAL CONSEQUENCES

Our IM work brought regional, national, and global recognition and prominence to the AND site. The protocols and the "culture" that we developed brought many invitations to showcase and share our progress as we developed mechanisms for data sharing, storage, and analysis that made data more readily available to researchers at the AND site as well as to others. In 1992, I travelled with three IM colleagues to Beijing, China, to work with the Chinese Academy of Science as they developed the Chinese Ecological Research Network, patterned after the LTER network. That cross-cultural experience provided a far deeper understanding that has served me well while working with international colleagues and students throughout my career. The professional friendships developed within the LTER IM community continue to this day.

My association with the AND program is responsible for a tipping point in my professional advancement in science and higher education administration. At an AND site visit dinner in the fall of 1993, the late NSF Program Officer, Tom Callahan, leaned across the table and asked if I had considered going to NSF? I had not, but after further discussion with my family and Mary Clutter, then NSF BIO Assistant Director, I did. I spent the 1994 calendar year at NSF in Washington, DC, as the first visiting Division Director of Biological Instrumentation and Resources, the precursor of the current Division of Biological Infrastructure. Interestingly, "biological infrastructure" was the label we used within the division to better describe our portfolio of supported activities. This was a profoundly positive and compelling experience with lasting, long-term impact on my life, both professionally and personally (Stafford, 1996).

CHALLENGES AND RECOMMENDATIONS

The LTER program is facing a real conundrum. Optimizing efforts to support the long-term research conducted on the scale of decades or centuries, is a nontrivial challenge, especially if the goal is to support hypothesis-driven inquiry, not just the accumulation of data and information. This requires a commitment to work *today* to support *tomorrow's* science, which will require committing resources to future studies that could otherwise be directed toward current investigations. This conundrum is exacerbated in today's research funding climate—with shrinking budgets, lower funding rates, and a general decline in public understanding and support for science. In my opinion, the expectation of continuing to collect long-term data in a funding context that requires the identification of new questions or conceptual themes is the Achilles heel of the LTER program. At the 2012 LTER All Scientists Meeting, Bob Robbins and I led a well-attended workshop on this challenging topic.

There has been, and will always be, a healthy tension between functioning at a site-versus network-level. Excellent site science is a prerequisite to receiving funding from NSF in general and from the LTER program in particular. Concomitantly, however, the expectation exists that sites contribute to a strong and vibrant network. This dichotomy becomes even more pronounced in the realm of data. Long-term, well-documented *regional* datasets, in addition to core *site* datasets, become one of the greatest assets of the network. This is where I believe the LTER network becomes far more than the mere sum of its parts.

Managing an LTER grant is not a straightforward proposition. There are many more moving parts than one would expect on first glance. By definition, an LTER grant is long-term and hence, should require a succession plan for the principal investigator. Successional planning is not a trivial undertaking but is essential for longevity of the site. I have seen several models. The most successful was at the AND site, where the first principal investigator was Jerry Franklin, followed in order by Fred Swanson, Mark Harmon, Barbara Bond, and currently Michael Nelson (an ethicist). Although these transitions were not easy, they were successful. I believe the reason for this was largely because individuals put their personal biases, disagreements, and opinions aside and focused on the larger intellectual goal, which was the successful longevity of the site. I would encourage the LTER network to give serious thought to how to catalyze the development of new principal investigators and senior research scientists. Thrusting junior scientists into administratively heavy jobs such as being a principal investigator of an LTER site, without an "apprenticeship" or adequate training, is less than optimal and will not necessarily be viewed as a positive step toward promotion, tenure, and advancement. Successional leadership plans should be carefully implemented across the network.

CONCLUSION

They say that timing is everything. In my case, it definitely was. I began my academic career at OSU in 1979, just as the AND LTER was launched. At that time, investments in data management were almost nonexistent. The fortuitous timing provided a blank canvas on which to design, test, and implement a data and information management system that helped establish standards and protocols across the LTER network. Today, data management is taken very seriously and plays a very significant role within the LTER network and in all NSF proposals. This essay chronicles my 30-plus-year journey to ensure that the vast wealth of research data and information from LTER sites is and will continue to be accessible for future generations of students and researchers, making sure that data management is implemented in a proactive and thoughtful manner!

In summary, I was very fortunate to begin my professional career in synchrony with the development of the AND site. Had I not, my professional career would have taken a very different trajectory! I encourage all graduate students and junior faculty to seek out the LTER site(s) nearest to them intellectually or geographically—or both. Establishing a professional network of colleagues early in one's career can be both extraordinarily helpful and enjoyable!

ACKNOWLEDGMENTS

I would like to acknowledge the many years of NSF funding of the H. J. Andrews Experimental Forest LTER site beginning with LTER at the site (LTER1), NSF Award DEB-8012162 (1980). This was followed by funding for LTER2: NSF Award DEB-8514325 (1985); LTER3: NSF Award DEB-9011663 (1990); LTER4: NSF Award DEB-9632921 (1996); LTER5: NSF Award DEB-0218088 (2002); LTER6: NSF Award DEB-0823380 (2008); and the current award, LTER7: NSF Award DEB-1440409 (2014). Although I was most active from 1980 to 2003, I have maintained my association with the AND. I have always felt that the core LTER grant was a magnet, attracting and leveraging the NSF investment. For example, I am currently collaborating with AND colleagues on BIO/DBI 1062572.

REFERENCES

Geier, M.G. (2007). *Necessary work: Discovering old forests, new outlooks, and community on the H. J. Andrews Experimental Forest, 1948-2000.* General Technical Report PNW-GTR-687. U.S. Department of Agriculture, Forest Service, Pacific Northwest Research Station, Portland, Oregon.

Stafford, S.G. (1996). Finding leadership opportunities in an era of dual-career families. *Bioscience,* **46**, 52–54.

Stafford, S.G. (1998). Issues and concepts of data management: The H.J. Andrews Forest Sciences data bank as a case study. In W.K. Michener, J.H. Porter, and S.G. Stafford, eds. *Data and Information Management in the Ecological Sciences: A Resource Guide,* pp. 1–5. Long-Term Ecological Research Network Office, Albuquerque, New Mexico.

Stafford, S.G., Alaback, P.B., Koerper, G.J., and Klopsch, M.W. (1984). Creation of a forest science data bank. *Journal of Forestry,* **82**, 432–433.

Stafford, S.G., Alaback, P.B., Waddell, K.L., and Slagle, R.L. (1986). Data management procedures in ecological research. In W.K. Michener, ed. *Research Data Management in the Ecological Sciences,* pp. 93–113. University of South Carolina Press, Columbia, South Carolina.

Stafford, S.G., Brunt, J.W., and Michener, W.K. (1994). Integration of scientific information management and environmental research. In W.K. Michener, J.W. Brunt, and S.G. Stafford, eds. *Environmental Information Management and Analysis: Ecosystem to the Global Scales,* pp. 3–19. Taylor & Francis, London.

Stafford, S.G., Spycher, G., and Klopsch, M.W. (1988). The evolution of the Forest Science data bank. *Journal of Forestry,* **86**, 50–51.

6 Science, Citizenship, and Humanities in the Ancient Forest of H. J. Andrews

Frederick J. Swanson

IN A NUTSHELL

1. The H. J. Andrews Experimental Forest Long-Term Ecological Research (LTER) program has nurtured a large, highly interdisciplinary community that has been a wonderful seedbed for emergence of ideas from our group, and for my own growth as a scientist, educator, collaborator, and communicator.
2. Collaborations for me as an individual and within the Andrews forest group have grown over the decades: research–land management since the 1950s, ecology–earth

sciences since the early 1970s, biophysical sciences–social sciences since the early 1990s, and humanities–arts–sciences over the past dozen years.
3. As a US Forest Service scientist in seamless collaboration with academic and land manager colleagues, the stable yet dynamic community that the LTER program fosters has served as a great platform for connecting science lessons with society through many means, ranging from development of regional conservation strategies and landscape management plans to storytelling. This is a practice of citizenship by individual scientists and by a science-based team.
4. The sustained learning that the LTER program has underwritten gives scientists a foundation for communicating findings from science and discussing their implications with the public, and the forest itself is a great stage for these conversations.

PERSONAL OVERVIEW

I have had a career of immersion in the International Biological Program (IBP) and in the LTER program since its inception. After completing graduate studies in geology in 1972, I had the good fortune to join the early stages of IBP in the Coniferous Forest Biome Project at the H. J. Andrews Experimental Forest (AND) in the Cascade Range of Oregon. Our team of forest and stream ecologists, and a few earth scientists, had the decade of the 1970s to coalesce, mature, and craft stories of the ecosystems of the Pacific Northwest. The Andrews forest was a wonderful place to do that. It has a complex, ancient forest with nearly 100-m tall trees and fast, cold, clear, mountain streams whose beauty and chill takes your breath away. The year 1980 was pivotal for the group in three ways. First, Jerry Franklin led a synthesis of our team's knowledge of old-growth forests, which set the stage for major transformation in public perception and policy toward federal forests a decade later and, incidentally, changed our lives. Second, Mount St. Helens produced a disturbance beyond imagination, so the team could test its green-forest ecosystem perspectives in this newly created gray landscape and hone its communications skills, because all the world was watching and curious about the eruption. Finally, 1980 was the year that the LTER program commenced.

The LTER program at AND began where IBP left off a few years earlier, but with a big shift in perspective: we researchers needed to take the long view and assume we would pursue it. Within IBP and the AND site, I studied disturbance processes (landslides, floods, wildfire, logging, forest roads), their interactions with ecosystems, and the implications for land management. I served as principal investigator at AND for two 6-year increments of funding from the National Science Foundation (NSF), and collaborated with colleagues at other LTER sites, especially those at Harvard Forest (New England) and Luquillo (Puerto Rico), and in International LTER programs, especially in Australia, Japan, and Taiwan. My career and research at AND have centered on work at disciplinary interfaces. The scope of disciplines addressed by AND has gradually grown. Earth science–life sciences and research–land management collaborations were going strong in the 1970s, biophysical sciences–social sciences interactions began in the context of the dispute about logging on federal forest lands in the Pacific Northwest ("the forest wars") circa 1990, and in the past decade we have engaged the humanities and arts in a program we call *long-term ecological reflection*s (Swanson, Goodrich, and Moore 2008). Currently, I am retired from my position as a Research Geologist with the US Forest Service but remain quite active in the Ecological Reflections program and also in the field of volcano ecology at Mount St. Helens and several volcanoes in southern Chile, where colleagues and I encourage the use of an approach like that of the LTER program.

APPROACH TO SCIENCE

According to plan, the LTER program has accomplished a great deal of long-term ecological research through long-term experimentation, retrospective studies, and other science approaches that take both the long and near views. But the LTER program has also achieved something much more important: it has fostered a culture of sharing ideas, data, and leadership tasks as well as supported strong networking across its science community of several thousand people spanning all career stages and based in several hundred institutions. The challenge of forming such a culture comes into sharp focus for me when I witness the strong interest scientists in other countries have to establish LTER-like science programs in their home countries. However, many of these efforts have faced serious difficulties posed by compartmentalized disciplinary science cultures and limited funding mechanisms. Development of a cooperative rather than competitive culture within the LTER program has been pushed by the NSF with the use of a carrot-and-stick approach. For example, NSF has supported All Scientists Meetings that bring the community together every 3 years for intensive networking and collaboration. On the other hand, NSF has required sharing of data with the threat of funding termination for failure to do so.

My 40-plus years of experience at the Andrews forest, including the component associated with the LTER program, convince me that a major component of the US environmental research portfolio should include the approach to science that AND and several other LTER sites have expressed: long-term, place-based research and monitoring that spans the spectrum from basic to applied research with a component of active land management that can be critically discussed with the public. In the case of AND, this is a form of adaptive management carried out in a partnership of academic scientists and land managers of the US Forest Service. This model has been very fruitful and influential. The LTER program has been a critical component of the approach, providing stable, core funding and the credibility that NSF brings to the science side of the partnership.

The LTER program has helped support my career-long commitment to one ecosystem, which deepens the sense of place and strengthens concern for the well-being of one's home ecosystem. At AND and other LTER sites, these feelings and the accumulated scientific knowledge have led to very important expressions of science citizenship, as lessons from long-term, site-based inquiry are synthesized and extended regionally and more broadly through mechanisms such as development of regional conservation strategies and approaches to environmental management. Examples of such regional extensions include the roles of AND science and scientists in the Northwest Forest Plan and the Wildlands and Woodlands approach at Harvard Forest in New England.

Increasingly, we recognize our science as a cultural enterprise. Our work as scientists and collaborations with those in the arts and humanities (Figure 6.1) can influence public perceptions of our bioregional icons: ancient forests; fast, cold rivers; and volcanoes in the case of the Pacific Northwest. During the heated public debate about the future of the region's native forests, the public was schooled in lessons about forest and stream ecology in the daily newspaper, and public appreciation and awe for these marvelously complex systems blossomed. A decade-long relationship with a creative nonfiction writer, Jon Luoma, resulted in half a dozen articles about various ecology research topics from AND appearing in outlets such as the science section of the *New York Times*. This relationship culminated in the highly readable book *The Hidden Forest: The Biography of an Ecosystem* (Luoma 1999). Thus, the patient approach to science that the LTER program engenders contributes to a patient approach to communications and the opportunity to build public understanding.

FIGURE 6.1 Log decomposition experiment in the J.J. Andrews Experimental Forest. Photo by Tom Iraci, US Forest Service. April 2005.

Being strongly place-based in a region of great conflict over the ecosystems and natural resources systems under study has encouraged the development of ever-widening collaborations, such as in those with social scientists and humanists. Interacting with people from these disciplines has stretched my thinking and liberated instincts that I had suppressed for several decades in the busy, focused pursuit of science and connections with land management. It is easy to think that science findings guide decisions about management of natural resources, but personal and societal values are at the root of such decisions, and social sciences and humanities help people to understand those values. The shift from the timber era of US Forest Service management (1950s–1980s) to the biodiversity era (since the mid-1990s) highlighted the importance of social sciences in understanding public perceptions and processes of governance. We need long-term social science research to proceed in tandem with biophysical science research in the LTER program. Critically, the humanities play a special role in deepening our engagement with these places where we practice science.

In addition to valuing the opportunity to be rooted in place, the LTER program has also given me opportunities to engage with other places and the scientists who work there. Consequently, ideas can be tested in diverse systems and social contexts, and I can bring ideas from those places and people back to my home turf. A year at Harvard Forest and participation in several proposals for the Luquillo (Puerto Rico) LTER afforded such opportunities.

ATTITUDES TOWARD TIME AND SPACE

As a geologist in a community dominated by ecologists with generally much shorter research time horizons, I typically find myself taking broader geographic and temporal views of ecosystems and issues than do many colleagues. Consequently, we must work together to develop a multiscale framework for collaborative undertakings. Early in my career, I made a table of geological and ecological phenomena at each of a series of time scales from days to tens of millions of years, and looked for intersecting geological-ecological phenomena at common times scales. This has been a useful device for communicating across disciplines. Broad system thinking and the sustained communities of the LTER program facilitate development of multiscale frameworks for doing science; it takes time for concepts and shared vocabulary to emerge in interdisciplinary work.

COLLABORATION

The continuity that the LTER program has provided for AND has revealed the richness of learning that can come from collaborations, especially seemingly unlikely ones. The study of the roles of deadwood in streams provides a rich example in the biophysical sciences; stream ecologists, forest ecologists, and geomorphologists needed to work together to understand the functions and dynamics of this system through floods, windstorms, and logging. Each field had avoided the topic because it was inconvenient and nontraditional, and there was not much wood in many study streams elsewhere because of past land use and other factors. In the wood-rich AND, it took all three perspectives to obtain the whole story. Although this may seem like an arcane research theme relegated to the big forests in only a small corner of the world, it turned out to have geographically broad significance to matters as diverse as carbon and nitrogen budgets for stream ecosystems and management of deadwood to sustain native aquatic species, including salmon.

Another intriguing example of collaboration comes from an ecosystem informatics program, which is an outgrowth of long experience in information management at AND. Mathematicians and computer scientists have valued the opportunity to join with environmental scientists to work on problems based in AND and within the interdisciplinary LTER community. Perhaps this has roots in the European mathematics community with a tradition of gatherings to ponder tough questions while walking in the Black Forest of Germany. Collaboration at the scale of the LTER network has been facilitated by a variety of mechanisms, such as ongoing working groups, triennial All Scientists Meetings featuring workshops convened by scientists around their favorite topics, and movement of scientists among sites as a result of career changes and visits.

APPLIED RESEARCH

Applied science and application of science to policy and management have taken many forms at AND. AND has been a seedbed for ideas that have influenced science, forest land management, and natural resource policy. The research–management partnership based at AND covers the full research–development–application spectrum with NSF-sponsored LTER at the basic research end, complemented by applied studies conducted jointly by researchers and US National Forest staff (Swanson et al. 2010). At the applied end of the spectrum, US National Forest partners implement innovative practices informed by science findings. Basic studies include characterization of old-growth forests, succession spanning nearly a millennium, and roles of deadwood in forests and streams. Applied studies concern exploration of alternative silviculture practices and use of interpretations of historic wildfire regimes to guide plans for future landscape management (Cissel, Swanson, and Weisberg 1999). The research–manager partnership involves two-way communications and frequent joint hosting of field tours to view long-term science and possible implications for new management approaches. This is a continuing, critical public discussion of the future of forests and forestry in the region.

COMMUNICATION

I am shy and especially dislike classroom teaching, so I have been fortunate to have had a US Forest Service research position rather than primarily being an educator, but telling stories in the forest is a great pleasure. AND and the Mount St. Helens blast zone provide amazing stages for communication with any person. The scenes are spectacular and visitors welcome stories of the place and help in making sense of all the unfamiliar complexity and immensity. With the great honor to work in such places comes a sense of responsibility to tell their stories and to facilitate access by others to find their own stories there—other scientists, artists, writers, and the public. The continuity that the LTER program provides for AND and the model that it provides for dealing with science and communications at Mount St. Helens have been critical to experiencing the evolution of stories and makes the stories richer.

MENTORING

In mentoring junior colleagues and students, I emphasize the values of the open culture and large, diverse LTER community that pushes networking beyond what is common in many discipline-based and professor–laboratory group opportunities. For example, I encourage researchers to take part in the LTER All Scientists Meetings, which are a model of open culture interactions where participants of all career stages share ideas and enthusiasm in workshops and informal interactions on dozens of topics of shared interest.

SKILL SET

As an earth historian (geologist), my standard tool kit is simply shovel, hammer, and hand lens for making observations of outcrops and landforms. Despite not being much of a tool user, some new technologies have benefited my work. For example, light detection and ranging (LIDAR) ground return images depicting topography at 2-m resolution have made it possible to interpret landforms that were very difficult to detect with the 80-ft contour maps I used in the early 1970s to study the vegetation-obscured Andrews forest landscape. The LTER program has impressed me with the importance of information management (IM) systems. These have proven to be very valuable for engagement with the humanities, a field quite unfamiliar with IM. As we developed the Ecological Reflections program, for example, we built a "database" of works from the writers in residence, including both their "raw data" (e.g., journal entries and draft poems) and finished products (published works). We have posted these cultural data online so that they will be available for enjoyment in the present and also for future analysis of our evolving cultural engagement with the land.

PERSONAL CONSEQUENCES

The LTER program strongly shaped the intellectual and social dimensions of my life, making possible the sustained engagement with one beautiful place, the Andrews forest, and also connecting me with the wide-ranging and very stimulating community of colleagues and friends across the network. A distinctive and recent personal consequence of this connection has been the chance to use the LTER program as a template for the Long-Term Ecological Reflections program (Swanson, Goodrich, and Moore 2008). In addition to IM, the Long-Term Ecological Reflections program benefits from infrastructure for science, created in large part through the LTER program and US Forest Service funding—field facilities, long-term experiments that are sources of inspiration, and a community of scholars. Since the Andrews Forest Ecological Reflections program began in 2002, more than 50 creative writers and philosophers have produced a rich collection of cultural data about the forest and human engagement there, including the science taking place. Now more than 20 programs of arts–humanities–science collaboration are underway at sites of long-term inquiry in the United States, many of them at LTER sites (www.ecologicalreflections.com).

CHALLENGES AND RECOMMENDATIONS

A challenging feature for the growth of the LTER program is its evolution beyond the familiar confines of organizational features internal to NSF and at the outer boundaries of NSF's scope. In some cases, there has been flexibility, in others not. Internal issues sometimes come down to turf and disciplinary traditions that make it difficult to assemble diverse teams within a single, large project. The scope of NSF has limits in terms of capacity to review proposals and projects suitable for funding. Growing collaboration among NSF and the National Endowments for the Arts and for the Humanities is a positive sign.

CONCLUSION

A feature of the LTER program that I greatly appreciate has been the incredible room for interdisciplinary work that it encourages. I can see my early instincts for interdisciplinarity in field research experiences as an undergraduate in Bermuda (sea water biogeochemistry, biology, geology) and as a graduate student in the Galápagos (volcanology, botany, seismology). The LTER program has been a very important training ground for me and many others in administrative and science leadership, where we researchers can gain experience leading small and large, formal and informal teams. These opportunities have been both more diverse and more rewarding than I can imagine in many academic and agency settings.

As I have learned over the years, leading an LTER program is very challenging because of the need to balance long-term experiments and observation programs while also attending to the issues of the day and including a component of pioneering work in the research portfolio. This tension gets more complicated as the record length grows, data management burdens increase, experiments age, and the science and societal contexts of the work evolve. The major impediment to keeping this work going has been that funding has not kept pace with the opportunities and sense of responsibility to do more. Also, as we have extended the disciplinary scope of our collaborations to social sciences and humanities, compartmentalization within NSF and between NSF and other federal agencies, such as National Endowments for the Arts and for the Humanities, has made it difficult to fund highly interdisciplinary work, especially on a sustained basis. There is no counterpart to the LTER program in the arts and humanities arenas.

In sum, I believe that the LTER program at both site and network scales is fulfilling its original mission to conduct long-term experiments, assemble long-term observations of gradual phenomena, be in a position to assess effects of periodic disturbance events, and curate and share the resulting data and stories. But an even greater payoff from more than three decades of research (and in some cases a decade of IBP before that and sometimes several decades of US Forest Service Experimental Forest life even before that) has been the creation of a science community with such an open, wide-ranging culture. At a personal level, AND has given me the opportunity to pursue socially relevant, intellectually challenging work in wild and beautiful landscapes with great colleagues. I hope the long-range outcome will be a community of scientists, land managers, artists, and humanists that is best able to respond to the mammoth environmental issues confronting society. I urge young scientists to engage with the LTER system in whatever ways best suit their circumstances: as outside observers seeking models of science culture, as active collaborators, and, perhaps, ultimately as science leaders.

ACKNOWLEDGMENTS

I am very grateful to the NSF and the US Forest Service Pacific Northwest Research Station for their many forms of support of this work at the H. J. Andrews Experimental Forest over the entire length of my time there.

REFERENCES

Cissel, J.H., Swanson, F.J., and Weisberg, P.J. (1999). Landscape management using historical fire regimes: Blue River, Oregon. *Ecological Applications*, **9**, 1217–1231.

Luoma, J.R. (1999). *The Hidden Forest: The Biography of an Ecosystem*. Henry Holt, New York (republished by Oregon State University Press, Corvallis, 2006).

Swanson, F.J., Eubanks, S., Adams, M.B., Brissette, J.C., and DeMuth, C. (2010). *Guide To Effective Research-Management Collaboration at Long-Term Environmental Research Sites*. General Technical Report PNW-GTR-821. U.S. Department of Agriculture, Forest Service, Pacific Northwest Research Station, Portland, Oregon.

Swanson, F.J., Goodrich, C., and Moore, K.D. (2008). Bridging boundaries: Scientists, creative writers, and the long view of the forest. *Frontiers in Ecology and the Environment*, **6**, 449–504.

PART THREE
ARCTIC (ARC) LTER SITE

7 Bridging Community and Ecosystem Ecology at the Arctic Long-Term Ecological Research Site via Collaborations

Laura Gough

IN A NUTSHELL

1. My research in the Long-Term Ecological Research (LTER) program helped to shape me into the ecologist that I am, working at the interface between communities and ecosystems on a variety of questions.
2. As a university educator and public speaker, I incorporate examples of LTER site-based empirical and theoretical research, as well as cross-site meta-analyses in my

teaching and presentations. My awareness of long-term research, in particular the response of North American ecosystems to global change, is heightened by my interactions within the LTER network.

3. Working in the LTER program has provided me with opportunities for collaborations both within the Arctic site and across the network. The LTER program has thus inadvertently provided the framework for all of my current and recently funded research projects.

4. These collaborations assisted in sustaining me through major life events, particularly having children, by helping me maintain my research productivity when my family required more of my time and attention.

PERSONAL OVERVIEW

Currently, I am a professor in the Department of Biology at the University of Texas at Arlington. I teach undergraduate and graduate courses in botany and ecology, and I also supervise MS and PhD students working in the tundra at the Arctic (ARC) LTER site and locally on urban ecology questions. I earned my PhD in plant biology from Louisiana State University and have been affiliated with ARC site since 1996, when I was hired as a postdoctoral scientist by Gus Shaver on a related grant. Since 1999, when I started my first faculty position, I have been an independently funded researcher affiliated with the ARC site, and for the past few years I have served as a member of the ARC Executive Committee. My research at ARC site is at the interface between the community and the ecosystem. My contributions to site-specific understanding have focused on the factors (abiotic and biotic) that control tundra plant species diversity, including the role of consumer species (Figure 7.1). In addition, I have been involved in a cross-site working group

FIGURE 7.1 Laura Gough assessing shrub height for a new experimental plot to examine how shrub vegetation affects associated arthropod and songbird communities near Toolik Field Station, Alaska, in May, 2010. (Photo courtesy of Laura Gough.)

in the LTER network (now called PDTNet: Productivity-Diversity-Traits Network) since 1996. Representing the ARC site in this group, one of my main contributions has been to emphasize the clonal growth of most of the plants in the herbaceous sites (grasslands, marshes, tundra) within the network, thus ensuring a better understanding of the interactions within these communities that may be affecting ecosystem responses to change.

APPROACH TO SCIENCE

During my first few years at the ARC site, I shifted from being an individual conducting research in plant community ecology in relative isolation to collaborating with theoretical and empirical ecologists working on much broader questions, mostly at the ecosystem level. I applied for a postdoctoral position based at ARC site almost on a whim. My undergraduate and dissertation research had focused on plant community ecology of coastal marshes, and I had little knowledge of the Arctic or of ecosystem ecology, beyond a graduate wetland soil biogeochemistry course. At the ARC site, I was suddenly in a research community comprising scientists from multiple institutions, many with extensive expertise in biogeochemistry, and with interests running the gamut from soils to plants to remote sensing to aquatic environments. Because the ARC program is based at Toolik Field Station, and not all researchers there are affiliated with the ARC program, I benefited from multiple interactions outside the LTER program as well, particularly with botanists who had extensive Arctic expertise.

My views of the importance of communities and particular species are continually challenged at the ARC site by my biogeochemist colleagues in a friendly and constructive way. The biotic interactions that I placed so much weight on as a student, particularly competition, were not considered the drivers of ecosystem carbon and nitrogen cycling in the tundra, and thus I was forced to justify my focus on species diversity and the factors that control community composition. This environment helped me to mature my own ideas that I have continued pursuing, as I have been able to demonstrate that biotic interactions can be important in Arctic tundra environments (e.g., Gough et al. 2012b). Simultaneously, because of my interactions with colleagues at the ARC site, my research fits into a broader context of element cycling and responses to global change.

In addition, participation in cross-LTER site activities has provided me with opportunities to ask larger-scale questions using data from multiple locations and meta-analysis techniques. For example, the LTER working group I attended in 1996 highlighted the kinds of data being collected across the LTER network and provided me with an "in" to access data from other sites. I learned about data access issues, deciphering metadata, and meta-analysis techniques. The workshop participants were open about sharing their data to look for cross-site trends, and this attitude encouraged my own support of data accessibility policies, partly because I was able to experience the power of such cross-site analyses in terms of detecting general patterns across biomes (e.g., Waide et al. 1999). Since then, my colleagues and I have published results from our cross-site meta-analyses focused on how plant traits influence the response of plant communities, productivity, and diversity to changes in the environment (e.g., Suding et al. 2005; Clark et al. 2007; Gough et al. 2012a). These activities also provided opportunities for new, smaller-scale collaborations. For example, I was able to successfully pursue National Science Foundation (NSF) funding with one of the other original workshop participants (John Moore) to broaden my research to include interactions among plants, soil communities, and food webs (e.g., Gough et al. 2012b).

Finally, although I do not incorporate a socioecological perspective into my research at the ARC site, my affiliation with the LTER program has affected my thinking on this topic. Because of my involvement in the LTER network, I know a great deal about urban research conducted at the two urban LTER sites. For several years I have been working with colleagues in North Texas to try to get local researchers to coalesce around topics related to urban ecology, and I supervise graduate students investigating related questions. I am not sure that I would have felt this was possible given the other demands on my time without my exposure to urban research through the LTER network.

ATTITUDES TOWARD TIME AND SPACE

One of the luxuries of conducting research at the ARC site is the ability to collect data on truly long-term experiments (some data 30-plus years old). These long-term data sets are particularly important in the Arctic, because environmental changes and biotic responses are relatively slow compared to those with warmer latitudes. The 6-year funding cycle of ARC program ensures that these experiments continue while my own, shorter-term grants provide the means for data collection. This framework therefore allows me to ask questions at multiple temporal scales simultaneously and to leverage long-term experiments at the ARC site when seeking my own funding.

In terms of space, working in the Arctic is challenging because topography over small and large spatial scales has dramatic effects on soil properties and microclimate—and thus the plant communities. Several researchers at the ARC site work at the landscape scale, both empirically and theoretically, which helps me put my plot-based research in a larger spatial context.

COLLABORATION

To ask large, societally relevant environmental research questions, such as those surrounding the impact of global change on Arctic tundra and its feedbacks to the climate system, requires a group of collaborators. The ARC program maintains a collegial environment, partly because of the long-term collaborations among the senior scientists and the atmosphere this creates for new members. For example, I can write proposals to fund my research team to collect data on extant experimental plots supported by the LTER program, but I do not have to pay for the maintenance myself. By piggy-backing on the infrastructure of the ARC site, but expanding the questions I ask beyond the initial framework of the LTER program, this collaborative structure represents a value-added investment for all.

These experiences in the LTER network have shown me the advantages and disadvantages of collaborative research. I have seen collaborations break down because of poor communication, unbalanced effort, and unwillingness to share data and credit for research. On the other hand, I have experienced the intellectual stimulation of thinking outside my own box, of learning something new about the system in which I am working, and of helping a colleague identify a possible mechanism to explain something she has documented by bringing in a community perspective. One reason the PDTNet group has been so productive is because the participants are (1) open and welcoming of new collaborations and (2) openly communicate regarding authorship plans and credit.

APPLIED RESEARCH

Because ARC site is remote, away from the direct influence of human activities, my research there has not led me to personally engage in "applied research," schoolyard activities, or other local outreach in Alaska. However, the very nature of studying the effects of climate change lends itself to outreach regarding the implications of combustion of fossil fuels, which I discuss in the next section.

COMMUNICATION

Although I do not think my activity at ARC site has directly changed my outreach or science communication in a place-based manner, when I speak to local environmental groups in Texas, I always couch my research within global change and talk about the Arctic as a part of Earth that is particularly vulnerable. For example, in a recent invited presentation to the Dallas chapter of the Native Plant Society of Texas, I spoke about the global carbon cycle, highlighted ARC research on this topic, and brought the theme "home" by describing North Texas native ecosystems and the ways in which prairies and savannahs contribute to the global carbon cycle. Because of my participation at ARC site, I can speak easily about both aquatic and terrestrial ecosystems, including about research projects far removed from my own that help us understand the Arctic system as a whole. Over the past few years, I have devoted more of my time to public outreach, but it is difficult to directly attribute that to the LTER program because I am now at a point in my career when I can shift some of my energy toward interaction with the public, an activity that I find very rewarding because of the opportunities for informal education.

MENTORING

I have been fortunate to have had outstanding mentors from ARC and other LTER sites. In particular, Gus Shaver, my postdoctoral mentor, taught me not only about ecosystem science, but about writing a budget, authoring a proposal review, and managing a field crew. He challenged me scientifically and provided me with logistical and moral support when I began pursuing my own funding. Along with those from ARC, my PDTNet colleagues (both peers and mentors, particularly Kay Gross from Kellogg Biological Station and Scott Collins from Sevilleta) have helped me through the years in which I was having children, provided advice when I needed to change institutions, counseled me when I was applying for promotion, and remain valuable colleagues within the LTER network.

Having had excellent mentoring myself provides me with a strong foundation to serve as a good mentor to junior scientists. I urge students and postdoctoral fellows in my laboratory to cultivate a "village" of mentors with different attitudes and professional and personal experiences. Junior scientists affiliated with the ARC site have many such opportunities. This expanded circle of mentors has been particularly important as my home institution has a small ecology program. I am able to recruit excellent laboratory members with the "carrot" of spending the summer at Toolik Field Station and interacting with an incredible group of scientists. My laboratory members also attend our annual spring planning meeting, providing an opportunity to network with other ARC scientists and develop new projects. My students then bring these experiences back to our home institution and enrich our entire graduate program with what they have learned as part of the LTER network.

SKILL SET

Working at the ARC site exposed me to collaborative research and provided an important education for me in the areas of biogeochemistry and ecosystem modeling. My scientific tool kit expanded substantially. For example, during my first summer as a postdoctoral fellow, I learned how to analyze soils for nitrogen and phosphorus availability at the field station under the tutelage of a senior research assistant in the ARC program. This knowledge of the subject matter and the required tools changed the questions that I chose to pursue. In addition, I learned meta-analysis techniques through my cross-site LTER collaboration as well.

PERSONAL CONSEQUENCES

I cannot overstate how important my research and collaborators at the ARC site were to keeping my research program going while I was having children. I was quite fortunate to have research funding through my own grants, but my involvement in ARC program meant that I had additional help with data collection, student supervision, and other aspects of fieldwork, when I could not be as involved or be in the field at all. I also had manuscripts in progress with my PDTNet collaborators who allowed my roles on various projects to shift to accommodate my growing family. My colleagues in the ARC program were patient as my involvement in the project slowed down, and they welcomed me back as I was able to take on administrative duties such as the Executive Committee 2 years ago. Perhaps this would have been possible with other collaborations that involve field station–based research, or if I were a member of a larger research group at my home institution. Regardless, my colleagues in the ARC program were crucial to maintaining career activity during a time when my family was my top priority.

CHALLENGES AND RECOMMENDATIONS

One of the challenges for the LTER network is how to be recognized as a valuable contributor to ecological knowledge without being resented as a private club by the broader ecological research community. I have colleagues who believe that science in the LTER program is not evaluated with the same rigor as are individual proposals and that the budgets are essentially preventing other, "better" science from being done. This can backfire on researchers in the LTER network if reviewers and panelists carry this thinking into their recommendations for funding. I completely agree that those of us associated with LTER sites benefit from the 6-year funding cycle in ways that scientists affiliated with other research stations and sites do not. I also benefit from my LTER affiliation in that my budgets have been reduced because I can get help with some logistics expenses from the ARC grant itself, such as maintenance of long-term experiments. However, the structure of the ARC program means that affiliated researchers, such as me, receive no direct research funding from the grant; I have had to acquire my own research grants to fund my work at the ARC site. This funding model is not used by all LTER sites, so the benefits (particularly financial) of being affiliated with an LTER program differ, and it is incumbent on us as individuals to make this clear in proposals and research products.

Perhaps the most important way that the LTER program could foster careers of junior scientists and educators is by continuing to develop a reputation as an open, welcoming, inclusive group. Much of this spirit of inclusiveness needs to happen at individual sites, but some of it can also be facilitated by the LTER network office. For example, recruiting "new blood" into the LTER program can occur via research opportunities for junior and more established scientists to join forces with current scientists in the LTER network to answer new questions. As many LTER sites transition to new leadership, these issues need to be directly addressed so that junior scientists can take advantage of the LTER program while embracing the challenges of collaborative research. There might be mechanisms by which intersite exchanges of personnel, particularly graduate students and postdoctoral fellows, could be facilitated so that junior scientists would be exposed to different site-based research and build their own networks of collaborators.

CONCLUSION

My experiences at the ARC site and in LTER Network have had a profound influence on my professional identity. In particular, collaborations outside my institution are crucial for me to maintain an active research program, given the resources at the University of Texas at Arlington. The ARC program provides me with an excellent recruiting tool for graduate students and a community of ecologists that keeps me focused and energized. For junior scientists at institutions that are similar to mine, these kinds of communities may mean the difference between success and failure, because of the difficulty of finding colleagues researching similar questions and collaborators in a smaller university or college. In addition, the data sets collected at LTER sites are a treasure trove for meta-analyses and opportunities for developing new collaborations with those who have collected the data. Collaborations in general, however, can also come with potential substantial costs. Junior scientists beginning their careers should discuss issues of collaboration with their mentors, and they should learn how to maintain open lines of communication to try to minimize conflicts regarding roles on the project, authorship, and credit. Because the LTER network is inherently a collaborative group, junior scientists should find many opportunities to learn how to manage collaborations successfully.

ACKNOWLEDGMENTS

As described in this essay, I am grateful to many of my LTER colleagues, including G. Shaver, S. T. Collins, A. Giblin, K. Gross, G. Kling, J. Moore, S. Pennings, E. Rastetter, and K. Suding. In particular from ARC, my scientific collaborations and personal interactions with S. Hobbie and N. Boelman have been invaluable. The Arctic LTER is funded by NSF grant DEB-1026843 to the Marine Biological Laboratory. Recent funding of my collaborative research projects based at ARC has also come from NSF (OPP 0908602 and 0909507).

REFERENCES

Clark, C.M., Cleland, E.E., Collins, S.L., Fargione, J.E., Gough, L., Gross, K.L., Pennings, S.C., Suding, K.N., and Grace, J.B. (2007). Environmental and plant community determinants of species loss following nitrogen enrichment. *Ecology Letters,* **10**, 596–607.

Gough, L., Gross, K.L., Cleland, E.E., Clark, C.M., Collins, S.L., Suding, K.N., Fargione, J.E., and Pennings, S.C. (2012a). Incorporating clonal growth form clarifies the role of plant height in response to nitrogen addition. *Oecologia,* **169**, 1053–1062.

Gough, L., Moore, J.C., Shaver, G.R., Simpson, R.T., and Johnson, D.R. (2012b). Above- and belowground responses of arctic tundra ecosystems to altered soil nutrients and mammalian herbivory. *Ecology,* **93**, 1683–1694.

Suding, K.N., Collins, S.L., Gough, L., Clark, C.M., Cleland, E.E., Gross, K.L., Milchunas, D.A., and Pennings, S.C. (2005). Functional and abundance based mechanisms explain diversity loss due to soil fertilization. *Proceedings of the National Academy of Sciences U.S.A.,* **102**, 4387–4392.

Waide, R.B., Willig, M.R., Mittelbach, G., Steiner, C., Gough, L., Dodson, S.I., Juday, G.P., and Parmenter, R. (1999). The relationship between productivity and species richness. *Annual Review of Ecology and Systematics,* **30**, 257–300.

8 Long-Term Ecological Research in the Arctic
Where Science Never Sleeps

John E. Hobbie

IN A NUTSHELL

1. When the Arctic (ARC) Long-Term Ecological Research (LTER) project began, I was an aquatic ecologist with experience in managing large projects in freshwaters and estuaries and a specialization in microbes. This project, which studies lakes, streams, and tundras, has greatly increased my breadth as an ecologist and allowed me to take part in terrestrial modeling, microbial studies in streams, and the role of soil mycorrhizal fungi in providing nutrients to many species of plants.
2. As a mentor to several postdoctoral fellows, my LTER research has enabled me to learn about other fields such as the application of molecular biology to microbial ecology.

3. The Arctic LTER project data, the long-term field experiments, and the facilities available at the University of Alaska field station brought me in contact with ecologists from many countries. One result of this association with experts was my coauthorship of a book on Arctic natural history aimed at communicating scientific knowledge to scientists and the general public unfamiliar with the Arctic (Huryn and Hobbie 2012).
4. I have always collaborated extensively with many scientists and encouraged collaboration as the best way to carry out ecosystem research. The Arctic LTER project brought many opportunities to broaden the scope of my collaboration to include terrestrial ecologists and microbiologists.

PERSONAL OVERVIEW

My PhD research was about year-round primary productivity of an Arctic lake but while on a postdoctoral fellowship at Uppsala University, Sweden, I switched to an emphasis on bacterial uptake kinetics in lakes. The techniques I helped develop in freshwater worked in the ocean and estuaries too (Hobbie and Williams 1984). In addition we developed the epifluorescence method for quantifying the abundance of planktonic bacteria. Our paper (Hobbie, Daley, and Jasper 1977) finally convinced oceanographers that bacteria are abundant (at 10^9 per liter) and important. Recently, I have used my understanding of kinetics of uptake to analyze microbial activity in the soil.

My Arctic expertise led to leadership of the aquatic part of the International Biological Program (IBP) at Barrow, Alaska, beginning in 1970. We (28 scientists, graduate students, and postdoctoral fellows) studied shallow ponds to quantify the carbon, nitrogen, and phosphorus cycles. In addition to nutrient cycling, we carried out whole-system experiments with fertilization and with crude oil. After the IBP, the aquatic ecologists took advantage of the newly opened pipeline support road to travel inland from the Arctic Ocean and study a foothills lake, Toolik Lake. In 1987, this location became the Arctic (ARC) site of the LTER program.

In the early 1980s, there were no coastal LTER projects. For this reason, a meeting was held at the Marine Biological Laboratory (MBL) at Woods Hole that laid out the need for long-term coastal research. From this meeting there developed a national study of coastal waters and estuaries that I coordinated; four projects were funded by the Ocean Studies Program of the National Science Foundation (NSF). This study, titled the Land-Margin Ecosystem Research (LMER), had the built-in problem that there was an open competition after the period of each project was completed. In contrast, the LTER projects are renewed if sufficient progress (e.g., publications and graduate student training) has taken place. This difference took away the long-term aspect of the LMER projects. Eventually, the LMER merged into the LTER program (several present LTER projects had their start as LMER sites; (Gholz, Marinelli, and Taylor, Chapter 3). This merger marks the start of the Ocean Studies contribution to the LTER program.

APPROACH TO SCIENCE

The LTER experience has influenced my ideas about how research should be carried out in several ways. First, I learned the importance of very good climate information. Summer climate data was collected at Toolik Lake from 1975 to 1987. By the time the LTER started in 1987, it was becoming clear that warming was occurring throughout most of Alaska and that the project had to invest in reliable instrumentation for winter data collection to build a more complete data set. One goal of the project is projecting the ecological changes into the future, and to do this scientists need accurate long-term climate data as well as experimental warming. They also need to know, if possible, how much climate change has already occurred at the LTER site.

The second idea that I gained from my participation in the LTER program is the importance of long-term manipulation experiments, some lasting more than 25 years. The changes seen in the first 9 years were not the same ones seen in later years. For example, the results from the stream fertilization experiment (extremely low amounts of phosphoric acid continuously added over each summer) changed after 9 years from an increase in the primary production of diatoms living on the rocks to a dramatic increase in moss cover in the stream; this changed the entire food web.

Third, it is very valuable to encourage scientists from outside the LTER program to sample long-term experiments and use the accumulated LTER data. The ARC site is one of a handful of Arctic research sites worldwide and is the only intensive US site. Many of the scientists from outside the LTER program who come to the Toolik Field Station (TFS) make use of the LTER data. A number also collect samples from the LTER experiments. The LTER project encourages these visits by paying for transportation and room and board at the TFS. One example is the research of a graduate student from the University of British Columbia who worked on the species and importance of mycorrhizal fungi (Deslippe et al. 2010). She chose to work at the ARC site because of the wealth of environmental data—much more than at any Canadian Arctic site. She discovered that there had been an unexpected switch in the species of ectomycorrhizal fungi in the long-term heated plots, a change from fungi that used mostly amino acids to species of fungi whose hyphae extended up to a meter from the plant roots and whose enzymes could break down very resistant organic molecules. This is one of the few examples of joining structure and ecological function for soil microbes.

Finally, I have learned more about the value of conducting syntheses. Putting together the LTER synthesis book for the ARC site triggered several questions that, while not exactly new, were very important. The synthesis came about at this time because each LTER site is expected to publish a synthesis volume summarizing past results and bringing in the scientific literature to explain and extend the results. One insight gained during the writing of the synthesis book was the finding that the 20 years of climate data showed no statistically significant trends in the mean annual air temperature (Cherry 2014). Yet, the temperature of deep permafrost was increasing over this period, and plant growth increased. One possible reason was that the aggregated temperature degree-days had increased over time—sort of a cumulative measure of the total heat affecting plant growth or microbial decomposition. We should have carried out analysis on this topic earlier, but the LTER synthesis forced us to realize the importance of this question.

ATTITUDES TOWARD TIME AND SPACE

Working at the ARC site has given me the opportunity to take part in modeling that predicted the future state of the Arctic ecosystem or of a region such as a large watershed. I have had to think about the time course of changes in vegetation and soil chemistry at the Toolik Lake ARC site and have also taken part in several mathematical modeling exercises. My previous modeling experiences were descriptions of carbon and nitrogen cycling, sort of a whole-system view of annual cycles. With the experiments and long-term observations at the ARC site, it has been possible to construct predictive models of the ecological effects of changes scientists expect to see over a century. For example, the effect of soil warming on nutrient release from the soil organic matter and what will happen to allocation of carbon resources in the vegetation (e.g., to leaves or to roots).

The basic observations about the natural system of the Arctic are provided by the ARC site. However, to use these observations in a mathematical model, the information from the experiments must also be incorporated. At the ARC site, we were able to fund both the observations and the upkeep and sampling of the long-term experiments. There is, however, an inherent conflict because the LTER funds go almost entirely to maintaining the experiments and making observations. There are no funds for a modeler and a postdoctoral fellow to construct a model. This conflict has been solved by individual scientists obtaining grants (mostly from NSF). This solution requires that the scientists involved are able to get funded. The same solution has been applied to the problem of how to bring in new questions and themes to the project. For example, it has become obvious that there are new thermokarst features forming on the landscapes around Toolik Lake. These arise when ice formed in permafrost melts and erosion channels or soil slumps occur. A geomorphologist was brought in, and a new project was eventually funded. Although this model of how to use the LTER long-term data in spatial and temporal modeling has worked for the ARC site until now, it may not be possible to continue in this way as funding becomes more and more difficult to obtain.

COLLABORATION

My experiences in the LTER program have only increased my belief in the value of collaboration. I have understood for a very long time how this valuable tool can be achieved, yet there has been a 30-year diminution of the size of regular ecosystem grants at NSF that has made it more and more difficult to carry out collaborative research. Being able to bring four or five scientists together on a grant is now a rarity and it is even difficult to bring two together because of funding limitations.

Working at the ARC site has widened my collaboration through our experiments with additions of nutrients to whole lakes and large lake chambers (limnocorrals). One big change that began through my involvement in research at the ARC site is my study of nutrients and mycorrhizae (fungi and plants). This entailed working with a mathematician, an isotope chemical ecologist, and a mycologist.

APPLIED RESEARCH

Before LTER funding at the ARC site began, back in the 1970s, we conducted a number of applied projects associated with the oil pipeline and the associated support road. For example, we investigated the impact of road dust and of an oil spill on tundra plants and on lake ecology. There has been no applied research as a part of the LTER program, although there are plans to communicate with Alaskan native communities about the projected changes in vegetation on the North Slope.

COMMUNICATION

The ARC site is located in a remote part of a remote state with the nearest villages accessible only by air. Barrow is 480 km to the northwest of Toolik. There is an outreach part of the project, but it is minimal in this particular situation. However, ARC has brought me in contact with the Arctic experts in natural history (i.e., ornithology, mammalogy, microbiology, plants), so Alex Huryn and I wrote a book (Huryn and Hobbie 2012) on the natural history of Arctic Alaska, a book in the LTER Network Series. This book is an attempt to communicate with the public that is interested in natural history, with the conservation community interested in wilderness, and with visitors to the North Slope. Groups that have bought multiple copies of the book include British Petroleum (Natural Resources Division) and guiding organizations that run rafting trips on the North Slope.

The ARC project also has a connection with the students in the Barrow public schools and transfers some funds to the village to organize a weekly science talk aimed at the native community. The students have visited the research sites (control and fertilized sites) that are set up for them at Barrow. Unfortunately, the science teachers at the Barrow schools all leave the village during the summer, so there is little opportunity for summer student research.

MENTORING

The MBL is not a degree-granting institution, but for the past 6 years it has had a graduate program in association with Brown University. However, many graduate students who have carried out research at ARC have been supported by research grants other than those from the LTER program. My experiences with the few postdoctoral fellows that the LTER program has paid for have all been very positive and worthwhile because our collaboration resulted in new scientific products such as mathematical models and new information on the molecular biology of aquatic microbes. The ARC project, however, has brought in a number of young scientists to work on the project by paying for travel to Alaska and for stays at the TFS, where they collected data that improved their proposals to NSF.

SKILL SET

When my involvement in the LTER program began in 1987, my scientific expertise was in limnology and in the ecology of aquatic microbes; now it has broadened to include the ecology of terrestrial microbes. This may not sound like much of a jump, but these are two very different fields; microbial ecologists of soils know almost nothing about the microbial ecology of water. This intellectual isolation is made even worse by the difficulty of making measurements in soil—any handling of soil samples is likely to destroy the soil structure and introduce sampling artifacts. My contention, expressed in several papers, is that there is a great deal of similarity in what bacteria are actually doing in both systems (Hobbie and Hobbie 2013).

In the ARC project, we have had the resources to bring to the Alaska site several scientists with stable isotope experience—this allows analysis of the changes in natural abundance of the stable isotopes that happen in undisturbed ecosystems. Through discussions with these experts, I analyzed the ^{15}N content of the plant leaves and the fruiting bodies of ectomycorrhizal fungi. The resulting paper (Hobbie and Hobbie 2005) showed that most of the nitrogen in the plants entered by way of the mycorrhizae rather than by direct uptake by roots from soil water.

PERSONAL CONSEQUENCES

Working with the LTER program has not led to any personal changes. When the ARC project began in 1987, I had already been carrying out large-project long-term research for 20 years in estuaries and in the Arctic. We had been testing ways to operate the ARC site for many years before its establishment as part of the LTER program in 1987.

However, I have very much enjoyed the opportunity to work with my son, who is a stable isotope expert based at the University of New Hampshire. This occurred because of the ^{15}N distribution in plants at the ARC site and the similarity to the isotopes in plants at his thesis research site, at Glacier Bay, Alaska.

In addition, the LTER meetings certainly have broadened my circle of scientific acquaintances. I have enjoyed the scientific contacts and discussions with outstanding scientists. These discussions and social events were not only enjoyable, but I have learned a lot. For example, just about everything I know about soil ecology has come from discussion and arguments, not from courses or textbooks.

CHALLENGES AND RECOMMENDATIONS

I have had a lot of good fortune in my scientific career, and it is difficult to identify impediments. My postdoctoral research led me into studies of the role of bacteria in lakes, estuaries, and oceans, a field desperately in need of new methods. I obtained a university position where I could teach mainly about limnology and oceanography, and then I moved to the (then) new institute at the MBL in Woods Hole. There I could interact with several top-flight ecosystem scientists and learn about modeling, stable isotopes, and forest ecology. The LTER program has added to my understanding of the importance of modeling and of large-scale, long-term experiments in ecology.

One opportunity that I believe has been missed within the LTER program is interchange about ways to operate among the leaders of the LTER projects. There is no standard way for such projects to be organized, write proposals, and write a synthesis book. This is fine, except that a number of times I have found out after the fact that there was probably a better way to do some task and I had not asked the right question. I am not asking for uniformity, just exposure to a wider range of ideas. These could have been aired at a meeting with a panel discussion focused on the question "What problems arose when LTER projects handled X?" For example, I have just finished coediting the Arctic LTER synthesis book, and I would have liked to know what problems other projects had with their books. I was surprised to learn that some other sites hired professionals to construct the index. I would have asked about costs and also about the gains and losses with a professional writer, such as the loss when the indexer does not understand an overarching theme and does not realize that certain paragraphs and figures should be included in references to this particular theme. Another question that has never been discussed is the makeup of the various proposals. I would have liked to have known what the NSF panels liked or disliked about a proposal or perhaps what the project leadership felt they should have done but could not because of time or space in the proposal.

CONCLUSION

Overall, the LTER program has had a good effect on my professional identity. This effect has occurred because I have had opportunities to carry out new kinds of research and work with new acquaintances at the project level. Because of my early experience with large research projects, I do not believe that my participation in the LTER program and projects has resulted in new leadership skills or a new realization of the importance of collaboration.

I would recommend that young scientists work at an LTER site. Collaboration with very good scientists, and a detailed knowledge of a single site is the best training.

ACKNOWLEDGMENTS

(AINA support)

My arctic research before the LTER project was supported by Air Force Cambridge Research Center through the Arctic Institute of North America, by the postdoctoral program of the National Institutes of Health, by the NSF funding for the US International Biological Program, and by the NSF Office of Polar Programs. In 1987, the NSF Arctic LTER funding began, which through the years has supported the funds for project management as well as for the collection and archiving of the multiyear data sets on natural occurrences and results of large-scale manipulation experiments at the Toolik Lake site (the present project is the Arctic LTER [NSF 1026843]). The many short-term research projects that have provided the scientific understanding of ecological processes and the development of models have been supported mainly by grants to individual scientists through NSF's Division of Environmental Biology and the NSF Office of Polar Programs. The support for the TFS, without which the LTER arctic project could not have functioned, now comes to the University of Alaska from the NSF Office of Polar Programs.

REFERENCES

Involvement of several NSF programs.

Cherry, J. E., Déry, S.J., Cheng, Y., Stieglitz, M., Jacobs, A.S., and Pan, F. (2014). Climate and hydrometeorology of the Toolik Lake region and the Kuparuk River basin: Past, present, and future. In J. E. Hobbie and G. W. Kling, eds. *Alaska's Changing Arctic*, pp. 21–60. Oxford, New York.

Deslippe, J.R., Hartmann, M., Mohn, W.W., and Simard, S.W. (2010). Long-term experimental manipulation of climate alters the ectomycorrhizal community of *Betula nana* in Arctic tundra. *Global Change Biology* (doi:10.1111/j.1365–2486.2010.02318.x)

Hobbie, J.E., Daley, R.J., and Jasper, J. (1977). Use of Nuclepore filters for counting bacteria by fluorescence microscopy. *Applied Environmental Microbiology*, **33**, 1225–1228.

Hobbie, J.E., and Hobbie, E.A. (2013). Microbes in nature are limited by carbon and energy: The starving-survival lifestyle in soil and consequences for estimating microbial rates. *Frontiers in Microbiology*, **4**, article 324. (doi:10.3389/fmicb.2013.00324)

Hobbie, J.E., and Hobbie, E.A. (2005). ^{15}N in symbiotic fungi and plants estimates nitrogen and carbon flux rates in Arctic tundra. *Ecology*, **87**, 816–822.

Hobbie, J.E., and Williams, P.J.LeB., eds. (1984). *Heterotrophic Activity in the Sea*. Plenum Press, New York.

Huryn, A.D., and Hobbie, J.E. (2012). *Land of Extremes: A Natural History of the Arctic North Slope of Alaska*. University of Alaska Press, Fairbanks.

Involvement of NSF research program

9 Forty Arctic Summers

Gaius R. Shaver

IN A NUTSHELL

1. I was committed to long-term, site-based, research long before the Arctic (ARC) Long-Term Ecological Research (LTER) site was established in 1987. Working with the LTER program since then has allowed me to reach my goals more easily than would have been possible otherwise.
2. Because of my deep involvement in research in the LTER program, most of the examples I use in teaching now come from LTER sites.
3. For the same reason, most of my communications with the public are about research in the LTER program.
4. I learned the value of collaboration as a graduate student, from my earliest mentors and collaborators. Being a part of the LTER program has helped me to develop a wide array of enjoyable, comfortable, and productive collaborations.
5. A message to students: be generous in all aspects of your research and professional life, because there is much more to be gained from generosity than there is to be lost.

PERSONAL OVERVIEW

I helped set up the ARC site of the LTER program in 1987 and have made it the focus of my scientific career for the past 27 years. My experience with integrated, site-based, multidisciplinary ecosystem research actually began in 1972, however, when as a graduate student I worked with the US Tundra Biome Study at Barrow, Alaska (Brown et al. 1980; Hobbie 1980). The Tundra Biome Study and its umbrella organization, the International Biological Program (IBP), ended officially in 1974, but the ideas developed and lessons learned from these programs were central to the later development of the LTER program (Coleman 2010). These lessons were central to the formation of my own professional worldview; key among them was the idea that long-term approaches, including long-term, whole-ecosystem experiments, were essential to understanding distribution, regulation, and change in populations, communities, and ecosystems everywhere. My dissertation research, on root growth at the Barrow site, benefited greatly from the interactions I had with the diverse group who worked there.

I finished my PhD in 1976, during a period when the need for a federally supported program of long-term, multidisciplinary, site-based ecological research was becoming increasingly clear. During the late 1970s, a number of excellent ecosystem research projects, several of them carryovers from the IBP days, were struggling to maintain long-term research with short-term funding. Some of them, such as the Hubbard Brook Ecosystem Study (Bormann and Likens 1981) had been under way since before the IBP era. During the late 1970s, Terry Chapin and I began our own long-term experiments in northern Alaska, not knowing whether we would be able to maintain them indefinitely (Shaver and Chapin 1995).

ARC received its first funding in December 1987. When we set up the field research in 1988 and 1989, it was designed largely as a continuation and expansion of existing research established at Toolik Lake over the previous decade. The logistics base, Toolik Field Station (now operated by the University of Alaska Fairbanks), was established in 1975 by John Hobbie and the Tundra Biome Aquatics Group, who wanted to shift their focus from shallow coastal ponds at Barrow to a large, deep lake. The Kuparuk River fertilization experiment (Peterson et al. 1993), now maintained by ARC, was started by this group in 1979 and was in its ninth year when ARC took over. Similarly, long-term fertilizer and transplant experiments in terrestrial ecology had been set up in 1979 to 1981 (Shaver, Fetcher, and Chapin 1986) and were incorporated into the ARC design. Since 1987, ARC research has been managed as a collaboration involving core monitoring, experimentation, and synthesis activities and a constantly changing array of separately funded projects, mostly from the National Science Foundation (NSF) Office of Polar Programs and NSF Ecosystem Studies Program.

The ARC program is now in its 27th year and I have been involved in every stage of its development, as a researcher, administrator, mentor of students and postdoctoral fellows, and currently as principal investigator (Figure 9.1). During this time, I have served on the LTER network Executive Committee for two terms, as well as on the Coordinating Committee, Science Council, and other LTER network committees. I have participated in four site reviews of other LTER sites and as a panelist in reviewing LTER proposals at NSF. I have participated in several multisite network synthesis activities. Finally, I have been able to view the LTER program from "the other side," as an NSF program officer from 1996 to 1998.

APPROACH TO SCIENCE

My experiences in the LTER program have generally confirmed and extended the lessons that I learned as a student about how ecological research is accomplished most effectively. It should be scale-aware and placed in a whole-system context for greatest

FIGURE 9.1 Gaius Shaver (left) describing the main research of the Arctic (ARC) Long-Term Ecological Research (LTER) program to two members of a site review team in 2013. Experimental plots are in the background, including fertilizer, greenhouse, and shade manipulations. (Photo courtesy of Gaius Shaver and the ARC LTER program.)

advancement of understanding. Furthermore, the greatest advances and the greatest opportunities often occur at the boundaries between disciplines, placing a premium on collaborations among researchers with different expertise. I must say, however, that the same trends in thinking have continued to develop outside of the LTER program as well. This is not an area where the LTER program has led the larger community of ecologists in a new direction so much as that the LTER program has helped greatly to facilitate a larger trend toward multidisciplinary, collaborative, whole-system approaches to scientific research. This larger trend includes changes in the way collaborative efforts are evaluated and rewarded in publications and in tenure and promotion actions. For example, several times a year I receive requests to review a young or mid-career faculty member for promotion; increasingly, these requests include explicit instructions about how to evaluate collaborative work.

At both the site and the network level, the LTER program has provided me with many specific research opportunities that probably would not have been available to me otherwise. By far the most important of these is simply the opportunity to maintain and observe changes in long-term experiments. I do realize, though, that I have not taken full advantage of all of the opportunities available within the LTER program. In particular, I have chosen not to participate actively in socioecological research. It has been interesting and enjoyable to discuss socioecological issues within the LTER network, and I think that I understand them better now, although I have no plans for a major shift in my personal research emphasis to incorporate social science.

ATTITUDES TOWARD TIME AND SPACE

I was interested in problems of scaling, in both time and space, before my association with the LTER program, and I am thankful that my association with it has allowed me to continue to develop these interests. Multiscale ecological research has been a core mission for the LTER program from its beginning in 1980. It is important to recognize, though, that this has also been a central issue over the past 30 to 40 years in a broad array of research on land-use change; climate change; and landscape, regional, and global ecology, most of it done outside the LTER network. It is not possible for me to distinguish the relative contributions of my experiences inside and outside the LTER program to the evolution of my thinking on these issues.

Many of those outside the LTER network wonder whether there is an inherent conflict in designing long-term research when funding for the research must be reviewed, renewed, and intellectually "refreshed" every few years. I have never seen this as a particularly difficult problem. The focal interest of a typical LTER site is broad enough and the core data collection is diverse enough that there is plenty of room for gradual evolution of core project goals from one funding cycle to the next. Frankly, it is a lot easier to do this now than in the days when projects with long-term goals had to be renewed and refreshed on a 3-year cycle.

COLLABORATION

Again, I was committed to collaborative research and convinced of its importance before I began working in the LTER program. My experiences within the LTER network have facilitated a greater array of collaborations than would have been otherwise available to me. At the ARC site, we have actively sought new collaborations with new investigators with new skills; every year we invite one to four new investigators to Toolik Lake to explore new collaborations, something we probably could not do without support from the LTER program.

The same trends toward increased collaboration have been occurring in all of science, not just environmental science, over the past 40 years. This trend is partly due to the many interesting and groundbreaking research discoveries that occur at the boundaries between disciplines. In addition, complex research requires multiple skills. In Arctic research, there have been numerous opportunities and pressures in support of international or Pan-Arctic collaborations, and I have benefited greatly from these as well. One of these collaborations is the International Tundra Experiment (ITEX), which has produced a series of meta-analyses of long-term observations of Arctic plant growth (e.g., Elmendorf et al. 2012). In the process, ITEX has developed its own comfortable culture of routine international collaboration. We have also participated in several productive LTER network-level syntheses.

The increase in collaborative research in all sciences has led to some problems in allocating proper credit to the involved scientists. These problems are now widely discussed as the system adjusts (e.g., in journal editorial policies and in promotion review policies).

APPLIED RESEARCH

The LTER program has had little impact on my contributions to applied science. Although the LTER program has provided opportunities for involvement in applied science, most of what I have done in this area is derived from other sources unique to the North Slope of Alaska and the Arctic. Early research on revegetation of development-related disturbances is an example (Gartner, Chapin, and Shaver 1986). Although ARC would like to be doing more applied research on Native Alaskan subsistence land use, the nearest permanent settlement, Anaktuvuk Pass, is 110 km from Toolik Lake and accessible only by air. This physical isolation makes it difficult to devise an applied research program that is well integrated with other research activities.

COMMUNICATION

My experiences in the LTER program have affected my teaching primarily because, as a result of being so heavily involved, most of the available personal examples that might be used in my teaching come from ARC research. Similarly, because the LTER Schoolyard Program and other network education activities have been most accessible to me, inevitably my work in education and public communication has been linked to the LTER program. Again, the remoteness of our site means that we cannot bring schoolchildren into the field. As a result, most of our efforts have focused on bringing journalists and K–12 teachers to Toolik Lake to work with us and learn about field research. They then take this experience with them to become more effective journalists and teachers of science.

MENTORING

Although my attitudes toward mentoring itself have evolved over the years, mostly as a result of simple experience with what works and what does not, I cannot identify anything in my attitudes or practices that I associate specifically with my experiences in the LTER program. At Toolik Lake, we have simply tried to optimize our involvement with students, postdoctoral fellows, and new investigators in the widest possible array of research, both collaborative and individual.

SKILL SET

The development of my scientific tool kit over the past 27 years is tightly linked to my choice of particular research activities and development of the skills needed to carry them out. However, I cannot say that I identify the development of these skills with something unique to the LTER program. The kind of research I do (terrestrial productivity, carbon and nutrient cycling) is actively pursued outside of LTER circles, and I believe my tool kit would have been equally influenced if I had chosen to be involved in different programs with different funding sources. Having a connection with the LTER program has simply made this easier.

PERSONAL CONSEQUENCES

The LTER program has been at the core of my professional life and has provided many, but by no means all, of my opportunities for new research, travel, collaboration, and friendship in science, including internationally. Of course, I am pleased to have developed strong friendships with many of the people I have worked with over this time. I feel good about the fact that the community we have at Toolik Lake is thriving and helping a continuous stream of young students to embark on their scientific careers. All of this is linked to the LTER program, but if I had been as deeply involved in a different research program, I am sure I would have been equally strongly affected, but in different ways.

The LTER program has provided the context for many of my professional career decisions and subsequent personal relationships. Most of these decisions have occurred as a result of particular choices about participation in one activity versus another; frequently, for example, I have linked my proposals for new research with ongoing ARC research because I thought it would increase the chances of funding success. These decisions have subsequently determined the trajectory of my research, taking me in different directions than I might have followed if I had simply proposed new research on the ideas that I was most excited about at the time.

CHALLENGES AND RECOMMENDATIONS

All federally funded research programs (and investigators as well) are under pressure now to increase the "broader impacts" of their research, including human resource development. The LTER program is doing a good job at this, with a variety of programs to increase participation at all levels. It could do much better, though. One key problem is that the LTER network is trying to do too much, spreading its resources too thinly where it would probably have greater impact by focusing more narrowly on fewer activities. This problem is especially compounded by pressures within the LTER network for all sites to participate in most or all of these activities. A second, even more important, problem is that the LTER program as a whole, including its programs in human resource development, are, in my view, increasingly isolated from parallel activities in both research and human resource development in the rest of the ecological community. The LTER network and its individual sites should all be reaching out more to people and programs not already linked to the LTER program, but they cannot, at least in part, because the LTER program is already trying to do too much with too few resources. This book, with its inward-looking focus on the LTER program alone and general lack of comparisons with similar programs and trends in biology and other sciences, is a fine example of this problem.

CONCLUSION

The LTER program and network have been outstandingly successful at promoting long-term, collaborative, site-based, ecosystem research while providing opportunities for researchers and students to develop their careers, participate in novel educational and mentoring opportunities, and explore applications of basic ecological science. Over the past 30 years, all of these trends have also been occurring outside of the LTER program, and indeed throughout all of the natural sciences. The LTER program has had an advantage, though, in that its founders at NSF took care to incorporate lessons learned during the IBP and the early days of the Ecosystems Studies Program as they designed the LTER program. Much of the success of the LTER program, thus, is due to the way it was designed. The wide range of LTER activities are not unique (other than in their specific topical focus), but they are of generally high quality and productivity, again because of the overall program design. Finally, it is essential to give due recognition to the early leaders, including Jerry Franklin, Jim Gosz, John Brooks, and Tom Callahan, who all worked hard to create a culture of openness and collaboration.

My participation in the LTER program has made it much easier for me to attain the career research goals that I had settled on before I was engaged in the program. These goals have always been focused on long-term change and regulation of plant populations, communities, and ecosystems using experimental approaches and based on deep, multidisciplinary understanding of well-studied sites. If I had tried to do this outside of the LTER program, it would have been much harder and I might have been forced to take a more narrow approach. I learned as a student the need for and importance of collaborative, multidisciplinary approaches to "big picture" and long-term ecological problems, and I am grateful that my participation in the LTER program has allowed me to continue working in this way.

If there is one lesson that I try to impart to aspiring ecologists and students, it is to "be generous—with your data, your time, your ideas, your equipment and supplies, and your friendship." By being generous there is much more to be gained than lost, while selfishness often leads to missed opportunities, limited productivity, and narrow horizons. I learned this lesson myself from my earliest mentors and colleagues (particular thanks to Hal Mooney, Dwight Billings, Terry Chapin, and John Hobbie), but applying this lesson within the LTER program is what has made my career most rewarding and enjoyable.

ACKNOWLEDGMENTS

Since 1975, the research of the Arctic LTER group, its predecessors, and its long-term collaborators has been supported by more than 50 grants from NSF. The most recent LTER award (of five since 1988) is DEB-1026843; most other awards (and long-term logistical support) have come from NSF's Office of Polar Programs and from the Division of Environmental Biology's Ecosystem Studies and Ecology programs. We are particularly grateful to multiple program officers at NSF who have provided consistent support and valuable advice over the past four decades. Finally, none of this would have happened without the leadership and generosity of J. Hobbie.

REFERENCES

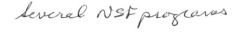
several NSF programs

Bormann, F.H., and Likens, G.E. (1981). *Pattern and Process in a Forested Ecosystem.* Springer-Verlag, New York.

Brown, J., Miller, P.C., Tieszen, L.L., and Bunnell, F.L., eds. (1980). *An Arctic Ecosystem: The Coastal Tundra at Barrow, Alaska.* US IBP Synthesis Series, #12. Dowden, Hutchinson & Ross, Stroudsburg, Pennsylvania.

Coleman, D.C. (2010). *Big Ecology: The Emergence of Ecosystem Science.* University of California Press, Berkeley.

Elmendorf, S.C., Henry, G.H.R., Hollister, R.D., et al. (2012). Plot-scale evidence of tundra vegetation change and links to recent summer warming. *Nature Climate Change*, **2**, 453–457.

Gartner, B. L., Chapin, F.S., III, and Shaver, G.R. (1986). Reproduction by seed in *Eriophorum vaginatum* in Alaskan tussock tundra. *Journal of Ecology*, **74**, 1–19.

Hobbie, J.E., ed. (1980). *Limnology of Tundra Ponds, Barrow, Alaska.* US IBP Synthesis Series, #13. Dowden, Hutchinson & Ross, Stroudsburg, Pennsylvania.

Peterson, B.J., Deegan, L., Helfrich, J., Hobbie, J.E., Hular, M., Moller, B., Ford, T.E., Hershey, A., Hiltner, A., Kipphut, G., Lock, M.A., Fiebig, D.M., McKinley, V., Miller, M.C., Vestal, J.R., Ventullo, R., and Volk, G. (1993). Biological responses of a tundra river to fertilization. *Ecology*, **74**, 653–672.

Shaver, G. R., Fetcher, N., and Chapin, F.S., III. (1986). Growth and flowering in *Eriophorum vaginatum*: Annual and latitudinal variation. *Ecology*, **67**, 1524–1525.

Shaver, G.R., and Chapin, F.S., III. (1995). Long-term responses to factorial NPK fertilizer treatment by Alaskan wet and moist tundra sedge species. *Ecography*, **18**, 259–275.

PART FOUR
BALTIMORE ECOSYSTEM STUDY (BES) LTER SITE

10 Of Fish and Platypus
If You Could Ask a Fish What It Feels Like to Swim

J. Morgan Grove

IN A NUTSHELL

1. I was in the first cohort of scientists who was specifically trained in long-term, socioecological research. My cohort may have a "disciplinary home," but we are less likely to be exclusive to a single discipline.
2. My research requires diverse approaches and skill sets that address the spatial, organizational, and temporal complexity of human ecosystems.
3. Donald Stokes identifies several categories of research, including (1) pure basic research, (2) pure applied research, and (3) use-inspired basic research. Most of my research and the research from Baltimore Ecosystem Study (BES) is use-inspired basic research, which is intended to advance both science and decision making.

4. Collaborative research is not for everyone. My collaborative research in BES is more like playing in a jazz ensemble than a regimented orchestra.
5. My participation in the BES Long Term Ecological Research (LTER) project has fostered lifelong friendships that are multigenerational.

PERSONAL OVERVIEW

I imagine that if I could ask a fish what it feels like to swim, it would be puzzled. It has never known any other way of being than to swim. Likewise, I am puzzled when asked what it is like to work on an LTER project. I have never known any other type of research program. Furthermore, the project I work on, the BES, is quite different from all but one of the other US LTER projects. It is an urban site and was designed to be interdisciplinary from its conception. Perhaps I am being asked what it is like to be a platypus. (The platypus is a strange mammal that has a duck-like bill, beaver tail, otter feet; that lays eggs; and that carries venom. When the platypus was first encountered by Europeans in the late 1790s, a pelt and sketch were sent back to Great Britain. British scientists thought initially that the evidence provided was a hoax.)

How did I come to have such a bizarre combination of traits and how does it feel? In this essay, I try to answer these questions by describing my professional training and experience working as a co–principal investigator and the lead for the social science research team for the BES for the past 17 years.

When I was in college, I chose to double-major in architecture and environmental science. I enjoyed both majors on their own but was fascinated with the combination of the two programs. My first introduction to urban ecology, and one of the books that has influenced me most throughout my career, was Anne Spirn's *The Granite Garden* (1984). Spirn described the dynamic relationships among buildings and ecologies in the city and how planning and design could create healthy and attractive urban environments. While planners and architects might see the environment as background to buildings, and ecologists might see buildings as background to the environment, Spirn's book presented the case for how good urban planning and design depended on seeing built form and ecology as inextricably connected, yin and yang, with neither view in the background.

My career plans changed unexpectedly in graduate school. My career goal had been to become an architect with a professional practice that focused on integrated designs of building and land. To develop the necessary skills, I had planned a two-master's degree approach: a graduate degree in environmental science, followed by a professional degree in architecture. My plans began to alter as I took courses in social forestry, watershed management, and ecology while attending the Yale School of Forestry and Environmental Studies with an internship in Baltimore, Maryland.

My internship in Baltimore was arranged in 1989 by my faculty advisor, William R. Burch, Jr., who had become friends with Ralph Jones, the Director of Baltimore City's Department of Recreation and Parks. Jones was intrigued with Burch's work in social forestry, which focused on the links between community development and natural resource management. Burch and his students had implemented social forestry programs internationally in rural areas. Jones wanted to explore the idea of applying social forestry programs in the United States to urban areas. The purpose of my summer internship was to prepare a report describing how Yale and the city department could work together to realize Jones' vision of creating capacity within the department to link environmental rehabilitation with neighborhood revitalization. Burch instructed me to use the perspective of social forestry programs for my assessment and report (Burch and Grove 1993).

Burch visited me several times during my first summer in Baltimore. I showed him the vacant lots and abandoned houses, the steep stream valleys, and the extensive forested lands in the city. We discussed the social ecology of those diverse lands, watershed management, and links to the Chesapeake Bay. We also discussed what we had learned from Herb Bormann and the Hubbard Brook LTER project. Bormann and Burch were close friends and colleagues, and I had taken several of Bormann's ecology classes. Considering these experiences, it is not surprising that Burch said to me during one of our walks that we really should start thinking about how this whole effort could be the first urban long-term study. But instead of studying forest clear-cuts such as Hubbard Brook and its impacts on water, we could study urban forest restoration to improve water quality and revitalize neighborhoods. This brief synopsis of my training and experience helps explain how I became a platypus and the diverse background I brought to the BES program.

I have never managed to leave Baltimore since my summer internship in 1989. I finished my MS degree in 1992 and my PhD in 1996. My dissertation, which focused on one of the urban watersheds in Baltimore City and County, is titled *The Relationship between Patterns and Processes of Social Stratification and Vegetation of an Urban–Rural Watershed*. Some aspects of my dissertation anticipated our work in the BES program. I developed a spatial-temporal, interdisciplinary approach to studying urban watersheds that combined the variable source area approach from hydrology, the shifting steady-state approach from ecology and Hubbard Brook, and the social area analysis approach from sociology. I called this an integrated, patch-dynamics approach, which built on the patch-dynamics work of Steward Pickett and others. I also used long-term, time-series data from 1970 to 1990 to test for temporal lags between neighborhood and vegetation change.

As the ink was drying on my doctoral defense forms, I began work as a postdoctoral fellow with the US Forest Service. Soon after I started that position, the National Science Foundation (NSF) announced a competition to establish two urban long-term research projects as part of the LTER network. During my work in Baltimore, Burch and I had connected with Steward Pickett, Mark McDonnell, Rich Pouyat, Peter Groffman, Mary Cadenasso, and others at the Cary Institute of Ecosystem Studies. They had developed a research project focused on the New York City metropolitan region: the Urban-Rural Gradient Ecology (URGE) project. Working with Pickett, Pouyat, Groffman, and Cadenasso, Bill Burch and I led the social science effort for the BES proposal. The result of the urban LTER competition was the selection of BES project, with Steward Pickett as the principal investigator, and the Central–Arizona Phoenix (CAP) LTER, led by Nancy Grimm and Chuck Redman.

APPROACH TO SCIENCE

My training at Yale and participation in the BES program fundamentally shaped my approach to environmental science. At Yale University, my training was in social forestry and based on an interdisciplinary orientation. The concept of society in social forestry signifies a broader agenda than growing trees. The goals of social forestry can be social, economic, and cultural: group formation and collective action, institutional development, and the establishment of sustainable social structures and values systems to mobilize and organize individuals and households. Social forestry solutions may incorporate both social and ecological concepts and data.

Activities at the BES site represent both a research project and a program. The research project is characterized by the scientists and staff, research questions and analyses, and production of knowledge. The research program is the platform for research, with study sites and other infrastructure for collecting data and access to long-term data (Grove et al. 2013). As a research project, BES has been a highly energized and collegial intellectual cauldron of diverse ideas and approaches. Because of this project environment, the perspectives and skills that I brought to BES have grown and flourished. I have been exposed to and learned so much about hydrology, biogeochemistry, ecology, meteorology, economics, and engineering—to name a few disciplines—while being challenged and encouraged to advance the potential contributions of anthropology, sociology, and geography.

The long-term focus of BES has required that I ask novel science questions and adopt new science approaches. Long-term research is more than collecting data for a long time. It means observing how the system functions under variable social, economic, and ecological conditions. For instance, what are the dynamics of Baltimore's urban ecology under changing conditions of formal and informal housing segregation, economic prosperity and decline, and wet and dry or hot and cold periods? It also means understanding slow and fast types of interacting social, economic, and ecological change and examining social, economic, and ecological inertias and path dependencies that accumulate and constrain the socioecological futures for the Baltimore region. As a research program, BES has given me access to a wealth and diversity of long-term socioecological data that would be extremely difficult to amass and manage if I were an individual researcher, working on a grant proposal for a 3-year project.

ATTITUDES TOWARD TIME AND SPACE

The organizing framework for the BES is an urban patch-dynamics approach. This approach is based on the idea that urban systems are spatially heterogeneous and can be described as patch mosaics. These mosaics can be classified using physical, biological, and social factors, or a mix of these types of factors. At BES, we think of patch dynamics in terms of spatial, organizational, and temporal complexity. Thus, my work is concerned with a multiscale approach to the spatial and temporal dynamics of urban ecological systems (Grove and Burch 1997). I am particularly interested in the socioecological dynamics of residential neighborhoods. There are about 276 neighborhoods in Baltimore City. We can still observe the legacies and path dependencies of segregation, mortgage red-lining, and disinvestment from the 1930s in the current conditions of some neighborhoods in terms of population density, land abandonment, race, and ecology.

COLLABORATION

My experience in the LTER program has affected me profoundly as a scientist. It has been an affirmation of my interdisciplinary training and provided me with the opportunity to pursue interdisciplinary research at a high level. This has occurred because I have been the member of a terrific ensemble of collaborators. I use the term *ensemble* purposefully because the participants in BES are much more like a jazz ensemble than a regimented orchestra. Our collaborations build lines of research based on riffs of listening, trust, and field walkabouts (visits to locations and sharing observations, questions, and interpretations about the history, dynamics, and future of a place from social, economic, and ecological perspectives). In these ways we play off of each other organically yet with direction.

My collaborative experiences in BES led me to believe that there is no single formula for interdisciplinary, environmental research. However, the fundamental criteria that apply to individual, disciplinary research apply equally to multi-investigator, interdisciplinary research. Interdisciplinary teams should always be able to describe how their research advances social or ecological theories, methods, or practices.

Collaborative research is not for everyone. Continuing the music metaphor, scientists who require solos or prefer duets might not do well in our BES jazz ensemble. Likewise, scientists who prefer to play in a symphony with clear instrument sections, sheet music, and notes to play may not do well either. Playing in an interdisciplinary ensemble takes time and commitment. It can be expensive in terms of time and energy, yet intellectually intriguing and rewarding. Honestly, who could imagine on their own that the relationships between urban stream water quality, anaerobic habitat for denitrifying bacteria in storm water detention ponds, poor maintenance, and lack of government inspections could be so interesting and potentially significant?

There is growing interest and capacity for socioecological research on a national and international basis in general, and in cities in particular. This growing interest has afforded me the opportunity to collaborate with scientists working in other cities in the United States and beyond. I find it interesting that Baltimore sometimes seems to have more in common with older European cities than with newer US cities such as Phoenix. For instance, because of its shrinking population and aging infrastructure, Baltimore shares many of the same types of challenges as cities such as Dresden, Germany, or Strasbourg, France. To promote mutual learning for urban sustainability and resilience, it makes great sense to me that we will eventually have an international network of urban long-term research sites associated with a diverse array of urban types.

APPLIED RESEARCH

My training in architecture, social forestry, watershed management, and ecology has provided me with the basis to comfortably exist among the domains of science-based professions and basic and applied research. In particular, social forestry is problem-oriented and addresses social and ecological concerns. Solutions occupy a middle-ground that neither blames humans for their destructive impacts nor places too much faith in technological solutions. Furthermore, social forestry can be both a science-based profession and a type of applied research, much like a medical doctor working in a research hospital (Burch and Grove 1993).

In his book *Pasteur's Quadrant,* Donald Stokes (2007) identifies several categories of research, including (1) pure basic research, (2) pure applied research, and (3) use-inspired basic research (the fourth quadrant has no label: no quest for fundamental understanding and no consideration of use). Stokes defines use-inspired basic research as science that is designed to both enhance fundamental understanding and address a practical issue. This quadrant is called "Pasteur's Quadrant" because biologist Louis Pasteur's work on immunology and vaccination both advanced fundamental understanding of biology and saved countless lives. In this quadrant, urban ecologists may work to advance scientific theories and methods as well as address practical problems. Although some of our work at BES can be located in each of these quadrants, most of our research is use-inspired basic research that advances both science and decision making.

I have found that there is a dynamic link between the "researchers" and the "users" in the use-inspired quadrant. One cycle of research and use inspires another cycle of research and use. Thus, decision-makers are an important type of collaborator (preceding section) and an important group with whom to closely communicate (following section) in ways that are more than "science delivery."

COMMUNICATION

My position in the US Forest Service's Research and Development Mission Area means that my communication activities rarely include students in a formal classroom setting. Most often, I work with various types of decision-makers from government, nongovernmental organizations, community groups, and businesses (Figure 10.1). These decision-makers usually have questions that are specific to a place, and that place is located in a mosaic of places, at certain scales, and in a particular time. Thus, my job is to provide a BES "way of thinking" to help them understand the spatial, scalar, and temporal context of their question and the range of possible solutions.

MENTORING

I am involved in what I call "colateral mentoring" because I am not formally responsible for training or advising students. What I mean by colateral is that there are students who work at BES and are advised by other researchers. In cases where I am working with that researcher, I may be asked to assist in advising the student. My assistance may include helping the student with research topics that complement existing and previous research, making connections with local partners and their interests, and identifying and obtaining data.

My overall frame for mentoring students is based on a question that I think all scientists and scientists-in-training should be able to answer: how does your research advance theory, methods, or practice? This question can be refined further in the case of BES and its focus. What is the interdisciplinary approach: what social and ecological variables are included? How does the research address different types of spatial, organizational, and temporal complexity? How can the research question be answered using existing and new data?

FIGURE 10.1 Morgan Grove (center background) leading a discussion at a community garden as part of activities in the Baltimore Ecosystem Study Long-Term Ecological Research program. (Photo courtesy of Olga Maltseva Brillman and Camden Partners.)

SKILL SET

My training in social forestry provided me with a diverse tool kit from the social and ecological sciences. Academic disciplines, both basic and applied, may preferentially use a limited set of methods, tools, and techniques. However, social forestry as a scientific practice tends not to favor one method, tool, or technique over another because the goals and constraints associated with many social forestry projects involve social, economic, cultural, and environmental drivers at multiple scales. Some of these methods include ethnographies and case studies, experimentation or direct manipulations, after-the-fact analyses, cross-sectional studies, and longitudinal studies to collect data. Specific tools include remote sensing, key informant and focus group surveys, observational studies, and analyses of "social scats" (administrative documents, records, and operational data from government agencies, private businesses, nonprofits, and neighborhood associations). Finally, these tools can be applied using a variety of techniques that include social area analyses, maps and geographic information systems, transects, point surveys, seasonal calendars, flow diagrams of critical resources, decision making trees, and network diagrams of organizational relationships. These empirical approaches from social forestry are an inclusive perspective toward social and ecological data (Grove and Hohmann 1992).

One challenge at BES has been to develop a data framework that facilitates our ability to relate the diverse methods, tools, and techniques from the range of disciplines involved in BES research. A key facet of our framework is to strategically combine extensive types of data such as remote sensing, census data, or administrative data with intensive types of data such as field plots or open-ended interviews. Extensive and intensive data can come from a variety of data sources, include physical, biological, and social variables, and can be associated with different scales and periods of time.

PERSONAL CONSEQUENCES

My involvement in BES and the LTER network has been a professional and personal journey for which I feel fortunate and grateful. I am amazed at how many parts of the United States I have seen and the diverse perspectives I have experienced. More importantly, I have developed lifelong friendships because we work on a long-term project, meet with scholars from other LTER sites at annual Science Council Meetings and triennial LTER All Scientists Meetings. Furthermore, these friendships are multigenerational, and I feel as if I am part of an extended family.

CHALLENGES AND RECOMMENDATIONS

A risk for the LTER network is that although it does unique science and is extremely productive, it is no longer a shiny new thing. Thus, as science administrators at NSF come and go, new ones may desire to roll out something new, not steward the old. In this case, a decision may be made to close down the LTER network and invest its funds in new projects. Thus, the LTER network faces two challenges: producing novel, cutting-edge science on new topics while retaining its commitment to creating knowledge about the long-term dynamics of ecological systems that is possible only because of its long-term funding. These challenges may be the same thing, but it can be difficult to convince new administrators of this reality.

A second challenge for the LTER network is long-term support for the study of LTER sites as socioecological systems. The initial rationale for the LTER network was that understanding the long-term dynamics of ecological systems required long-term funding. However, this same rationale has not been applied to support interdisciplinary research. Internally, the LTER network has been supportive of interdisciplinary research. Although the training of most of the senior scientists in the LTER program has had a disciplinary focus, these scientists have encouraged interdisciplinary research. Currently, most NSF support for interdisciplinary research on LTER sites comes from NSF programs external to the LTER program. The support from these external NSF programs occurs on 3- to 5-year funding cycles. Thus, long-term direction and support for interdisciplinary research is needed on a widespread basis in the LTER network.

Finally, today it may be necessary to ask whether two urban LTER sites are sufficient (Polsky et al. 2014). When the BES and CAP sites were funded in 1997, the question was simply whether they would succeed as long-term socioecological research projects. It really was an experiment in science. Today, CAP and BES are models of success in terms of advances in theories, methods, and practice, as well as students trained. Many other countries look to the United States and its two urban sites as models for their own investments in urban socioecological research. Thus, the questions have changed. Are Baltimore and Phoenix sufficiently representative of all cases of urbanization in the United States and are they sufficient for the needed production of knowledge and training of students in urban long term research and practice?

CONCLUSION

The LTER program has had a dominant influence on my professional identity. It has given me the opportunity to take full advantage of my interests and passions, my training, and my abilities. It has enabled me to grow and contribute in ways that are unimaginable outside of the LTER program.

My advice and encouragement to students is that it is OK to be a platypus. One has to have good training in theory and methods, but one does not have to be defined by a single discipline. Expertise in theory and methods may be associated with a particular discipline. But it is silly to think, particularly in the case of socioecological systems, that any one discipline is the keeper of all knowledge and able to explain all things. Be the best platypus that you can be.

ACKNOWLEDGMENTS

This chapter was supported by the US Forest Service and funding from the NSF LTER program. This material is based on work supported by the NSF grant DEB-1027188. Any opinions, findings, and conclusions or recommendations expressed in this material are those of the author and do not necessarily reflect the views of the US Forest Service or the NSF.

REFERENCES

Burch, W.R., Jr., and Grove, J.M. (1993). People, trees and participation on the urban frontier. *Unasylva*, **44**, 19–27.

Grove, J.M. and Burch, W.R., Jr. (1997). A social ecology approach and applications of urban ecosystem and landscape analyses: A case study of Baltimore, Maryland. *Journal of Urban Ecosystems*, **1**, 259–275.

Grove, J.M., and Hohmann, M. (1992). GIS and social forestry. *Journal of Forestry*, **90**, 10–15.

Grove, J.M., Pickett, S.T.A., Whitmer, A., and Cadenasso, M. (2013). Building an urban LTSER: The case of the Baltimore Ecosystem Study and the D.C./B.C. ULTRA-Ex Project. In: S.J. Singh, M. Chertow, M. Mirtl, and M. Schmid, eds. Long Term Socio-Ecological Research: Studies in Society-Nature Across Spatial and Temporal Scales, pp. 369–408, Springer-Verlag.

Polsky, C., Grove, J.M, Knudson, C., Groffman, P.M., Bettez, N., Cavender-Bares, J., Hall, S., Heffernan, J., Hobbie, S., Larson, K., Morse, J., Neill, C., Nelson, K., Ogden, L., O'Neill-Dunne, J., Pataki, D., Steele, M., and Chowdhury, R.R. (2014). Assessing the homogenization of urban land management with an application to US residential lawncare. *Proceedings of the National Academy of Sciences U.S.A.*, **111**, 4432–4437. (doi:10.1073/pnas.1323995111)

Spirn, A.W. (1984). *The Granite Garden: Urban Nature and Human Design.* Basic Books, New York.

Stokes, D.E. 1997. *Pasteur's Quadrant—Basic Science and Technological Innovation*, Brookings Institution Press, Washington, DC.

11 Long-Term Ecological Research on the Urban Frontier

Benefiting from Baltimore

Steward T. A. Pickett

IN A NUTSHELL

1. The Long-Term Ecological Research (LTER) program has made me a more effective scientist because I have had to learn about disciplines that are very distant from my own, and it has helped me see the relevance of my own interests in the context of rapidly changing systems in which human agency is inescapable.
2. Being a part of the Baltimore Ecosystem Study (BES) site has extended my educational activities to primary and secondary school situations. It has been both an eye opener and personally very rewarding to interact in city classrooms and after-school programs.

3. I have found myself in demand as a public speaker as a result of serving as leader of one of the two urban LTER programs. My communication skills and strategies have been greatly improved as a result.
4. Collaboration has taught me to listen more effectively and to emphasize dialogue rather than exposition. Multidisciplinary urban field trips are powerful tools for joint research and for communication with people in the community.

PERSONAL OVERVIEW

My role in the LTER network has been as principal investigator of the BES site from its inception in 1997. Before involvement in the LTER program, I conducted urban ecological research in metropolitan New York. My interests beyond urban studies include vegetation dynamics, natural disturbance, and landscape ecology. At the time that my involvement in the LTER program began, I became part of a multidisciplinary and international team conducting a 10-year study of the linkages between rivers and upland savannas in Kruger National Park, South Africa. In the LTER network, I have been a member of the committee on scientific initiatives and the Science Council. I have also contributed to cross-site integration through workshops at the LTER network's triennial All Scientists Meetings and to cross-site activities such as comparison of disturbance across the network (Peters et al. 2011).

I hold a BS and a PhD in botany, specializing in plant ecology. I am currently Distinguished Senior Scientist at the Cary Institute for Ecosystem Studies, a flexible position that has allowed me to explore the cross-disciplinary and synthetic approaches required to lead an urban LTER program. My site-specific and network-wide contributions have promoted a socioecological approach to research.

APPROACH TO SCIENCE

My experiences in the LTER program have expanded my scientific repertoire, and allowed me to develop latent interests. Although my approach before the funding of the BES project was collaborative, the extent of collaborations has vastly expanded in managing urban, socioecological research. For example, I now routinely interact with physical scientists, with social scientists of various kinds (e.g., geographers, sociologists, anthropologists, and economists), with urban designers (e.g., planners, architects, and landscape architects), and with environmental ethicists.

The interdisciplinary LTER project has deepened my understanding of theory and developed my sensitivity to the role of metaphor across disciplines, with members of the public and with professional colleagues in urban and regional policy or management. Furthermore, my penchant for synthesis has been more fully developed in leading the BES project. This skill has been honed in composing annual project reports and in writing overview papers about BES research. Synthesis has also been required in preparing three BES proposals that have unified the ideas, methods, and motivations of diverse specialties.

Comparison and place-based approaches are both required for successful research in the LTER program. For example, to understand and work in Baltimore, I have had to appreciate the social and environmental history of the place and understand the changing policy effects on Baltimore's regional ecology. But it has been important *not* to assume that all urban systems are like Baltimore. Interactions with the Central Arizona–Phoenix site; several of the Urban Long-Term Research Areas–Exploratory (ULTRA-Ex) cities; other urban regions with substantial research efforts, such as Chicago, New York, and

Beijing; and several of the French Zones Ateliers cities have been important to understanding the generality or uniqueness of Baltimore's socioecological structure and functioning (Childers et al. 2014).

ATTITUDES TOWARD TIME AND SPACE

Long times and extensive spaces are key to successful urban ecology. At the BES site, we have ongoing conversations across disciplines about how to maintain core, long-term data collection while experimenting with new questions and theoretical motivations in the context of the evolving policy and environment of Baltimore. I have learned to constantly probe new conceptual areas, especially those that might facilitate interdisciplinarity, and to explore how those relate to core data while sometimes requiring new data streams. It requires effort to avoid the pitfalls of short-term efforts that satisfy the need for novelty.

Urban research has heightened my spatial awareness. The consideration of rapid changes in urbanization worldwide, and the extent and speed of connections between urban areas with each other, as well as with so-called rural or wild lands, have become an unavoidable part of studying the ecology of Baltimore. These complexities are important parts of the larger theory of urban systems (Boone et al. 2014).

COLLABORATION

Even before becoming involved in the LTER network, I was a proponent of collaborative research. That is one reason why I accepted a position at the Cary Institute, which was founded to promote collaborative research. That spirit stimulated my first involvement with urban ecosystem research, led by Mark McDonnell (McDonnell and Pickett 1990), who was then at the Cary Institute. All of this was founded on experiences in graduate school in the laboratory of Fakhri Bazzaz, who often said that a crucial part of his ability to effectively train graduate students was frequent and well-nurtured collaboration with other laboratories. Continued experience at Rutgers, especially interdisciplinary interactions and collaborations with Andrew "Pete" Vayda and his students in human ecology, prepared me for the discipline-hopping required by an urban LTER program.

Helping to establish and to lead the BES site has been an exercise in collaboration. However, there have been familiar challenges: contrasting disciplinary reward systems experienced by collaborators, misunderstandings due to how different disciplines use the same words, time constraints, incompatible or unexamined mental models, and spatially dispersed locations of the collaborators. However, focusing a diverse community of collaborators on the same place, especially more specific locations such as stream sampling networks, study plots, and shared neighborhoods, was important for collaboration. Regular interdisciplinary meetings and the occasional interdisciplinary field trip have been important collaborative tools.

Another tool has been a meeting style that ensures that all voices are heard, and where there is periodic stock-taking by the whole group. A process of democratic input, followed by strategic winnowing and search for cross-disciplinary connections, characterizes the meeting style I have helped the BES program develop.

Although international interactions have been important to me since early in my career, my experiences in the LTER program have demanded increased international interactions. My early overseas interactions were with the International Association of Vegetation Science, and through Vayda's group, with researchers in Indonesia. Further international comparisons were through Moshe Shachak and his work in the Negev

Desert. However, since the establishment of the BES project, my opportunities for international collaboration have burgeoned. Particularly important have been interactions with members of the French Zones Ateliers working in urban systems, with Chinese urban ecology colleagues, and more recently, with a team working on urban-like rural settlements in South Africa (McHale et al. 2013). These international collaborations have been important for (1) putting the study in Baltimore in perspective, (2) sharing methodologies, and (3) contributing to a truly international urban ecological science.

The costs and benefits of collaboration in the LTER network have to be balanced. Collaboration takes time and personal energy. It also takes intellectual energy to understand the mindsets and the different ways that various specialties understand the same or similar words. It requires patience and a willingness to abandon control of beloved concepts. Sometimes, collaborations do not bear fruit, and so a "plan B" has to be percolating in the background. Not all collaborative investments yield a scientific profit, but those that do result in new research questions, new research approaches, creative new data sets, and novel relevance to stakeholders.

APPLIED RESEARCH

Applied research is a key to success in an urban setting. Communities, agencies, and policy makers, all of whom are formal or informal partners, must see some relevance in urban long-term ecological research. I have developed positive relationships with nongovernmental organizations such as Baltimore's Parks & People Foundation, in addition to community groups in some of the neighborhoods in which we work, and key environmental managers and policy makers in Baltimore City, Baltimore County, and the State of Maryland. These relationships have helped refine scientific questions, gain access to research sites, expose important background information, and place our insights in the hands of those who have need for (and interest in) them. Urban community forestry and neighborhood revitalization are applied management areas that I have had to become conversant in as a result of membership in the LTER program. In all these relationships, it has been important for me to learn how to negotiate across different scales of action, motivations, and professional time frames. However, the successes can be notable. Research at the BES site on riparian ecosystem function and on urban tree canopy effects and restoration have affected policy in the Chesapeake Bay and the State of Maryland.

COMMUNICATION

Being the leader of one of two urban sites in the LTER network has thrust me into a number of spotlights. It has proven to be a bully pulpit, and I have had to meet the demands of increased public communication. My public lecturing has increased. I have had to learn new techniques and adopt new presentation technologies. Radio interviews have become a familiar venue for me.

My position does not involve regular classroom lectures. However, several times a year I receive invitations to speak in the classes of colleagues elsewhere. I have also participated in distributed, urban socioecological graduate classes. My educational experience now also extends to participation in urban design and landscape architecture studios. Although these have mostly been in collaboration with Brian McGrath, now at Parsons School of Design–The New School, I have participated in studios and design critiques at several other architecture schools (McGrath et al. 2007; Pickett, Cadenasso, and McGrath. 2013).

I have had many opportunities to interact with student and teachers at elementary, middle, and high school levels. The most enjoyable of these have been on the school grounds or in the neighborhoods, although we bring students and teachers to some of our meetings as well. Again, there is a new language I have had to learn to do this, and I have had to adopt new sensitivities to the concerns of lay people and youth in the city and suburbs.

Perhaps the most significant change in "outreach" as a result of my experiences at the BES site has been to realize that one-directional information transfer is far less successful than two-way dialogue with those interested in ecological knowledge about cities and urban regions. We quickly shifted from a "delivery model" of science in Baltimore to a model of community engagement through continual dialogue, in which our research community could learn as well as teach. Hence, abandoning literal outreach was one of my first and most important lessons from the urban LTER site.

MENTORING

My mentoring now includes advice and examples for activities and concerns that were not included before my involvement with the LTER program. For example, I have shared advice on interacting with members of the public during research in the field, writing and speaking for the general public, sensitivity to social assumptions about various publics, and preparing products for the public along with the usual scientific publications. Furthermore, I have been able to share experiences and lessons learned about cross-disciplinary interactions. Frank discussion about both costs and benefits for collaboration and cross-disciplinary research are now part of my mentoring strategy. Although it is difficult to always put in practice, it has become clear to me that consistent mentoring of young students over the long term is very desirable for the diversification of the scientific community. The BES site also mentors teachers through summer institutes, and the Research Experiences for Teachers Program at the National Science Foundation (NSF). We work with teachers to develop, promulgate, and support curricula that involve understanding urban ecosystem dynamics (Figure 11.1).

SKILL SET

The LTER program has exposed me to geographic information systems, although it is more efficient to rely on those more specialized in this technology. We developed a novel urban land cover classification that relied on this technology (Cadenasso, Pickett, and Schwarz 2007). Much of my field research, when I am able to accomplish any, brings familiar methods to unfamiliar field locations. New technologies of plot marking and explorations of electronic data gathering are results of the urban experience. Within the BES project, of course, there are many new technologies for automation of stream, soil, and atmospheric measurements that are being employed. Although I do not personally operate any of these, I have to be familiar with them as a part of my synthesis of the project.

PERSONAL CONSEQUENCES

Involvement in research at the BES site and in the broader LTER network has demanded a great deal of rewarding travel, including the promotion of international–LTER interactions. As a "poster child" for research at the BES site and consequently for urban ecology in general, I have had the pleasure to be exposed to urbanized regions and their

FIGURE 11.1 Students gathering data on plant species composition in urban areas as part of the Baltimore Ecosystem Study Long-Term Ecological Research program. (Photo courtesy of Steward Pickett.)

cultural contexts in Europe, Latin America, and Asia. I have developed warm friendships and professional relationships within the United States and overseas. As an African American, I have a lifetime's experience of cross-cultural interaction and understanding, and practice in code-switching. These experiences have been useful in urban ecology and in the growing social networks that I inhabit.

I believe that participation in the LTER program, and my role as a leader of one of the two urban sites in particular, has been a significant boost to my career and a platform for unimagined exposure to new professional perspectives and scholarly areas. In particular, this role has generated involvement with urban design, with sustainability and resilience science, and with scholarship on environmental justice. The LTER program has also thrust me into situations where it is necessary to correct the misconceptions about ecological science that exist in other professions, disciplines, and in the public.

CHALLENGES AND RECOMMENDATIONS

The training and networking that exist at the individual sites and collectively are already quite successful in improving the research capacities of students, postdoctoral fellows, and scientists. Although there may be creative and effective ways to improve these, I think that there are several more pressing needs. First, it is important to enhance the capacity of individuals and LTER sites in education, communication, and outreach. For example, there are currently good programs, such as the Schoolyard Ecology effort, that enhance the role of education at LTER sites. There might be improvements to achieve through integrating the outcomes and lessons learned from these efforts with site and network scientific programs. Second, I think that capacity for communication with various publics (yes, plural) can be enhanced. Capacity would certainly include training in strategies and techniques for summarizing and presenting messages in various contexts and to various audiences. But equally important for improving communication capacity would be access to experts who can help produce graphics, stories, videos, and so on. These things are rarely part of the training and resources available to researchers and may be useful for education as well, perhaps improving the linkages between LTER education and research.

The issue of recruitment is a thorny one. Two issues come to mind. First, there seems to be a perception outside the LTER program that it is a closed system. Second, because the LTER program has fixed resources, there is an allocation problem between existing efforts versus beginning new initiatives and attracting new participants. To deal with the first, it might be useful to highlight the existing stories of successful inclusion of new persons and institutions into LTER sites and cross-site efforts. For example, cross-site efforts sometimes extend beyond the LTER network. Second, the opportunity for explaining the success of leveraging additional support can be perhaps better exploited to indicate the openness of the LTER community and the desire for inclusion.

CONCLUSION

In summary, my career has benefited immensely from involvement in the LTER program. Indeed, urban ecology as a modern science in the United States would hardly exist without the involvement of the LTER program. Some of my early experiences in synthesis, collaboration, and interdisciplinary work preceded my involvement in the LTER program. However, all of these early penchants have been developed in completely novel and intensive ways as a result of the commitment of the LTER program to urban ecology and its openness to understanding socioecological systems.

Intraregional comparisons have been important to the BES project, and these have required interacting with various disciplines, from hydrology to sociology. Carefully framed comparisons have been an important substitute for experimentation in those situations where ethical and practical considerations have proved constraining.

The LTER program has influenced my thoughts about the process of environmental research. It has confirmed that the most effective environmental research combines biogeophysical perspectives with social perspectives. Although it is possible to answer environmental research questions from a biogeophysical perspective, understanding the origins of environmental impacts as well as their potential solutions requires including social perspectives.

I encourage young investigators to explore the social–environmental connections in research and in publications. I believe that much of the excitement in ecology resides on the interdisciplinary frontiers required to understand socioecological systems. Read broadly, but more importantly, talk with experts in other fields, because it is easy to misinterpret texts read in isolation. In fact, the early history of urban ecology as represented by the "Chicago School" had suffered from this error, according to Light (2009). Go out in the field with people from other disciplines and contrasting approaches, and engage them in conversation about the places and situations you visit. Write short essays jointly to clarify issues of shared concern. These can find a life on blogs, in curricula, in discussion groups, and may serve as useful finger exercises for the symphonic syntheses to come. Engage policy and management, but always do so from the most substantive empirical and clearest conceptual base. Do not expect metaphor alone to be a permanent bridge. Most of all, it is OK to have fun doing this. LTER programs are a great preparation for a career.

This advice to new investigators would be disingenuous without simultaneously encouraging my senior colleagues who review younger colleagues' job applications, proposals, and promotion packets to recognize the benefits of interdisciplinary research along with the burdens that it involves. Interdisciplinary environmental research is costly and demands much time, and it should be rewarded. Finally, elders in LTER programs are and should continue to be welcoming of new partners of any career stage.

ACKNOWLEDGMENTS

I am grateful to my many colleagues at the BES site for their patience as I have worked to learn enough about their specialties to promote integration in the project. I am especially pleased to acknowledge the open and collegial approach of social scientists M. Grove, C. Boone, G. Buckley, L. Ogden, and K. Dow, as well as urban designers B. McGrath and V. Marshall. BES LTER is supported by NSF award DEB-1027188. The complementary activities of the Urban Sustainability Research Coordination Network (RCN 1140070) has also been instrumental in my experience as I have explored the connection of ecology with other disciplines and urban professions. The intellectual partnership with D. Childers, Codirector of the RCN, has facilitated this growth as well.

REFERENCES

Boone, C.G., Redman, C.L., Blanco, H., Haase, D., Koch, J., Lwasa, S., Nagendra, H., Pauleit, S., Pickett, S.T.A., Seto, K.C., and Yokohari, M. (2014). Group 4: Reconceptualizing urban land use. In K.C. Seto and A. Reenberg, eds. In press. *Rethinking Urban Land Use in a Global Era*. MIT Press, Cambridge.

Cadenasso, M.L., Pickett, S.T.A., and Schwarz, K. (2007). Spatial heterogeneity in urban ecosystems: Reconceptualizing land cover and a framework for classification. *Frontiers in Ecology and Environment,* **5**, 80–88.

Childers, D.L., Pickett, S.T.A., Grove, J.M., Ogden, L., and Whitmer, A. (2014). Advancing urban sustainability theory and action: Challenges and opportunities. *Landscape and Urban Planning,* **125**, 320–325.

Light, J. S. (2009). *The Nature of Cities: Ecological Visions and the American Urban Professions 1920-1960*. Johns Hopkins University Press, Baltimore.

McDonnell, M.J., and Pickett, S.T.A. (1990). Ecosystem structure and function along urban-rural gradients: An unexploited opportunity for ecology. *Ecology,* **71**, 1232–1237.

McGrath, B.P., Marshall, V., Cadenasso, M.L., Pickett, S.T.A., Plunz, R., and Towers, J., eds. (2007). *Designing Patch Dynamics*. Columbia University Graduate School of Architecture, Preservation and Planning, New York.

McHale, M.R., Bunn, D.N., Pickett, S.T.A., and Twine, W. (2013). Urban ecology in a developing world: How advanced socio-ecological theory needs Africa. *Frontiers in Ecology and Environment*. (doi:10.1890/120157)

Peters, D.P.C., Lugo, A.E., Chapin, III, F.S., Pickett, S.T.A., Duniway, M., Rocha, A.V., Swanson, F.J., Laney, C., and Jones, J. (2011). Cross-system comparisons elucidate disturbance complexities and generalities. *Ecosphere,* **2**, art 81.

Pickett, S.T.A., Cadenasso, M. L., and McGrath, B., eds. (2013). *Resilience in Ecology and Urban Design: Linking Theory and Practice for Sustainable Cities*. Springer, New York.

PART FIVE
CEDAR CREEK ECOSYSTEM SCIENCE RESERVE (CDR) LTER SITE

12 Beneficiary of a Changed Paradigm
Perspectives of a "Next-Generation" Scientist

Elizabeth T. Borer

IN A NUTSHELL

1. As a scientist, the Long-Term Ecological Research (LTER) program has deeply influenced my approach to scientific inquiry by creating an environment of effective collaboration and long-term evaluations of ecosystems. The increasing emphasis on data management and sharing has shaped both the philosophy and implementation of my scientific projects.
2. I have become a highly collaborative scientist because of my experiences with the effectiveness of collaborative inquiry, put in place by initiatives including the LTER

program and institutes such as National Center for Ecological Analysis and Synthesis (NCEAS).

PERSONAL OVERVIEW

I have been involved in the LTER program since I began my first faculty position at Oregon State University in 2004. Although my primary site affiliation is now Cedar Creek Ecosystem Science Reserve (CDR), I have ongoing experiments and collaborations spanning nine LTER sites (Borer et al. 2014b). I am a community ecologist with work that bridges into ecosystems. My research focuses on quantifying the consequences of global changes (e.g., nitrogen deposition, species invasions and extinctions) for interactions among species, including host–pathogen, plant–herbivore, and plant–plant interactions, and the resulting consequences for ecosystem functions. Since 2007, I have been the lead principal investigator of the Nutrient Network (NutNet; www.nutnet.org), a global scientific cooperative of more than 100 scientists performing identically replicated experiments at more than 75 sites in 17 countries on 6 continents to examine the interactive effects of herbivory and multiple nutrients on controlling critical processes and functions in the world's grasslands (Borer et al. 2014a). I am currently an associate professor in the Ecology, Evolution, and Behavior Department at the University of Minnesota and serve as senior personnel on the ongoing National Science Foundation (NSF) grant supporting CDR.

APPROACH TO SCIENCE

My LTER site affiliation is not entirely clear in my own mind, even though I am listed as a scientist at CDR. Although I have ongoing projects at LTER sites, primarily at CDR, I do not consider myself a site-based researcher in the LTER program. Yet the success of the LTER program in shifting perspectives has been so great that, until I sat down to think about my answers to the survey questions, I did not even realize how much I have relied on the cultural shifts that the LTER program put into motion to gain the perspectives I currently have on ecological research.

Each of us is a product of our time. I was a graduate student when the first ecological synthesis center, the NCEAS, was just opening its doors and beginning to define ecological synthesis. At that same time, the National Ecological Observatory Network (NEON), a highly instrumented observational platform, was a new idea just under consideration, and meta-analysis and large collaborative groups were still highly suspect in the ecological literature. Two urban LTER sites were in the process of being funded, but there was debate about whether studying urban ecosystems could produce "real" and "important" ecological insights. These debates, new tools, and new ideas were exciting to me and shaped my approach to science. During this same time, the LTER program was already well established, with some LTER sites just reaching 15 years of continuous research and funding. The value of a long-term research focus became folded into my scientific approach.

Midway into my career, I see myself as philosophically a "next-generation" scientist from the LTER program—standing on the shoulders of the incredible infrastructure and reenvisioning of ecological science spanning collaboration, data management and sharing, and long-term evaluations of ecosystem structure and functioning put in place by the LTER program. I have adopted each of these elements established by the LTER program in my own scientific approach. For example, since the year I started a faculty position (and felt I could realistically develop my own long-term studies), I have planned and maintained decade-scale projects and have resampled decades-old, abandoned experiments

(Borer et al. 2009a) to complement the inference from my shorter-term studies. However, I have extended this long-term approach to concurrently document patterns and relationships that emerge across sites and ecosystems. Whereas the LTER program has created an amazing research infrastructure that has revealed the importance of long-term, place-based work, my research has increasingly focused on identifying the scales (e.g., taxonomic, functional, spatial, and temporal) at which there is generality in the structure and functioning of ecological systems, regardless of location (Borer et al. 2014a).

Before I became involved in research at LTER sites, I naively thought of parallel research using standard methods at many sites as a central component of science in the LTER program and embraced the importance of such an approach for furthering our understanding of the factors determining the composition and function of ecological systems. However, although LTER sites are often referred to as a national network of research sites, this seems primarily administrative. As I became more involved in cross-site research, I discovered that the use of standardized methods to make measurements at multiple sites was rare. For this reason, comparative or synthetic research involving data from multiple LTER sites usually require the use of meta-analytical approaches to gain insights into the role of location and conditions in ecological processes and responses across the diverse ecosystems represented within the LTER network. My initial naiveté and later realization that relatively little standardized data were being collected across the LTER sites was part of the motivation that led me to start large-scale research projects replicated across large spatial scales. In one of my ongoing studies, for example, my collaborators and I specifically targeted LTER sites for replicated data collection to complement and benefit from the long-term, data-rich LTER site-level work (Borer et al. 2014a).

The focus of my research is on quantifying the effects of global changes on the composition and function of ecological communities. Part of this research focus includes understanding the role of location in determining ecological function and discovering which ecosystem responses to perturbations are similar, regardless of location or conditions, and which responses depend on local conditions. NutNet is a study that I participated in envisioning and that I have co-coordinated since 2005; it is explicitly organized around multisite, global-scale replication (Borer et al. 2014a). This study is a collaborative, distributed experiment that uses standard protocols to characterize the species diversity and composition of herb-dominated plant communities. NutNet arose from the vision of a group of seven junior faculty and postdoctoral fellows who created a new approach to ecological research. We combined the highly collaborative nature of NCEAS, the long-term focus of the LTER program, and the spatial replication and standard protocols of NEON, overlaying these on an experimental framework designed to sort out ecologically general from site-specific responses. This project's central theme is coordinated observation and measurement of the responses of grassland ecosystems to experimental manipulations of nutrients and vertebrate herbivores, and it is currently being replicated at more than 75 sites worldwide. We have developed data sharing and management for this network using work that originated in the LTER program. Initially, I did not realize that incorporation of nine LTER sites into NutNet would be a rare use of this network, but at least at present, it is one of only a few experimental studies to span this many LTER sites. Thus, this distributed, multisite approach represents an amalgam of many different approaches to the study of ecological systems, including those in the LTER network.

ATTITUDES TOWARD TIME AND SPACE

My involvement with the LTER program has affected my scientific approach by highlighting the importance of long-term research and the need for collaborative research networks. First, while the time scales of PhD projects and 3-year federal research grants create a focus on time scales of a few months to a few years, the LTER program has clearly demonstrated that ecological responses to perturbations in the first few years can be entirely different from changes after a decade or longer, leading to limited, or incorrect, long-term predictions stemming from short-term studies. Publications from LTER sites demonstrating the importance of timescale in ecological responses were highly influential for me as I developed a research vision early in my career, and they continue to influence how I think about change in ecological systems.

COLLABORATION

Collaborative science, a hallmark of the LTER program, can lead to important and novel insights that would not have arisen with solo efforts. Collaboration enriches understanding of the complex environments under study at LTER sites and also allows studies to be replicated across many sites. I share a deep appreciation for the collaborative nature of science in the LTER program. This approach is among the greatest draws of the LTER program for me, because it enriches my understanding of ecological systems. At CDR, for example, each of the core experiments has two to eight investigators involved in planning treatments and studying responses. Often, several undergraduates, graduate students, and postdoctoral fellows are involved, providing a rich diversity of hypotheses, observations, and interpretations. In addition, each investigator is involved in multiple projects, thereby creating a strong network of interactions. The varied perspectives in this collaborative network never fail to provide me with novel insights into my study questions and a greater depth of understanding about the systems my collaborators and I jointly seek to understand.

APPLIED SCIENCE

My research at LTER sites has not increased my interactions with stakeholders in agriculture, natural resources, or restoration in the United States or abroad.

COMMUNICATION

Although I regularly bring research, specimens, and scientific ideas to K–12 classrooms, this has not been via LTER programs. However, since arriving at CDR a few years ago, I have begun planning with the CDR education coordinator, Mary Spivey, to communicate some of my research questions and findings from work at CDR to K–12 students and teachers via the Cedar Creek Schoolyard Programs, which reach an impressive number of people.

MENTORING

In addition to scientific collaborations, the social aspects of research in the LTER program include some of my most valued teaching and training opportunities. The depth of interactions that are possible when doing fieldwork with undergraduates makes it one of my favorite venues for teaching, and much of my current field work is done at CDR. Collaborative science, such as projects within the spheres of the LTER program and NCEAS, has led to increasingly lengthy authorship lines on ecological publications. This is not new territory; genomics, medicine, and physics are well into the era of highly collaborative projects and publications. There is a diversity of opinions about collaborative science in ecology, but when I review tenure packets, articles, or proposals, I am not at all put off by long author lines or multi-institution collaborations. Establishing clear individual contributions and areas of expertise by leading papers and grants is critical, but participating in collaborations can be an extremely valuable and enriching experience for ecological science and for scientists at each career stage.

SKILL SET

In my own research, NutNet in particular, I emulate the data management and sharing policies developed within the LTER program (Borer et al. 2009b, 2014a). The LTER best practices for data and metadata standards, data integration and preservation, as well as database management and querying have laid the groundwork for researchers across the field of ecology to effectively document, store, access, and preserve valuable ecological data (Michener et al. 2009, 2011). Decades of effort on ecological information management within the LTER have allowed me to expand my own tool kit and have percolated out to enrich the field of ecological research with well-articulated guidance for effective data management.

PERSONAL CONSEQUENCES

Ecological science is a social enterprise, and my approach to answering questions is often highly collaborative, in part because I enjoy the people in my field as scientists and as people. I often become friends with my collaborators—because of our common interests, passion for our shared work, and the development of mutual respect and trust that arises through collaboration. In addition, because of the similar demands of our jobs as academic scientists regardless of institution or country, we often face a similar set of choices as we attempt to maintain work–life balance. Intuition and academic research both suggest that these factors are key predictors of friendships. Although my friendships formed within my profession extend beyond my interactions with personnel in the LTER program, they certainly include these people.

CHALLENGES AND RECOMMENDATIONS

There are opportunities for increasing participation and recruiting "new blood" into the LTER network, and this has been a priority at CDR and the H. J. Andrews Experimental Forest LTER sites, the two sites where I have had the greatest involvement. The LTER program is intended to generate deep insights from long-term study of a research site through many lenses. This approach is undeniably valuable as one approach to the science of ecology. However, LTER sites, in my mind, are the people and ideas as much as the place. For many sites, the same researchers have been involved since the outset—or even before—and their scientific contributions are associated with both their name and the name of the site. However, the questions of this survey and format of this book reflect the perspective that there *is* a site-based experience in the LTER program. For some this certainly reflects their experience, but for many in the field of ecology, this assumption risks being alienating, creating "insiders" and "outsiders"—those who perceive themselves as having an "LTER experience" and those who do not. Perhaps this stems from the goals of the LTER program—collaborative, long-term, site-based research may foster tight-knit groups of long-term collaborators at a site. Compounding this is that LTER sites that have been funded for 20 or 30 years have accumulated an astounding corpus of literature, making it challenging for new investigators (graduate students, postdoctoral fellows, or faculty new to a site) to carve out an intellectual niche. Yet, despite all of this, I have felt very welcomed as a researcher at several LTER sites, so this need not be an impediment.

The LTER program is, in fact, facilitating very broad participation through its extremely high standards for data quality, management, and access at the site-scale (Michener et al. 2011). My experience with data from LTER sites is that it is highly accessible and extremely well-managed, providing the larger ecological community with a valuable data resource for novel insights and discovery. The data-sharing and integration for novel insights create a body of knowledge about each LTER site that is qualitatively different from other field sites where I have worked. The management and provisioning of high-quality data by LTER sites is a substantial service to the greater scientific community that also serves to increase participation by "non-LTER" scientists. Further evidence of the breadth of participation fostered by data access at LTER sites is that many publications using site-level data from the LTER program do not include as authors the scientists from the LTER sites who envisioned and collected these data.

Beyond the perception of "clubbiness," the more challenging impediment to catalyzing participation is funding for both research and training. Enhancing the development of young researchers and recruiting "new blood" will not occur without sufficient funds. Although this is not a severe current impediment because of the ongoing NSF commitment to the LTER program, it is a point on which to remain vigilant—the long-term viability of this program depends on continuous, long-term funding to support both the science and the training of scientists and communicators at each site.

CONCLUSION

Because my research focus draws strongly on cross-site comparisons and has never been fully nested within the LTER program or any single LTER site, addressing questions about how my "LTER experience" has shaped me as a scientist has been challenging. Although I work at LTER sites and have become deeply involved in creating a vision for CDR, in particular, I still do not define myself as an "LTER researcher." This perspective is likely shared by many who have been influenced by the LTER program. Yet I value my work at CDR and other LTER sites, my interactions with a diverse group of scientists, and the integrative, long-term, site-based ecological understanding that has arisen from decades of research at this and other LTER sites. The extensive and well-managed data, the long-term focus, and the integrative understanding each provide a significant draw for me to continue my work at these sites. I am committed to expanding knowledge and understanding of ecology through work at CDR, and after nearly 4 years of work and collaboration there, I am gaining momentum in my work at this site.

ACKNOWLEDGMENTS

Thanks to my many colleagues and friends at Cedar Creek LTER and other LTER sites and Nutrient Network collaborators around the world. Special thanks to F. Swanson and E. Seabloom, who read and commented on this essay. This work was supported by the NSF Research Coordination Network (DEB-1042132) and Long-Term Ecological Research (DEB-1234162 to Cedar Creek LTER) programs, the Institute on the Environment (DG-0001-13), and the University of Minnesota.

REFERENCES

Borer, E.T., Harpole, W.S., Adler, P.B., Lind, E.M., Orrock, J.L., Seabloom, E.W., and Smith, M.D. (2014a). Finding generality in ecology: A model for globally distributed experiments. *Methods in Ecology and Evoution,* **5,** 65–73.

Borer, E.T., Mitchell, C.E., Power, A G., and Seabloom, E.W. (2009a). Consumers indirectly increase infection risk in grassland food webs. *Proceedings of the National Academy of Science,* U.S.A. **106,** 503–506.

Borer, E.T., Seabloom, E.W., Gruner, D.S., Harpole, W.S., Hillebrand, H., Lind, E.M., Adler, P.B., Alberti, J., Anderson, T.M., Bakker, J.D., Biederman, L., Blumenthal, D., Brown, C.S., Brudvig, L.A., Buckley, Y.M., Cadotte, M., Chu, C., Cleland, E.E., Crawley, M.J., Daleo, P., Damschen, E.I., Davies, K.F., DeCrappeo, N.M., Du, G., Firn, J., Hautier, Y., Heckman, R.W., Hector, A., HilleRisLambers, J., Iribarne, O., Klein, J.A., Knops, J.M.H., La Pierre, K.J., Leakey, A.D.B., Li, W., MacDougall, A.S., McCulley, R.L., Melbourne, B.A., Mitchell, C.E., Moore, J.L., Mortensen, B., O'Halloran, L.R., Orrock, J.L., Pascual, J., Prober, S.M., Pyke, D.A., Risch, A.C., Schuetz, M., Smith, M.D., Stevens, C.J., Sullivan, L.L., Williams, R.J., Wragg, P.D., Wright, J.P., and Yang, L.H. (2014b). Herbivores and nutrients control grassland plant diversity via light limitation. *Nature,* **508,** 517–520.

Borer, E.T., Seabloom, E.W., Jones, M.B., and Schildhauer, M. (2009b). Some simple guidelines for effective data management. *Bulletin of the Ecological Society of America,* **90,** 205–214.

Michener, W.K., Brunt, J.W., Helly, J.J., Kirchner, R.B., and Stafford, S.G. (1997). Nongeospatial metadata for the ecological sciences. *Ecological Applications,* **7,** 330–342.

Michener, W.K., Porter, J., Servilla, M., and Vanderbilt, K. (2011). Long term ecological research and information management. *Ecological Informatics,* **6,** 13–24.

13 Listening to Nature and Letting Data Be "Trump"

David Tilman

IN A NUTSHELL

1. My long-term research, which has focused on major ecological mysteries and questions, has provided many unexpected insights into the processes, mechanisms, and feedbacks that determine the structure and functioning of the grassland ecosystems. Many of these insights have emerged from exploring these questions with a combination of well-replicated field experiments, long-term observations, and predictions of theory.
2. A major source of research creativity has been my instinct to pay the deepest attention to any rigorous results that fall outside the realm of current paradigms, concepts, or theoretical predictions, including the predictions of my own theories. I refer to this as "listening to nature" and letting data be "trump." It is when current

ideas fail and "things fall apart" that new hypotheses are generated that are so crucial for the advancement of science.
3. My teaching builds on this approach: trying to have each lecture explore a mystery or paradox, including those with which I am currently grappling, and challenging my students to propose solutions. Perhaps because they are not saturated with the current paradigms of ecology, students and members of the public frequently respond to ecological mysteries with great creativity.
4. I believe that the amazing privilege of having public support for my research obliges me to communicate my findings of relevance to society through public talks, testimony to legislative committees, interviews with the media, and discussion with business leaders.

PERSONAL OVERVIEW

In 1981, when we were writing the initial Cedar Creek Long-Term Ecological Research (LTER) proposal, I was a 32-year-old, 6-year-post-PhD associate professor at the University of Minnesota. I had spent most of my career doing mathematical theory and laboratory studies of resource competition between freshwater algae. I had started doing nutrient-addition field experiments in the Cedar Creek grasslands only 3 years earlier and had just finished writing a book on resource competition (Tilman 1982). Hutchinson's (1961) "paradox of the plankton"—the search for the forces and processes that allowed so many competing species to coexist with each other, whether algae in lakes, herbaceous species in grasslands, or trees in tropical forests—was intriguing. The possibility that simple but mechanistic theories of interspecific interaction might not just solve Hutchinson's diversity paradox but also help to transform ecology into a more experimental, mechanistic, and predictive science was especially fascinating. At the encouragement of David Parmalee, the Director of Cedar Creek, and in collaboration with the mammalian ecologist John Tester, I assembled and led the team that wrote an LTER proposal titled "Micro and Macro Views of Succession, Productivity and Dynamics in Temperate Ecosystems." Leading an LTER program for the past 32 years has immensely broadened my interests, changed my career in ways that I never could have foreseen, and led me to conclude that long-term, site-based research that combines experiments, observations, and theory is the bedrock of our discipline.

APPROACH TO SCIENCE

My long-term work at Cedar Creek taught me, time and again, how much more rapidly science can progress when a major question is simultaneously addressed with new theory, experiments, and long-term observations. Each of these three approaches informs and shapes the others. I rarely explore a topic with theory unless inspired by unexpected results of my experiments or observations; my experiments often test a novel prediction of theory. This cyclic approach, which is reminiscent of simple conceptualizations of the scientific process, proved to be particularly powerful when doing long-term research at the same site, perhaps because it eliminates the time lags that occur when some individuals do observations or experiments and others do theory.

Some of the higher-impact work that I have been involved in at the Cedar Creek Ecosystem Science Reserve (CDR) occurred because I learned to "listen to nature," by which I mean that I became especially sensitive to and excited by any results that fell outside the realm of then-current paradigms, concepts, or theoretical predictions. Instead of dismissing such results as "wrong," I paid particular attention to them. No matter how seemingly solid and accepted an idea might be, no matter how rigorous the related theory might be and how believable its predictions, and no matter who had done the work or proposed the theory, myself included, data were trump. If, after thorough and rigorous analyses, data rejected an existing idea and suggested some set of alternative possibilities, the data always won the debate and set the research agenda on a new trajectory.

A dramatic instance of this happened when I started analyzing data on plant abundances and net primary productivity before, during, and after a major drought that occurred in 1987–1988. I was puzzled by the results, which seemed to suggest that greater diversity led to greater ecosystem stability. I enlisted the collaboration of John Downing, whose statistical skills greatly exceeded my own, in a search for answers. Despite dozens of different analyses that Downing and I subsequently did, we consistently found that plant communities that were more diverse had primary productivity that was more stable, which is to say, more resistant to drought and more resilient when drought ended (Tilman and Downing 1994). The possibility that higher diversity led to greater community or ecosystem stability contradicted two then-current views that I knew well because I taught them in my classes: (1) species composition was the major driver of community and ecosystem processes and (2) higher species diversity was not only of little relevance to stability but also likely made it more difficult for ecosystems to be stable.

When, in 1993, analyses of our large data set offered a compelling case against those views, the data won the debate. We began setting up a completely randomized and replicated biodiversity experiment, the first of its kind. National Science Foundation (NSF) funding through the LTER program, as well as existing unfettered funds from the Andrew Mellon Foundation allowed us to immediately begin this major new initiative. We knew that such an experiment would be the only way to address the questions and possibilities that our analyses had suggested. This experiment, and others that followed around the world, were pivotal in resolving the ensuing biodiversity debate. Long-term funding allowed the experiment to reveal many unexpected findings. The most important of these, I believe, is that the effects of biodiversity increase through time and grow to be as great as, or greater than, those of nitrogen, water, herbivory, and others ecological forces long-known to be of importance (Tilman, Reich, and Isbell 2012).

ATTITUDES TOWARD SPACE AND TIME

Casual but long-term observations in several successional grassland fields that I started visiting regularly in 1980 led to my interest in spatial ecology. I initially imagined that interspecific competition caused the spatial patterning in the locations and abundances of various perennial plant species. However, a decade of observation, combined with results of competition experiments, suggested something quite different: that the best competitors were poor dispersers and were absent from many locales within a field because of their poor dispersal ability and not their competitive ability. Some dominant competitors took 10 to 20 years to reach a successional field, and at least as long to spread across the field. The same factors caused the short-term effects of some experimental manipulations to differ from the long-term results because the species favored by a treatment, such as an herbivore exclosure, often had to migrate into a plot before their responses occurred. The net effect of these and other results were greatly increased interest in spatial processes and temporal dynamics.

COLLABORATION

Collaboration was essential in planning the biodiversity experiment, as it was for subsequent activities, such as the biodiversity, CO_2, and nitrogen deposition experiments (Reich et al. 2001). Collaborative planning is slow and at times frustrating. It must be done with equal sensitivity to the needs of each team member and to the major goals of the experiment. Particular team members have a strong sense of how an experiment should be designed to address the issues of greatest interest to them but less insight into how that might affect the work of others. One way to address such issues is to expand the experimental design to accommodate all wishes, but the resultant design is almost always too large, complex, and grandiose to be practical. Our discussions of the biodiversity experiment, for instance, extended over several months before we agreed on design solutions that seemed to be a workable and wise compromise. Even then, designs have had to be changed on the fly. For instance, we compromised on having a 13 × 13-m plot size in our biodiversity experiment, hoping that it would be large enough to capture effects of treatments on insect communities but small enough for us to maintain desired plant species compositions via weeding. Within a few years, it was clear that weeding several hundred plots of that size was unaffordable, and we reduced plot size to 9 × 9 m. Fortunately, that size was sufficient for our work on insects. We later found that even the 3 × 3-m plots in our small biodiversity experiment were sufficient to detect effects of plant diversity and plant community composition on insect communities.

In 1981, when we were writing our first LTER proposal, most ecologists, myself included, had a very different view of collaboration than we do today. Collaboration was uncommon in ecology then. The tradition at that time was that each ecologist was a free agent, the sole proprietor of his or her own individualistic research enterprise. Moreover, at that time a long-term experiment ran for 2 or 3 years. To cover the diversity of topics in NSF's request for LTER proposals, our team originally included seven faculty investigators. We had presented ourselves, in our proposal, as a coherent team. However, when on receiving funding we met to formalize and implement our research plans, many of the coinvestigators thought that each investigator should individually determine what research to do at Cedar Creek and should have full and sole discretion on use of "their" slice of the grant. This, of course, was how scientific free agents did science, but I believed it could threaten our ability to achieve meaningful long-term research and to collaborate. I struggled to find a solution to this dilemma, deciding that we would use our funds from the LTER program to support a suite of shared long-term experimental and site-based observational studies. Some members of our initial team immediately lost interest in the LTER program. When our research began in the spring of 1982, it consisted of each of us pursuing our individual ideas but doing our work, as much as possible, within the same sites or experiments.

Our individualistic perspectives gradually broadened as our team interacted and learned from each other. For instance, I initially focused on how plant traits and interspecific competition for nitrogen affected plant community composition and diversity along a nitrogen addition gradient. A mystery soon arose: why were nitrogen-fixing legumes so rare in grasslands in which soil nitrogen was clearly the major limiting soil nutrient? Various herbivore exclosure experiments kept suggesting the same answer: many legumes were rare because they were preferentially consumed by deer and small mammals (Ritchie, Tilman, and Knops 1998). Indeed, such herbivory seemed to keep fast-growing legumes at such low densities that it took them many years to disperse from the few sites where they occurred into herbivore-exclosure plots. Eliminating herbivory led, eventually, to great increases in these species of legumes, and then to increases in soil carbon and nitrogen (Knops, Richie, and Tilman 2000). These and other collaborations broadened my perspective from an initial focus on nitrogen competition to include dispersal dynamics (Tilman 1994), herbivory, disease dynamics (Mitchell, Tilman, and Groth 2002), and feedback effects of herbivory and plant community composition on soil carbon and nitrogen.

Ecological linkages led to intellectual linkages, and we began discovering other interactions and feedback effects that were structuring grassland ecosystems. My collaborative network grew as questions arose that demanded skills and knowledge that only a diverse team could provide. I was on a path toward multidisciplinary collaborative research, a path that would broaden through the decades to include the application of our findings to economic, environmental, and policy issues; public communication through engagement with the media; and providing advice to state and federal agencies, legislators, and political leaders.

APPLIED RESEARCH

My career also branched out from its initial focus on mechanisms of interactions among species and the causes of the earth's high biodiversity, to issues such as the environmental impacts of agriculture and of biofuels. The latter two interests grew out of my interests in biodiversity. The agricultural land clearing that is used in the poorer nations of the world to meet the globally accelerating demand for food is a major threat to biodiversity and a major contributor to global greenhouse gas emissions. Demands for biofuels also threaten biodiversity and, unless done properly, can escalate greenhouse gas emissions.

I cannot say that my activities in the agriculture and biofuels arenas were a direct result of the LTER program, although it is clear that long-term research is uniquely able to provide scientific knowledge of relevance to many major environmental problems. Rather, I suggest that the existence of the LTER program at the same time that humans were recognized as dominating global ecosystems (Vitousek et al. 1997) led to a synergistic interaction that allowed me and many other scientists in the LTER network to have the skills and long-term data needed to address emerging environmental issues.

COMMUNICATION

It is just as difficult to disentangle cause and effect in a career as in ecosystems. As our research at the CDR LTER site became more relevant to major environmental problems, I grew more interested in communicating our findings with the media, governmental and corporate leaders, and the public. I was fortunate to be chosen to be a Pew Scholar in Conservation Biology, which provided extensive (and for me, at least, much needed) training in media communication. The LTER Schoolyard Program initiated an interest, now shared by most CDR researchers, in passing on the love of science and concern for the environment to the next generation.

MENTORING

When I was working on my PhD at the University of Michigan in the early 1970s, many of my fellow graduate students and I experienced the open, ardent, and few-holds-barred discussion of science practiced by the faculty of the zoology department. This, to me, was a hallmark of science as individualistic intellectual free agency. I quickly learned, however, that it was not the way to lead an LTER program. Leadership is much more about mentoring, respecting, listening, and establishing trust and communication. As I saw in the best deans and provosts I knew at the University of Minnesota, leaders work for those whom they lead, not vice versa. The success of the CDR program has depended as much on what we achieved as a team as on our individual research efforts. As the university hired ecologists, I would take my new colleagues on tours of the Cedar Creek site, encourage them to work there, invite them to add their skills to ongoing projects and, as possible, allocate funds from the LTER program to support them. I make no claims of great mentoring expertise; my early mentors, especially Peter and Susan Kilham, were better mentors than I. I have greatly benefited from the help and guidance of many others, including Steve

Hubbell, John Vandermeer, Dan Janzen, Bob May, and Paul Ehrlich. I have used them as models and tried to pay my debt forward by encouraging and perhaps inspiring the faculty, students, and staff of the CDR.

SKILL SET

Massive increases in computing power since the LTER program began have drastically changed the skills needed to do ecological research. A single class in probability and statistics was once sufficient. I have since gained skills in experimental design, general linear models, nonparametric methods, advanced multiple regression, and spatial statistics; I regularly use Mathematica. Computers linked to various instruments gather most data. A wide range of computer and statistical skills has become essential to my science.

PERSONAL CONSEQUENCES

Over the past three decades I have learned much from scholars from other LTER sites and now count many of them as friends. Annual meetings that set aside time for informal conversations have rewarded me with insights into the similarities and differences in the controls of community ecology and ecosystem processes across the wide diversity of terrestrial, aquatic, and marine sites that compose the LTER network.

CHALLENGES AND RECOMMENDATIONS

Those who initiated the LTER program more than three decades ago correctly envisioned that major advances in ecology could come from combining a diverse team of ecologists, long-term site-based research, and the rigors of NSF peer-reviewed funding. Successful LTER sites are evolving entities that build on past findings by continually opening new areas of inquiry. They are data- and history-rich, which make them attractive to others who wish to join in site-based work, or to use data collected at one or more sites as the basis of their own research. Therein lie three major challenges confronting ecology and the future of the LTER program.

These challenges are to (1) fully open all ecological data for analysis and synthesis by teams of scholars whether or not they are a part of the current LTER network; (2) establish a universal ethic of data sharing that maximizes the rate that knowledge advances by appropriately awarding, encouraging, and supporting all aspects of the intellectual process; and (3) greatly increase the number of sites and scholars involved in long-term studies in evolution, behavior, and ecology.

Unlike most other data gathered by NSF grants, the vast majority of data collected with funds from the LTER program are now online, well documented, and being used by scientists who work outside of the LTER network. Indeed, ecology has a growing number of young scholars for whom "other people's data" are the foundation of most of the papers that they write. This is an important advance, one that merits full NSF support via grants focused on this type of synthetic research. However, until ethical and practical issues are resolved, its long-term viability is uncertain. Neither NSF, scholarly journals, nor scientific societies have grappled with the ethical implications of the massive and immediate data openness that the US Congress and NSF now require.

There are two equally compelling sides to this issue. Let us first consider the perspective of individuals who pose novel questions, design the critical experimental or observational studies, obtain the funding, and dedicate their time and energy to the success of their endeavors. Is it appropriate for someone to search the Web, download data, do their own analyses, and publish a paper before those who did the work have published on it? Or, should a single paper by the originators make the data fair game for the data hunters, even when the originators had planned, from the start, to explore the same issue? If so, what would motivate someone to dedicate themselves to fieldwork?

Conversely, science is built on the findings of many others. These prior findings are much more useful to the enterprise of science when the underlying data are available for all to use. Moreover, essentially all scientific research done at academic institutions is publically supported, and thus should belong to the public, which is the well-founded logical basis of federal law and of NSF's data access policy.

An ethic is needed that resolves these conflicts by appropriately rewarding all roles in the scientific process. A simple yet powerful ethic is this: if a paper could not have been written without the intellectual efforts of someone, it should not be published without them as an author. Journals, scholarly societies, and academic institutions need to establish rules of ethics that both appropriately acknowledge all those whose ideas, data, analyses, and writing create a scientific paper, and that sanction those who deviate from the rules.

What should NSF do? To maximize the benefits of long-term research in ecology, NSF must go beyond merely funding LTER sites. The immense data sets gathered across all the LTER sites since the early 1980s harbor numerous as-yet undiscovered insights and advances. However, NSF support goes only to those working at an LTER site. Competitive funds are also needed to support analyses done by a diverse array of other scholars. Individuals and teams should have peer-reviewed funds available to support the mining and interpretation of data from all sources so that the discipline, as a whole, can better address major ecological issues and questions. As discussed earlier, NSF should also expand the opportunity to do long-term science across the full sweep of the environmental sciences.

What should you do as a young scholar? The impact of your research depends first and foremost on the issues that you tackle. Ask big questions and pursue them wholeheartedly. The grandest challenges, I feel, are at the interface of the frontiers of science and society. Build a network of collaborators whose skills and interests complement yours. And, take advantage of the power of combining long-term observations, experiments and theory as you dedicate your career to the questions worthy of your efforts.

CONCLUSION

There is immense power in long-term studies of a site when done by a team of researchers who dedicate their careers to the site, and whose creativity and productivity are honed by the pressures of peer-reviewed proposals. This power is magnified when all data are archived and made freely available for use by others. However, even this will not achieve the true potential of long-term ecological research. Scientific advances require a confluence of the results of observations, experiments, and theory, and major advances occur when such a confluence is found to apply across a broad range of systems. A wide diversity of aquatic and marine ecosystems and their scholars are not yet represented in the LTER network, and scholars working in many types of terrestrial ecosystems are underrepresented. Knowledge of fundamental importance to science, and to the environmental problems that society faces, will continue to accrue at too slow a pace until NSF broadens the LTER network to include more ecosystems and scholars; to embrace evolution, behavior, and population biology as fully as the current LTER program embraces ecosystem ecology; and to fund the efforts of non-site-based data synthesizers and theoreticians.

The majority of any successes that I have been fortunate to have in my career have occurred because of NSF funding, especially through the LTER program, to which I am deeply indebted. I urge young scholars of ecology, behavior, and evolution to find a system that intrigues them and to dedicate their lives to its ever-fuller understanding. I urge those at NSF to broaden the scope of long-term research and its funding. Ecology is, after all, the study of the most complex system on earth, the 10 million–piece puzzle called life, of which humans are a part, and on which we depend.

ACKNOWLEDGMENTS

I would like to thank the NSF (DEB-1234162) for supporting this research for the past 30 years.

REFERENCES

Huntly, N., and Inouye, R.S. (1987). Small mammal populations of an old-field chronosequence: Successional patterns and associations with vegetation. *Journal of Mammalogy*, **68**, 739–745.

Hutchinson, G.E. (1961). Paradox of the plankton. *The American Naturalist*, **95**, 137–145.

Knops, J.M.H., Ritchie, M.E., and Tilman, D. (2000). Selective herbivory on a nitrogen fixing legume (*Lathyrus venosus*) influences productivity and ecosystem nitrogen pools in an oak savanna. *Ecoscience*, **7**, 166–174.

Mitchell, C., Tilman, D., and Groth, J.V. (2002). Effects of grassland plant species diversity, abundance, and composition on foliar fungal disease. *Ecology*, **83**, 1713–1726.

Reich, P.B., Knops, J., Tilman, D., Craine, J., Ellsworth, D., Tjoelker, M., Lee, T., Wedin, D., Naeem, S., Bahauddin, D., Hendrey, G., Jose, S., Wrage, K., Goth, J., and Bengston, W. (2001). Plant diversity enhances ecosystem responses to elevated CO_2 and nitrogen deposition. *Nature*, **410**, 809–812.

Ritchie, M.E., Tilman, D., and Knops, J.M.H. (1998). Herbivore effects on plant and nitrogen dynamics in oak savanna. *Ecology*, **79**, 165–177.

Tilman, D. (1994). Competition and biodiversity in spatially structured habitats. *Ecology*, **75**, 2–16.

Tilman, D. (1982). *Resource Competition and Community Structure*. Monographs in Population Biology, Princeton University Press, Princeton.

Tilman, D., and Downing J.A. (1994). Biodiversity and stability in grasslands. *Nature*, **367**, 363–365.

Tilman, D., Reich, P.B., and Isbell, F. (2012). Biodiversity impacts ecosystem productivity as much as resource disturbance or herbivory. *Proceedings of the National Academy of Sciences U.S.A.*, **109**, 10394–10397.

Vitousek, P.M., Mooney, H.A., Lubchenco, J., and Melillo, J.M. (1997). Human domination of earth's ecosystems. *Science*, **277**, 494–499.

PART SIX
CENTRAL ARIZONA–PHOENIX (CAP) LTER SITE

14 The Socializing of an Ecosystem Ecologist

Interdisciplinarity from a Career Spent in the Long-Term Ecological Research Network

Daniel L. Childers

IN A NUTSHELL

1. The broad interdisciplinarity of my science and my worldview are direct products of my career spent in the Long-Term Ecological Research (LTER) program.
2. I attribute the holistic systems approaches that I use in my teaching and mentoring to my career spent in the LTER program.

3. I am able to converse with a broad array of collaborators and practitioners because of my career spent in the LTER program.
4. My career is rich with interdisciplinary collaborations and partnerships thanks to the LTER program.
5. My life is rich with friends that I have met throughout my career spent in the LTER program.

PERSONAL OVERVIEW

There are probably few mid-career scientists who have spent virtually all of their careers associated with the LTER network. As one of these few, I view this as a tremendous asset. My experiences in the LTER program began in 1983 with the North Inlet Program (NIN), where my master's research, advised by the late Hank McKellar, involved modeling salt marsh ecosystem dynamics. After completing my PhD at Louisiana State University (LSU) in 1989, I returned to the NIN for a 3-year postdoctoral fellowship with Fred Sklar at the Baruch Marine Laboratory. I worked with Fred on another of his National Science Foundation (NSF) grants, but there was considerable overlap between that research and the work being done at NIN.

When the NSF released a solicitation for new coastal LTER sites in 1998, I was an assistant professor at Florida International University (FIU) in Miami. We gathered a core group of Everglades colleagues and answered this solicitation with a proposal to study coastal ecosystem dynamics in the Florida Everglades. Our proposal was successful, and by early 2000 the new Florida Coastal Everglades LTER program (FCE) was off and running. I directed FCE from its inception until I left FIU in 2008 for Arizona State University (ASU).

On arriving at ASU in 2008, I immediately became involved with the Central Arizona–Phoenix (CAP) LTER program. I was excited about my move to ASU and the new School of Sustainability because I felt as if it were a rare mid-career opportunity to change the trajectory of, and perhaps even the impact of, my career. Recently, I was fortunate to have the opportunity to lead CAP for 2 years, from 2010 to 2012. I quickly learned that it is one thing to build and lead an LTER program, as I had done with FCE, but that taking over leadership of a long-standing and fully functioning LTER program (CAP) is something altogether different!

My career path thus far has been marked by a significant expansion of my worldview and the scope of my science. This is largely because my career has been so rich in LTER experiences. I am now codirector of the Urban Sustainability Research Coordination Network (RCN) that NSF funded in Fall 2011. I lead this RCN with Steward Pickett, who is director of the Baltimore Ecosystem Study LTER program. My many LTER experiences are largely responsible for my growth from being a classically trained wetland ecosystem ecologist into an interdisciplinary scientist who studies many systems from a diversity of perspectives.

APPROACH TO SCIENCE

Bob Waide, director of the LTER network office and informal "keeper of LTER network histories," once told me that he believes that I am the only individual who has been director of two different LTER programs. If true, I find this fact to be humbling, but more importantly it reminds me that I cannot easily attribute some fraction of my career development to my LTER experience. These experiences have been too continuous, long term, entwined, and all encompassing. My captivation with, and commitment to, holistic systems science began with my interactions as an undergraduate with the late Bill Odum at the University of Virginia; grew through my graduate and postgraduate work with Hank McKellar, John Day, and Fred Sklar; and truly matured through my involvement with the FCE and CAP programs. Holistic systems approaches to science and research, which have always been a hallmark of LTER research, are essentially all I have ever known and practiced.

Hank McKellar, my MS advisor at the University of South Carolina (USC), was an H. T. Odum student. After working with Hank at NIN, I took a 4-year hiatus from the LTER network to earn my PhD at LSU, where I worked with John Day—also an H. T. Odum student. By the time I returned to USC's Baruch Marine Laboratory for a postdoctoral fellowship in 1989, I was fully indoctrinated in the "Odumesque" systems approach to science and was thus well prepared for yet more LTER-based ecosystem research. Interestingly, NSF developed the LTER program in the late 1970s as a follow-up to the International Biological Program (IBP) of the 1950s and 1960s, with a systems view that was strongly influenced by the works of H. T. Odum and his brother Eugene.

Around the turn of the century, my approach to science began to broaden considerably and become much more interdisciplinary. This transition was strongly influenced by my involvement, as an investigator with a large, multiyear planning grant that the LTER network received from the NSF in 2003. Under the leadership of Scott Collins, then at NSF, this broad-scoped effort ultimately led to an expanded thinking of LTER science beyond traditional ecology to include human actions, decisions, motivations, and interactions with their environment (Collins et al. 2011). Key to this transition was the active involvement of several energetic social science leaders, including Morgan Grove and Ted Gragson, who helped the LTER network see the many advantages of a more interdisciplinary approach to LTER science. Thanks to my close involvement with the Network Planning Grant, I had already been "drinking the Kool-Aid" of socioecological research when it came time to write FCE's first renewal proposal in 2005. In that process, I urged our research team to expand our boundaries beyond Everglades National Park—conceptually, geographically, and disciplinarily—to include the city of Miami, its residents, and its activities. After 2006, Laura Ogden's fine leadership of our new human dimensions group broadened the FCE program into urban ecosystem research and truly interdisciplinary science by including a cadre of excellent social scientists in the FCE endeavor.

Not surprisingly, the two urban LTER programs have consistently provided leadership on, and examples of, the power, success, and rewards of such interdisciplinary research. As one of only two urban LTER programs in the network, CAP has always taken a very interdisciplinary approach to studying urban ecosystems as complex adaptive socioecological systems. I was a full convert to the value of both cross-disciplinary collaborations and urban ecosystem science by the time I began working with and leading the CAP program and coleading the Urban Sustainability RCN.

ATTITUDES TOWARD TIME AND SPACE

It is difficult to articulate how my attitudes toward time and space have broadened, independent of my continuing evolution into a truly interdisciplinary LTER scientist. I feel certain that my LTER experiences are the source of my initiation into interdisciplinary socioecological approaches and my interest in the coproduction of research questions, approaches, and analyses with colleagues from many other disciplines. This significant expansion of my worldview and scientific attitudes began with the LTER planning grant, grew considerably as we added a human dimensions and urban component to the FCE research agenda in 2006, and has recently truly blossomed through leading the Urban Sustainability RCN with Steward Pickett. The growth of this RCN has focused on interdisciplinary and cross-generational membership and on including urban practitioners and decision-makers. Our intercity comparative approach includes many LTER sites, numerous LTER scientists, and scientists from around the world (Pickett et al. 2013; Childers et al. 2014). Our network goals include not only developing a better understanding of urban ecosystems but also applying that knowledge to making cities more sustainable. In short, my attitudes toward space have expanded to include a global view of human-dominated ecosystems, and my attitudes toward time now include the multigenerational views that characterize sustainability.

COLLABORATION

I have always made the point to my students that ecosystem science by its very nature is collaborative. Notably, ecosystem ecologists seldom publish single-authored papers, and when they do they are typically reviews. As my research and worldview has become more interdisciplinary, the diversity, extent, and richness of my collaborations have exploded. Today, my colleagues are more likely to be social scientists, economists, planners, designers, humanists, and even artists. Helping to lead the Network Planning Grant introduced me to the new vocabularies and approaches used by social scientists, and I now comfortably articulate in those vocabularies. I understand colleagues from many other disciplines. This evolution to coproducing truly interdisciplinary urban systems science was not easy, and I often feel as though my learning curve today is as steep as it was during my graduate school years.

LTER programs are large, highly collaborative endeavors that are constantly growing in scope, disciplinary expertise, and number of people involved. This happens at rates that far exceed the small budget increment that NSF gives each program every 6 years when renewal proposals are funded. One might ask how these interpersonal collaborations begin and evolve. In the case of FCE, we have always been coproducing scientific knowledge about the south Florida ecosystem with key agency scientists and decision makers. Our original proposal was coproduced with colleagues at the South Florida Water Management District, Everglades National Park, the US Geological Service, and beyond. These collaborations have always been easy because many of us were already colleagues and friends.

In my leadership of CAP, from 2010 to 2012, I recognized that I faced several collaborative challenges, particularly given that I had only been on the ASU faculty for 2 years at that point. A key strategy I used to meet some of these challenges was to spend my first semester as director of CAP meeting with CAP researchers. I scheduled half-hour meetings with more than 50 CAP researchers, meeting in their offices or via telephone at their leisure. Although I asked everyone the same general questions, mostly I listened and took copious notes. Those meetings and notes remain invaluable to me, and a serendipitous outcome of this time-consuming process was that it allowed me to quickly build considerable trust and confidence among CAP scientists. As any leader of large, collaborative, diverse groups knows, this is priceless.

APPLIED RESEARCH

Theoretical and basic science is the hallmark of the NSF, and the results of this should form a foundation for a better society. Many researchers worry about their scientific objectivity coming into question and are loath to be involved with applied research. This is a realm in which I can no longer operate. I am on the faculty of the first School of Sustainability in the United States, of which the "knowledge to action" and "science for societal good" concepts are key tenets. Sustainability science invokes various strategies for coproducing scientific knowledge with the managers and policy-makers who need it to make informed decisions. As I noted above, long before I knew about this concept of coproduction, we were practicing it at FCE—beginning with our original proposal. From working with us on research questions and sampling designs to supporting FCE research in many ways, our agency colleagues have always been part of the LTER science at FCE. At CAP, we work closely with city managers and planners to ensure a seamless, efficient interface between research products and decision-making. There is also considerable interaction between CAP LTER and the Decision Center for a Desert City, an NSF-funded Decision-Making Under Uncertainty center at ASU. Both FCE and CAP are "hubs" that integrate a wide diversity of socioecological research, but key "spokes in the wheel" are resource and city managers who directly use LTER knowledge. Notably, several long-standing LTER programs are based on US Forest Service land and include government scientists and managers. Although there may be some among us who remain uncomfortable with applied science, LTER research is relevant to many aspects of American society today.

COMMUNICATION

The interdisciplinary expansion of my science and career has required me to learn entirely new vocabularies, particularly in the social and sustainability sciences. With this came new abilities to communicate with a broad range of collaborators. I have been able to parlay these new communication skills into my lectures and interactions with students, and the LTER program deserves considerable credit for these pedagogical advances. Through the years I have also mentored high school students in LTER research experiences. Both at FCE and at CAP, I have hosted high school students in my laboratory, many funded through the NSF's Research Assistantships for High School Students Program. One of these students, who recently finished his undergraduate degree, even published his high school research in a peer-reviewed journal as first author!

MENTORING

Today I am prouder and more vocal than ever that H. T. Odum is my "academic grandfather." Holistic systems approaches to science and research, which have always been a hallmark of LTER research, are essentially all I have ever known and practiced. And I have always incorporated this systems perspective into my undergraduate and graduate teaching as well as my student mentoring. Even when my graduate students use methods and tools with which they are most familiar (e.g., microbial ecology, metagenomics), they formulate their research questions through a holistic, systems science lens. In my 20 years as an academic, I have advised more than two dozen MS and PhD students. Most of them conducted their thesis and dissertation research within the FCE and CAP LTER programs, and the work of those I mentored at FIU before FCE was funded contributed to our original LTER proposal. I am quite proud that several of my students have taken leadership roles in the FCE and CAP student groups, and in the LTER network-wide student group. Most of my graduate students have gone on to academic or environmentally related careers and have maintained contact with the LTER network. Finally, I have mentored numerous undergraduate students in LTER research at both FCE and CAP. Many of them have gone on to LTER-related graduate programs.

SKILL SET

Perhaps the most important growth in my skill set that I directly attribute to my LTER experience is with leadership abilities and interpersonal management skills. While at the Baruch Marine Laboratory for my postdoctoral fellowship, I witnessed the unfortunate defunding and discontinuation of the NIN program by the NSF in 1991. The lessons I learned from watching this devastating process, and the postfunding ramifications, were invaluable, particularly when I found myself building and directing a new LTER program only a few years later. Although my research skill sets have expanded considerably with the "interdisciplinization" of my career, it is my leadership and management capacities that have strengthened the most as a result of my LTER experiences.

PERSONAL CONSEQUENCES

By their very nature, LTER programs are large, highly collaborative endeavors that are constantly growing in scope, disciplinary expertise, and number of people. This growth typically far exceeds the small budget increment that NSF gives each program every 6 years at renewal. How do these interpersonal collaborations begin, evolve, and grow? In the case of FCE, we began coproducing scientific knowledge about the south Florida ecosystem with key agency scientists because many of them were already our colleagues and friends. In fact, I would hazard a guess that a large proportion of interpersonal collaborations that take place at any LTER site either began as friendships, or strengthened as friendships grew. In my case, I am very lucky to have a large group of collaborators whom I also consider friends; some of these fine people are, in fact, very close friends. Virtually all of my lasting and fulfilling collaborations have been LTER-based. I have the LTER network to thank for my [current] success as an ecosystem ecologist, sustainability scientist, interdisciplinary researcher, collaborator, and leader. But what I treasure most about my long-term involvement with the LTER program is the many rich and fulfilling friendships that I have developed, and come to depend on, over the years.

CHALLENGES AND RECOMMENDATIONS

Looking back on my LTER-rich career, most of the major challenges that I have faced have been direct products of poor, counterproductive, even destructive decisions by university administrators. In all cases I was able to "weather the storm" and I am now [almost certainly] stronger, more agile, and savvier. During my 30-plus years of involvement with the LTER network, I have seen it weather numerous challenges. Many of these have been budget-related, but we have had challenges within the network as well (e.g., the network self-planning process in the mid-2000s was a challenge for many). The NSF has the challenge of managing this large, long-term research program in the most consistent and predictable way possible. The rapid turnover of NSF program officers who oversee the LTER program, each of whom has their own ideas of what the network should be, does not seem to be a viable strategy for this challenge. All said, however, the positives of being associated with LTER far outweigh the challenges, and I will continue to recommend that my students and colleagues engage with or stay engaged with the LTER network as much as possible.

CONCLUSION

I honestly cannot envision what my career, my scientific impact, or my life would look like without the LTER program. I am proud that many of my students who also got their start in LTER science remain connected, and I am hopeful that my future students will do the same. In the nearly 35 years that the LTER program has been in existence, it has had major impacts on society and the environment. My 30-plus years with the LTER program have had a similar influence on me.

ACKNOWLEDGMENTS

Over the years, support has been provided to the author by the NSF through the CAP LTER program (grant 1027188), the FCE LTER program (grants 1237517, 0620409, and 9910514), and the Urban Sustainability RCN (grant 1140070). Beyond this research support, there are simply far too many people to thank in this small space!

REFERENCES

Childers, D.L., Pickett, S.T.A., Grove, J.M., Ogden, L., and Whitmer, A. (2014). Advancing urban sustainability theory and action: Challenges and opportunities. *Landscape and Urban Planning,* **125**, 320–328.

Collins, S.L., Carpenter, S.R., Childers, D.L., Gragson, T.L., Grimm, N.B., Grove, J. M., Harlan, S.L., Knapp, A.K., Kofinas, G.P., Magnuson, J.J., McDowell, W.H., Melack, J.M., Ogden, L.A., Ornstein, D., Robertson, G.P., Smith, M.D., Swinton, S.R., and Whitmer, A., (2011). An integrated conceptual framework for socio-ecological research. *Frontiers in Ecology and the Environment,* **96**, 351–357.

Pickett, S.T.A., Boone, C.G., Cadenasso, M.L., Childers, D.L., Ogden, L.A., McHale, M., and Grove, J.M. (2013). Ecological science and transformation to the sustainable city. *Cities,* **32**, S10–S20.

15 An Urban Ecological Journey

Nancy B. Grimm

IN A NUTSHELL

1. A long-term approach is definitive for my career, which has evolved at a single place over more than 30 years. But the Long-Term Research Ecology (LTER) program, and especially its urban research, has broadened my thinking far beyond the boundaries of the ecosystem science tradition in which I was trained.
2. I have added to my expectations of students that they learn collaboration, use a diversity of approaches, explore existing data, and document and archive their own data. I anticipate that they will find careers in a broader diversity of areas than academia.
3. The urban research in the LTER program has provided an incentive for me to work on communicating with the public, educators, students, and practitioners. I am still learning but am much more motivated than previously to reach out to these communities.
4. Collaboration is standard practice for ecosystem science but the LTER program has expanded the types of scientists with whom I collaborate as well as the extent of my external collaborations.
5. My decision to lead the Central Arizona–Phoenix (CAP) LTER project was therefore life-changing in extending the horizons of my science, mentoring, collaborations, and outreach.

PERSONAL OVERVIEW

Since 1997, when the CAP program began, I have been involved in the LTER program. I was the original principal investigator, and Charles Redman and I were codirectors from 1997 to 2010. In 2010, after successfully renewing the CAP project, I took a 2-year hiatus to work at the National Science Foundation (NSF). I returned in 2012 and am currently the principal investigator and sole director. This has been my only involvement in the LTER program throughout my career, although as an undergraduate, I conducted research at what was to become the H. J. Andrews Experimental Forest site.

I am trained as a stream ecologist and biogeochemist, and I have been at Arizona State University (ASU) for my entire graduate and postgraduate career. Currently I am a professor, having moved through ranks, first as a non-tenure track research faculty member, then as an "academic professional," and finally as an associate and then full professor. I lead somewhat of a double life, scientifically. My research prior to involvement in the LTER program was focused on desert stream ecosystems—their biogeochemistry, spatial heterogeneity, and responses to disturbance. My work on these subjects has continued and developed even while the urban research has flourished.

I was certainly aware of the LTER program and was perhaps its biggest fan before becoming involved in the program as a researcher, contributing to LTER panels and site visits (interestingly, since I have become involved in the LTER network, I have not been asked to do any of these things). Beginning in 1997, I have engaged in network activities including *BioScience* special issues, several working groups, and cross-site proposals. My contributions lie primarily in advancing socioecological understanding as applied to cities (Grimm et al. 2000; Grimm et al. 2008), although I also have made some inroads toward understanding urban biogeochemistry and stream ecology (Grimm et al. 2005).

APPROACH TO SCIENCE

My experiences in urban ecology have substantially broadened my perspective on ecology. An important aspect of this more inclusive perspective is the recognition that human-dominated ecosystems are appropriate subjects for ecological study but that an ecosystem approach (in which I was trained) is necessary, but not sufficient, for a full understanding of urban ecological systems. My experiences in the LTER program also have increased my contact and association with scientists in several other disciplines and forced me to think outside the box of the ecological thinking that I learned as a graduate and undergraduate student. From this experience, considering the diverse approaches and motivations of other scholars of the urban milieu has enriched my approach to science. For example, environmental engineers are much more motivated by the quest for solutions to environmental problems than by seeking answers to interesting questions, which I would maintain is a primary motivator for ecologists. Thus, the general problem of water quality in Phoenix is approached by one of my colleagues, Paul Westerhoff, as identifying the compounds that result in taste and odor problems in the waters that supply the area's household taps. These differences led us to begin examining persistent organic pollutants in various urban waters. Awareness by urban geographers and sociologists of disparities in the access of different segments of the population to amenities, or their disproportionate exposure to hazards, has led us to several investigations of environmental justice in the region. For example, thanks to the work of sociologist Sharon Harlan and her ecologist and climatologist collaborators, not only do we understand where in the city heat waves are most severe, but we know which people are most affected by them (Harlan et al. 2006). We have extensive spatial maps of soil chemistry, including metal pollutants, which we now can view in the context of both pollutant origins and consequences for local residents, thanks to the work of geochemists Xiaoding Zhuo and Everett Shock and social science colleagues such as Chris Boone (Zhuo, Boone, and Shock 2012). Finally, work in my group has blended the perspectives of engineering, management, and ecohydrology

FIGURE 15.1 One of the largest "green" stormwater infrastructure systems within the Central Arizona–Phoenix Long-Term Ecological Research site is Indian Bend Wash in Scottsdale, Arizona (left, showing the green belt looking north from the air). (Photo courtesy of Nancy Grimm.)

in understanding how stormwater dynamics are influenced by land use, human behavior, infrastructure design, and ecohydrologic connectivity (Figure 15.1).

I believe that environmental research should be performed in whatever way is most appropriate to answering the questions that we ask about systems. For example, I am not an experimentalist to the exclusion of observation; I am not wedded to a hypothetico-deductive approach if that does not serve the advancement of understanding. In the case of urban socioecological research, despite the fact that we have found an interdisciplinary perspective to be essential to our understanding, there is still meaningful ecological research (without any attendant social science) that can be done in cities.

We have continually refined our research questions to focus on the *reciprocal* interactions between human and ecological phenomena (Grimm et al. 2013). This approach is the crux of a coupled human–ecological system. Although we could certainly ask strictly ecological questions with our ecologist "blinders" on, the answers we got would suffer from limited spatial or temporal applicability. I am growing more aware, as well, of the sometimes hidden interactions between the technological sphere and the socioecological sphere. Consider a roadway with the intended function to transport people and goods from one place to another. Such a system may also act as a corridor for the movement of some species and a barrier for others. As another example, recent extreme events such as Superstorm Sandy illustrate the value not just of the protection that intact ecosystems can provide but of blending ecosystem components into designed systems in a hybrid approach to coastal management. Thus, such concepts as "ecosystem services" and even "ecosystem function" confront a different reality when we consider the built environment. Even more novel interdisciplinary interactions will be necessary to uncover some aspects of urban ecosystem functioning that are intimately tied to technology.

ATTITUDES TOWARD TIME AND SPACE

Because I have spent nearly my entire career, including graduate school, at the same place, I have nurtured a long-term perspective in nearly everything I do. My stream research has been ongoing since 1978, with only a brief hiatus. Taking on the research within the CAP project thus was not a huge stretch, even though the subject matter was new to me: my stream–watershed research had already begun to emphasize questions of how interannual variability (via disturbance regimes) and spatial heterogeneity influenced ecosystem pattern and process. What we found in the CAP region was a spatially heterogeneous ecosystem, at least at one scale, but with relatively muted temporal variability because of human management.

A funding context that requires the identification of new questions or conceptual themes is standard operating procedure for our science, and there is no reason that an exception should be made for long-term research. The nature of science is to continually ask new questions and identify new theory, models, or conceptual frameworks in which to place them. The questions jump out from careful study of the system. For me, the biggest challenge is to ensure that we are collecting data that are relevant to the big changes and big challenges of this complex socioecosystem. I worry most when it seems that those big changes and challenges have little to do with ecological phenomena. A carbon budget that is overwhelmingly dominated by fuel imports and carbon dioxide emission from the transportation sector, for example, might suggest to some that attention to carbon sequestration by urban trees is misplaced.

COLLABORATION

Collaboration is generally characteristic of an ecosystem approach, for the simple reason that, to effectively study whole ecosystems, there is too much for one person (plus one or two graduate students) to do. Collaboration is a skill that must be learned; it can occur within disciplines or it can be interdisciplinary. In the CAP program, we have very explicitly and conscientiously incorporated training in collaboration into aspects of our program over the years. In the Integrative Graduate Education and Training Program (IGERT) in Urban Ecology, we require one chapter of each IGERT fellow's dissertation to be collaborative (with someone other than his or her major professor).

In my own case, I have always collaborated with others, especially through the Lotic Intersite Nitrogen Experiment, or LINX, a multisite, NSF-funded research project focused on stream nitrogen dynamics. But my experiences in the LTER program have changed the way in which I have collaborated. Most importantly, aside from the interdisciplinary approaches, I believe that the availability of funding for working groups has greatly increased my collaboration with scientists in other locations and at other LTER sites. It has made me think more about how to export some of the work we are doing to other locations both nationally and internationally. I have also greatly benefited from the National Center for Ecological Analysis and Synthesis, in finding new ways to collaborate on synthesizing ecological and other information.

APPLIED RESEARCH

Rather than being an explicit consequence of my experiences in the LTER program, my rising interest in "applied" research, or more accurately, translating knowledge into action, resulted from a conscious decision to enter into federal service for 2 years, as a NSF program director and then a senior scientist with the National Climate Assessment (NCA). However, the decision to go to Washington, DC, was certainly influenced by my experiences within the LTER program, and my position as a scientist in the LTER network was an asset to my NSF and NCA activities. Those experiences shaped many of my current opinions about applied science (it needs a closer communication with basic science and vice versa) and especially about actionable science (doing science that will provide answers to critical questions about sustainability).

Locally, we have tried to keep our partnerships with local conservation and management entities active and mutually beneficial. Because CAP is an urban LTER site, the research we are doing has obvious relevance for municipal governments, nongovernmental organizations, and even local businesses. But the real task is to elevate the level of attention to whole-system challenges (social, technological, economic, and ecological), beyond the obvious and pressing social issues any city must face (e.g., crime, traffic congestion, growth).

COMMUNICATION

Communicating science is a critical aspect of our long-term value as a project within the university and local community. Communication to the public or the media is not something that comes to me easily; I tend to shy away from reporters' calls, although I recognize how important they can be. Fortunately, we have an excellent communications team within the project and in our home institute. Being part of the LTER program has helped in my teaching, particularly in bringing real-world examples from multiple ecosystem types to my teaching of general ecology.

The two urban LTER projects seemed from the outset to be ideal test beds for the development of schoolyard programs that would affect a large number of children. I think that we have been extremely successful in bringing ecology to local classrooms in a relevant way, through our work with teachers. Our schools are participating in CAP events such as the annual poster symposium. Students present the results of their research projects, based on protocols developed with the award-winning Ecology Explorers Program, alongside CAP researchers. My own personal involvement with the Ecology Explorers program, unfortunately, has been minor.

There is a great need for scientists to exchange information with practitioners about the types of actions that can be taken in the face of climate change impacts (i.e., climate adaptation). I do not mean this to be a prescriptive exchange. Our role as scientists is to illuminate the potential consequences of a course of action and communicate the uncertainty associated with our judgment. But what has struck me increasingly in recent years is how little scientists actually think about what kinds of questions practitioners might want the answers to in the first place! Without an understanding of the needs of practitioners firmly in mind at the outset of any investigation that purports to be relevant to decision-making or management, the opportunities for communication, translation of the science, and application of our findings are often missed. I would like to see a more explicit recognition of the value to both the science and society of the role played by individuals who cross that divide.

MENTORING

Undoubtedly, one of the biggest challenges in leading the large, interdisciplinary CAP program is supporting scientists at diverse career stages who come from disciplines with widely varying expectations of scientific productivity. We have been extremely lucky, because as we have evolved as a project, adding ever-increasing numbers of disciplinary perspectives. At the same time, the institution at which we are based has evolved in parallel: replacing disciplinary departments with interdisciplinary schools and creating institutes for the purpose of fostering new, transdisciplinary approaches to thorny issues such as sustainability. In concert with this change, even the most recalcitrant academic units have had to reconsider their tenure and promotion criteria and have largely valued interdisciplinary, collaborative research to a greater extent than previously.

My approach to mentoring graduate students has changed as a result of my experiences in the LTER program. Even though the urban IGERT program has ended, I continue to incorporate the idea of learning about collaboration in my training of graduate students. I also emphasize training and experience in a breadth of research approaches: empirical data collection, modeling, comparative ecosystem ecology, experiments, and analysis of existing, long-term data. The last is a skill that I believe all ecology students need to acquire in the era of big data and open science. Learning how to navigate databases, evaluate data collected by others for quality, and develop new statistical techniques to answer research questions with these data is becoming part of the identity of a new kind of ecologist. Of course, I do not really know how to do these things, but I encourage my students to develop these skills and I learn from them in the process.

One other aspect of training students has definitely changed. I no longer expect or want my students to follow a narrow path toward academia. Through my experiences in the LTER program (along with experiences in government), I have developed a much greater appreciation for skills of translation, communication, and more action-oriented agendas, and I understand that today's students are more motivated in this direction, as well. For example, there is a need to train "translators" of ecological understanding that can cross the knowledge–action gap, and I would like to encourage my students to pursue that kind of training if they are so inclined. Beyond that example, my students have moved on to American Association for the Advancement of Science fellowships in the federal government, and to positions in the private sector.

SKILL SET

Most of the skills I acquired as a consequence of my association with the LTER program have to do with leadership. I now know how to run a 50-minute meeting, and I have picked up some skills in running workshops intended to evince creative ideas from individual participants. I know how to pull together disparate bits to assemble a large, interdisciplinary proposal. And along the way, I have learned a great deal from my social science colleagues about the theories, biases, interests, and habits of that group of scholars that help me to integrate their work into the overall project.

PERSONAL CONSEQUENCES

I was a reluctant LTER principal investigator. When the urban competition was first announced, Chuck Redman, who had just taken on the directorship of ASU's Center for Environmental Studies, decided to bring together people from all over the university to talk "urban ecology." I do not even know how Chuck found me, but he did, and he managed to keep a large group of scholars from multiple disciplines coming to meetings to plan a proposal for the urban LTER competition. Those were fun times—having absolutely no belief in the fantasy that we could get this thing funded but having great big intellectual discussions about how one might do such a thing. Although Chuck was the pure inspiration behind our first proposal, when it came time to decide on the leadership he declined, because as an archaeologist he would have been unknown to the ecological research community. So there I was. Chuck and I were codirectors for the first 13 years of the CAP project, and I can say without hesitation that I did it because of the force of Chuck's intellect and commitment. Almost instantly I saw leadership qualities in him that I wanted to emulate. And I am still learning from Chuck, still admiring his style, the loyalty he inspires, and the way he creates a really wonderful environment of intellectual stimulation and fun.

Before we embarked on this journey, I knew maybe five people outside my own department. By the time we submitted the proposal, I knew at least 20 people from different departments, many pretty well. The relationships I have formed and been able to maintain across the university have been an outstanding aspect of being the principal investigator of the CAP project.

I also have made great friends in the wider LTER program. It is truly a wonderful community: very open, very diverse, and very committed to the long-term research approach. There is, of course, the annual travel to Science Council meetings and the triennial All Scientists Meetings. But way beyond that, my travel schedule is full and I credit (blame) the LTER program for that. I suppose one additional personal consequence deserves mention: the fact that I married the guy who used to be the NSF program director for the LTER program and is now the chair of its Science Council. Today, as a two-LTER, two-state, two-institution couple, our lives are intertwined with the LTER program in ways that perhaps few others have experienced.

CHALLENGES AND RECOMMENDATIONS

Many LTER projects go through an ontogeny of increasingly committed budgets and reduced flexibility as they proceed from their first through their second, third, and fourth renewals. The CAP project, in its third grant cycle, is no exception, although we have aspired to retain the ability to change directions as needed. One strategy we have adopted is to hold most resources centrally, supporting technical crews, instrumentation, database expertise, and facilities that are available to all researchers and accepting requests for summer salary, travel, and graduate- and undergraduate-student support on an annual basis.

Since the beginning of the CAP project, we have maintained open doors. We are always looking for new participants and creative ways to support and inspire graduate students in the traditions of the LTER program. I believe that herein lies the future of the program, not just for the CAP project, but for the entire network. As a network, opportunities to conduct synthesis research that involve graduate students and postdoctoral fellows will be most effective in promoting the continued use of the LTER approach and philosophy.

A second area that I believe we must advance, probably way beyond the current investment, is the documentation, access, and discovery of data from not only our data collection efforts but from other networks and long-term studies. We need to train students in methods of harvesting and using existing data to test exciting hypotheses derived from long-term observation.

These opportunities, although they represent an exciting potential future for the LTER program, also require continued support from the network and, more importantly, from NSF. The LTER program becomes vulnerable to an ever-present risk that it will be seen as a program past its useful life when compared to the shiny new penny of such programs as NEON. However, LTER scientists can and must continue to highlight the contributions they are making to ecological and socioecological knowledge and the ways in which the program is adapting and growing in the face of novel theoretical and empirical discoveries.

CONCLUSION

No other single event in my career has had a greater impact on my development as a scientist than the decision to lead the CAP project. There are several reasons for this, which I have explained in some detail in this essay. Being a leader of this urban LTER program has, deservedly or not, placed me in a position of leadership for an emerging field—urban ecology—and has afforded me many opportunities to synthesize and communicate my own views on what the field is about and where it is headed. Ultimately, I think that the LTER program has changed my professional identity, because there are probably many more people who know me as an urban ecologist than as a stream biogeochemist.

My strong assertion about the importance of the CAP project to my trajectory as a scientist raises the question of whether it is the nature of the subject matter (urban socioecological systems) or the LTER program that has had the greater impact. I think both play a role. Certainly, if I had come to direct an LTER project focused on desert ecosystems, this tremendous expansion in my view of what constitutes proper subject matter for ecology would not have occurred. Yet, the reputation of the LTER program itself—and the fact that it sanctioned the establishment of two urban LTERs—had a profound role in raising the status of urban ecology as a worthy topic for investigation.

ACKNOWLEDGMENTS

Many individuals—too many to name—have helped to shape and challenge my views of ecology, urban ecosystems, and social–ecological–technological systems, and I am grateful to all of them: ecologists, life scientists, physical scientists, social scientists, and engineers I have worked with, reviewers who trashed my proposals or papers anonymously, students and postdoctoral fellows who expanded my thinking and inspired me to set an example. The opportunity to enter the LTER world was made possible especially by C. Redman and S. Collins. The NSF is acknowledged for funding for my urban research from 1997 to present under grants 9714833, 0423704, and 1026865 (LTER); 9987612 and 0504248 (IGERT); and 0514382 and 0918457 (Ecosystems).

REFERENCES

Grimm, N., Redman, C., Boone, C., Childers, D., Harlan, S, and Turner, B.L., II. (2013). Viewing the urban socio-ecological system through a sustainability lens: Lessons and prospects from the Central Arizona–Phoenix LTER Programme. In S. J. Singh, H. Haberl, M. Chertow, M. Mirtl, and M. Schmid, eds. *Long Term Socio-Ecological Research*, pp. 217–246. Springer, Netherlands.

Grimm, N.B., Faeth, S.H., Golubiewski, N.E., Redman, C.L., Wu, J., Bai, X., and Briggs, J.M. (2008). Global change and the ecology of cities. *Science*, **319**, 756–760.

Grimm, N.B., Grove, J.M., Pickett, S.T.A., and Redman, C.L. (2000). Integrated approaches to long-term studies of urban ecological systems. *BioScience*, **50**, 571–584.

Grimm, N.B., Sheibley, R.W., Crenshaw, C.L., Dahm, C.N., Roach, W.J., and Zeglin, L.H. (2005). N retention and transformation in urban streams. *Journal of the North American Benthological Society*, **24**, 626–642.

Harlan, S.L., Brazel, A.J., Prashad, L., Stefanov, W.L., and Larsen, L. (2006). Neighborhood microclimates and vulnerability to heat stress. *Social Science & Medicine*, **63**, 2847–2863.

Zhuo, X., Boone, C.G., and Shock, E.L. (2012). Soil lead distribution and environmental justics in the Phoenix metropolitan region. *Environmental Justice*, **5**, 206–213.

PART SEVEN
COWEETA (CWT) LTER SITE

16 An Anthropologist Joins the Long-Term Ecological Research Network

Ted L. Gragson

IN A NUTSHELL

1. Environmental science has no room for theoretical or methodological hegemony, and questions cannot be asked in the absence of purposeful design.
2. Education must simultaneously engage students in thinking and doing, ideally in collaboration.
3. Communication is a two-way process in which scientists are challenged to be credible and legitimate in conveying salient results to diverse audiences.

4. Collaboration is about leveraging individual skills toward a common purpose, which can only succeed when trust exists between investigators.

PERSONAL OVERVIEW

I was trained as an ecological anthropologist with an emphasis on behavioral and ecosystem ecology at the University of Montana and the Pennsylvania State University. I have conducted archaeological, behavioral, cultural, and historical research throughout the western and the southeastern United States, as well as in several countries in lowland South America, the Dominican Republic, and southern France. Currently, I am professor and head of the Department of Anthropology at the University of Georgia. In 1997, I was invited to join the Long-Term Ecological Research (LTER) program at the Coweeta site (CWT). CWT is based in the eastern deciduous forest of the southern Appalachian Mountains, and I was brought in to collaborate on regionalization of what had been exclusively a site-based project. Just prior to joining CWT, I had been involved for several years in regional conservation activities in Paraguay and Bolivia. Since 2002, I have served as principal investigator of CWT, leading the successful grant renewal efforts in 2002 and 2008. I recently completed the 2014 renewal effort, which was successful. There has been a dramatic shift over the period of my involvement in the LTER program in attitudes within the network to regionalization and participation by scientists from disciplines other than ecology (Gragson and Grove 2006; Robertson et al. 2012). Several colleagues and I have helped to foster this shift through our involvement on the LTER Social Science Standing Committee (1998–present), leadership in the LTER planning activities (2004–2007), and service on the LTER Executive Board (2008–2011).

APPROACH TO SCIENCE

My experiences in the LTER program have influenced my ideas about the nature and conduct of environmental research. I professionally reside within a Department of Anthropology, of which I have been a member since 1992. I was hired when the department made a strategic decision to redirect the graduate program toward environmental and ecological anthropology. These combined experiences have crystallized for me the notion that there is no room for theoretical or methodological hegemony in the practice of environmental science. Individual theories and particular methods need to be wedded to questions to which we seek answers; they are not merely for philosophical consideration with no view to their design for a purpose or utility. As an evaluator and reviewer of diverse kinds of effort, my experiences in the LTER program have sharpened the filters that I bring to discriminating unsubstantiated claims from those that represent honest attempts to build an argument that includes all the elements that rhetorical analysis identifies as necessary for a productive communicative exchange. I believe that research is kept alive by our ever-increasing capacity to ask questions about processes that unfold over long time intervals. The place where we do this is ultimately merely a vehicle for advancing knowledge.

I was trained as an ecological anthropologist and have long been motivated by research on the human–environment relationship (Gragson 1993)—what is now more commonly labeled socioecological research. A point of initial interest for me in joining CWT was the potential it presented for working collaboratively to answer questions about this relation in ways that neither existing data sources nor collaborations made possible in the areas where I was conducting research at the time (i.e., western Paraguay, eastern Bolivia). Although working alone in a data desert has its own appeal, working as a group toward a common objective with the data legacy of prior research has a profound effect on the ability to answer the same questions as well as the scale at which the results can be applied. There are several renowned longitudinal studies from various areas that have examined discrete socioecological situations. However, LTER sites are unique in (1) the range and depth of observations that are or can be made; (2) the potential for synergy between fundamentally different theoretical and methodological approaches that are or can be brought to bear on a problem; and (3) the cultural and infrastructural facilities that exist at an LTER site to ensure that past gains, most notably data and results, are readily available for future researchers to use. My approach to environmental science within the LTER program has led me to seek comparable opportunities elsewhere. The most rewarding has been International Long-Term Ecological Research (ILTER) in southern France, which involved investigating the fundamental socioecological question of how montane landscapes were transformed and soils evolved as a function of agropastoral activities across the Holocene.

ATTITUDES TOWARD TIME AND SPACE

A large proportion of research in the LTER program continues to be carried out at the plot scale or the reach scale in the context of what are effectively human exclosures (e.g., experimental forests, ranges, and reserves). When the spatiotemporal boundaries of inquiry have expanded beyond such exclosures, there has been a tendency to identify residual effects of human activity, and then to offer simplistic and deterministic accounts about the modal behavior of humanity. It would be just as stultifying to believe that sap flux observations on one tree can explain how a forest functions. In my experience, the opportunity presented by research supported by the LTER program is to bring theory and method to bear on questions that encompass yet transcend the plot and the reach, thus challenging us to understand the spatiotemporal implications of moving across levels, scales, and systems. Integrative research that is place-based, long-term, cross-scale, and comparative can lead researchers to answer questions in ways that transcend conventional understanding and knowledge (Collins et al. 2011).

COLLABORATION

My research experiences prior to involvement in the LTER program had been collaborative, although significantly more limited in scope, number of participants, and duration than in the LTER program. Although these prior experiences primed me for collaborative research, my experiences within the LTER program since 1997 have profoundly influenced my research. One of the greatest opportunities has been facilitating "intercultural communication" between different disciplines as principle investigator of CWT. We now take for granted that socioecological research in the LTER program is difficult, in part, because social scientists and ecological scientists have different ways of thinking, talking, writing about, and doing research (Gragson and Grove 2006). However, there are just as many kinds of social scientists as there are kinds of ecological scientists, and the opportunity is to find common ground on how to collaborate in answering a question that no single kind of scientist could ever answer alone. Anthropologists are often viewed as "Lone Ranger" researchers—going to distant places that they write about in isolation (Gragson 2013). Although I have performed this kind of solitary research, the possibilities of collaborative research opened by my experiences in the LTER program with members of other disciplines have been so rewarding that nearly every research undertaking I now participate in involves collaboration. Much of my research prior to involvement in the LTER program had been international, but my involvement with the LTER program created novel opportunities for collaboration in Europe not previously available to me. Probably the most significant effect of my involvement in activities sponsored by the LTER program has been to turn my anthropological interests onto the United States. More so than many disciplines, the typical focus of anthropologists on distant locations often leads to a failure to contribute to an understanding about their own backyard.

APPLIED RESEARCH

My experiences in the LTER program have affected my contributions to applied science, although perhaps more by reference to what I see as a general disconnect across the LTER network between research and practice. Prior to engaging in research as part of the LTER program, I had been involved for several years in international conservation and development, working with various governmental and nongovernmental agencies. Despite an expressed interest across the LTER program in contributing to applied research, the research is overwhelmingly "blue sky" in nature, and its ultimate translation into the realms of applied and usable science is more frequently only a token of its potential. Basic science may become applied science that contributes to new industries and technologies, as well as usable science contributing to policy design and the solution of problems (Stokes 1997; Dilling and Lemos 2011). However, issues with the training of future researchers, the pressures on individuals within academia, and the fact that most funding for site activities is earmarked as "research," not "application," are ongoing barriers to moving results between the realms of science. A particular challenge for research in the LTER program is that its long-term nature means it often unfolds on a very different time scale than what meets the needs of application. However, the comprehensive understanding possible from place-based, long-term, and integrative research that does justice to the spatiotemporal implications of moving across levels, scales, and systems is without parallel, if issues of collaboration and communication are managed properly.

COMMUNICATION

I see three outcomes in my professional academic life from my involvement with the LTER program.

First, my classes have evolved to engage students simultaneously in collaboratively thinking and doing (Figure 16.1). For example, based on previous collaborations with colleagues from several LTER sites, we designed and taught a distributed graduate seminar on long-term, socioecological methods. The virtual course combined research examples presented by researchers in the LTER program, followed by students conducting and presenting projects that had two requirements: (1) the research had to be a collaboration between two or more individuals and (2) it had to use existing data available in an LTER site data repository. In one offering, we had registered students from four LTER sites, researchers from seven LTER sites, and class auditors from three continents; it was truly exciting.

Second, without my previous experiences in the LTER program, my administration of the Department of Anthropology unit would have unfolded very differently than it has. The forces and incentives in academia promote individuality at the expense of the collective enterprise, yet it is ultimately the products of collective effort such as training the next generation of researchers that justify the existence of the academy. Drawing from my work with the LTER program, I try in my capacity as head of a medium-size department within a large, land-grant university to promote engagement in collaborative

FIGURE 16.1 Kristen Cecala directs a crew of researchers and graduate students from the Coweeta Long-Term Ecological Research program, who are sampling salamanders during a synoptic sampling project of 57 streams across 12 watersheds that represent distinct land uses. (Photo courtesy of Craig Stickney.)

education and research among faculty members, staff members, and students. By some of the same standards used at an LTER site to evaluate success (e.g., publications, leveraged funding, awards, and graduations), combined with the sense of fulfillment and collegiality among departmental members, we are succeeding in being something more than the sum of our individual parts.

Third, my experiences in the LTER program have transformed my thoughts on what "outreach" is as well as how to conduct it. The rigors of academic training leave many faculty members unprepared for involvement in service and outreach activities, which is compounded by the disincentives in academia for faculty members and students to participate in such activities. This situation produces specialists who are authorized for involvement in service and outreach activities yet lack access to the individuals with the knowledge that requires translation into a useful form. The Coweeta Listening Project was established in 2011 in partial response to this situation to engage scientists in the LTER program and the public in southern Appalachia in coproducing usable knowledge. The members participate in diverse boundary-spanning activities to achieve credibility, salience, and legitimacy in diverse settings that now include a syndicated newspaper column published in several western North Carolina counties within the CWT project area.

MENTORING

My experiences in the LTER program have provided numerous concrete examples of the costs and benefits of collaboration that contribute to my mentoring of students and faculty members, and in particular, how I view an individual's place within a developmental trajectory. We can all hope to provide the definitive answer with our research to some key question that the world has been waiting for (i.e., the solution to a current environmental problem). Experiences in the LTER program, however, have helped counter in myself the scientific hubris that gives rise to such hope, which we are all prone to at some point in our careers. Research in the LTER program is long-term precisely because we ask questions about processes that unfold over long time intervals. Accumulating the evidence with which to answer such questions can take a long time, which sometimes results in forgetting the objective is to understand the long-term process itself, not merely to collect a large mass of data. Experiences in the LTER program have also enabled me to help others understand what they can do at given points in their career as a function of their knowledge, capacity, and position. If individuals can understand that their knowledge and capacity to address fundamental environmental questions increases over time, then they can both overcome the frustrations with personal limitations at any moment in time as well as find environmental research compelling as they progress through their careers.

I have served as a department head for 4 years, during which time I have led four successful faculty searches for ecological anthropologists and successfully shepherded the dossiers of four additional ecological anthropologists through the promotion and tenure process. This gives me a solid vantage point for evaluating the progress of colleagues and students, several of whom are directly involved in LTER or LTER-like research. It would be a disservice to the department to evaluate all professional aspects of individuals within the discipline through the lens of my experiences in the LTER program. However, in reviewing articles, proposals, dissertations, and the accomplishments of individuals, I seek to value the diversity of theoretical and methodological approaches necessary to guide purposeful and meaningful activities toward full understanding of complex phenomena in space and time. My appreciation for diversity comes from my own collaborations within the LTER program, as well as from working beside or observing colleagues engaged in productive collaborations in the program.

SKILL SET

When I joined CWT, I was fairly literate in statistical and geographic analysis. Nevertheless, my skill set has decidedly increased through my experiences, particularly in the areas of computer-intensive and Bayesian approaches to numerical analysis (Gragson and Bolstad 2007) and in the requirements and approaches necessary to managing large, heterogeneous data sets. What I have personally enjoyed through involvement in research in the LTER program is coming to appreciate that I do not need to learn all the skills required for research because I can collaborate with others and we can leverage our individual skills toward a common purpose. Although such a position stems from having a common objective in mind, it succeeds only by building trusting relationships with other investigators.

PERSONAL CONSEQUENCES

From an early age, I was raised overseas in what would now be called a bicultural setting, and then worked for many years in lowland South America among culturally and linguistically diverse Native American groups in very remote settings. My experience with the LTER program has afforded travel opportunities to me that have expanded my network of colleagues and friends. An encounter with an individual at an ILTER meeting, for example, led to collaboration in research and education that has now lasted for more than a decade. More than a dozen students have received bachelor's, master's, and doctoral degrees or participated in postdoctoral activities; numerous faculty and student exchanges to participate in field research have taken place; and foreign study and degree programs have been created. My experiences in the LTER program have literally put me in contact with hundreds of individuals who I would otherwise never have met, several of whom have become close friends over time. More than anything, I know I can contact anyone within the large LTER community for help by referring to our points of common interest in place-based, long-term, cross-scale, and comparative research. At one level, the LTER community is simply another ethnic group with its own tribal organization.

By virtue of my connection to the LTER program, I have been invited to deliver presentations, serve on panels and boards, and provide academic and research leadership in situations that I find personally and professionally rewarding. There is no reason to think that the types of personal and professional outlets that my colleagues have who have followed a strict disciplinary pathway are any less rewarding, but I do know that mine are notably different. I have been brought into contact with individuals, groups, and ways of approaching socioecological problems as a consequence of my association with CWT that I would never have experienced, had I followed the course I had been trained to believe was mine on graduating from college.

CHALLENGES AND RECOMMENDATIONS

The LTER program operates like a grassroots organization: despite the absence of a central authority, individual sites share common ideals about the overall purpose of research. Sites can nevertheless develop independently of other sites by virtue of the institutions, personnel, infrastructure, and other circumstances unique to each site. Nonetheless, there is a near absence of LTER network-wide resources, guidelines, or training opportunities to build the capacity at sites for increasing their effectiveness in communicating with the public or engaging in service and outreach. These activities at best are seen as necessary evils, and at worst they are ignored completely. Their momentary burden, however, is ultimately critical to spanning the boundaries to achieve credibility, salience, and legitimacy among those who we eventually hope will care about or benefit from our research; or whose children will become the next generation of researchers.

CONCLUSION

Being trained as an ecological anthropologist, I had been steeped in the place-based holistic tradition of the discipline. Before joining CWT, I had carried out extensive and diverse research both domestically and across Latin America, some of which had been collaborative in nature if smaller in scale and scope than my research in the LTER program. My background and experiences were certainly relevant to engaging in long-term ecological science, and particularly the CWT regionalization effort. However, I had no idea how transformative joining CWT and the LTER network would be on my subsequent professional life. My identity at its core is that of an anthropologist interested in analyzing and understanding place-based human experiences in relation to their environment. Doing so within the context of CWT, however, has made it possible for me to carry out such research in ways that social scientists frequently discuss in the abstract as important yet seldom have the opportunity to carry out in practice. Developing place-based, holistic knowledge depends on the ability to understand many things about a situation that, without the advantages of collaboration with investigators from divergent disciplines, is simply impossible for any one individual to achieve.

ACKNOWLEDGMENTS

Material is based in part on work supported by the National Science Foundation (NSF) under grants DEB-1440485, DEB-0823293, and DEB-0218001. Any opinions, findings, conclusions, or recommendations expressed in the material are those of the author and do not necessarily reflect the views of the NSF.

REFERENCES

Collins, S.L., Carpenter, S.R., Swinton, S.M., Orenstein, D.E., Childers, D.L., Gragson, T.L., Grimm, N.B., Grove, J.M., Harlan, S.L., Kaye, J.P., Knapp, A.K., Kofinas, G.P., Magnuson, J.J., McDowell, W.J., Melack, J.M., Ogden, L.A., Robertson, G.P., Smith, M.D., and Whitmer, A.C. (2011). An integrated conceptual framework for long-term socioecological research. *Frontiers in Ecology and the Environment* **9**, 351–357.

Dilling, L., and Lemos, M.C. (2011). Creating usable science: Opportunities and constraints for climate knowledge use and their implications for science policy. *Global Environmental Change* **21**, 680–689.

Gragson, T.L. (1993). Human foraging in lowland South America: Pattern and process of resource procurement. *Research in Economic Anthropology* **14**, 107–138.

Gragson, T.L. (2013). The contribution of anthropology to concepts guiding LTSER research. In S. J. Singh, H. Haberl, M. Chertow, M. Mirtl, and M. Schmid, eds. *Long Term Socio-ecological Research: Studies in Society-nature Interactions Across Spatial and Temporal Scales*, pp. 189–214. Berlin, Springer.

Gragson, T.L., and Bolstad, P.V. (2007). A local analysis of early 18th century Cherokee settlement. *Social Science History* **31**, 435–468.

Gragson, T.L., and Grove, J.M. (2006). Social science in the context of the long term ecological research program. *Society and Natural Resources* **19**, 93–100.

Robertson, P., Brokaw, N., Collins, S., Ducklow, H., Foster, D., Gragson, T., Gries, C., Hamilton, S., McGuire, D., Moore, J., Stanley, E., Waide, R., and Williams, M. (2012). Long term ecological research in a human dominated world. *BioScience* **62**, 342–353.

Stokes, D.E. (1997). *Pasteur's Quadrant: Basic Science and Technological Innovation*. Brookings Institution Press, Washington, DC.

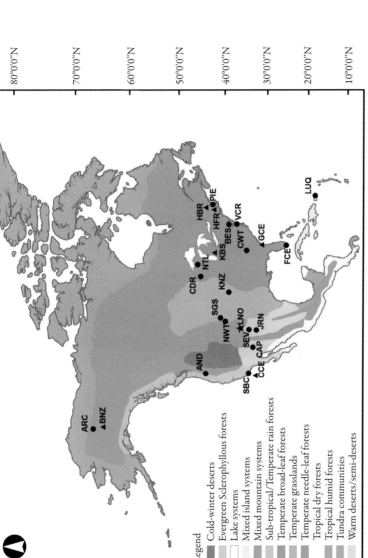

FIGURE 1.2 The Long-Term Ecological Research (LTER) network currently comprises 26 sites and a network office (LNO). Most of the sites (23) are disturbed across a diversity of biomes in North America (United States and Puerto Rico) and appear as circles (sites represented by essays) or triangles (sites not represented by essays) on the map. Two sites in Antarctica (McMurdo Dry Valleys site, MCM; Palmer Antarctica site, PAL) and one site in French Polynesia (Moorea Coral Reef site, MCR) are not represented in the graphic, although PAL and MCM are represented by essays in the book. Major biomes (Udvardy 1975) are represented by colors on the map; the LNO appears as a star. Three-letter acronyms identify particular LTER sites: AND, H. J. Andrews Experimental Forest site; ARC, Arctic site; BES, Baltimore Ecosystem Study site; BNZ, Bonanza Creek site; CCE, California Current Ecosystem site; CDR, Cedar Creek Ecosystem Science Reserve site; CP, Central Arizona–Phoenix site; CWT, Coweeta site; FCE, Florida Coastal Everglades site; GCE, Georgia Coastal Ecosystems site; HFR, Harvard Forest site; HBR, Hubbard Brook site; JRN, Jornada Basin site; KBS, Kellogg Biological Station site; KNZ, Konza Prairie site; LUQ, Luquillo site; MCM, McMurdo Dry Valleys site; MCR, Moorea Coral Reserve site; NWT, Niwot Ridge site; NTL, North Temperate Lakes site; PIE, Plum Island Ecosystem site; SBC, Santa Barbara Coastal site; SEV, Sevilleta site; SGS, Shortgrass Steppe site; and VCR, Virginia Coastal Reserve site. In 2016, funding for the LNO ended and the network communications office became operational at the National Center for Ecological Analysis and Synthesis in California (near SBC).

FIGURE 1.3 A tributary of McRae Creek within a stand of old-growth forest at the H.J. Andrews Experimental Forest Long-Term Ecological Research site. (Photo courtesy of Lina DeGregorio.)

FIGURE 1.4 Toolik Lake and the surrounding landscape at the Arctic Long-Term Ecological Research site. (Photo courtesy of James Laundre.)

FIGURE 1.5 West Franklin Street neighborhood and landscape at the Baltimore Ecosystem Study (BES) Long-Term Ecological Research site. (Photo courtesy of Steward Pickett, the BES LTER program.)

FIGURE 1.6 Aerial photograph of experimental manipulation plots and the surrounding landscape at the Cedar Creek Ecosystem Science Reserve Long-Term Ecological Research site. (Photo courtesy of David Tilman.)

FIGURE 1.7 The Phoenix metropolitan area (background), situated within the Sonoran Desert (foreground), represents the landscape of the Central Arizona–Phoenix (CAP) Long-Term Ecological Research (LTER) site. (Photo courtesy of Eyal Shochat and the CAP LTER program.)

FIGURE 1.8 Image depicting a typical landuse (foreground) and the broader landscape (background) in the Caler Creek Watershed that are typical of the Coweeta Long-Term Ecological Research site. (Photo courtesy of Ted Gragson.)

FIGURE 1.9 Aerial view of the coastal landscape that represents the Florida Coastal Everglades Long-Term Ecological Research site. (Photo courtesy of Stephen Davis.)

FIGURE 1.10 Creosote shrubland is one of the five major plant communities of the Chihuahuan Desert that compose the Jornada Basin Long-Term Ecological Research site. (Photo courtesy of Debra Peters.)

FIGURE 1.11 Landscape image of the Konza Prairie Long-Term Ecological Research site, with a pair of bison in the foreground. (Photo courtesy of Edward Raynor.)

FIGURE 1.12 Landscape image of the forests of the Luquillo Mountains, including suburbanizing areas in the lowlands, that represent the Luquillo Long-Term Ecological Research site. (Photo courtesy of Jerry Baurer and Oxford University Press.)

FIGURE 1.13 Staghorn coral represent dominant components of the underwater landscape that is typical of the Moorea Coastal Reserve Long-Term Ecological Research site. (Photo courtesy of Melissa Holbrook.)

FIGURE 1.14 Image of scientists collecting climatic data under harsh conditions typical of the Niwot Ridge Long-Term Ecological Research site. (Photo courtesy of John Marr and the National Science Foundation.)

FIGURE 1.15 Aerial photograph of the aquatic and terrestrial components of the landscape that represents the North Temperate Lakes Long-Term Ecological Research site. (Photo courtesy of Carl Bowser and Silver Pixel Images.)

FIGURE 1.16 Sea ice and glacier-covered mountains in Marguerite Bay (Andelaide Island) represent typical land- and sea-scapes of the Palmer Antarctica Long-Term Ecological Research site. (Photo courtesy of Hugh Ducklow.)

FIGURE 1.17 Tidal creek and salt marsh habitats are typical of the landscape at the Plum Island Ecosystem Long-Term Ecological Research site. (Photo courtesy of David Johnson.)

FIGURE 1.18 Giant kelp forest and fish compose typical seascapes of the Santa Barbara Coastal Long-Term Ecological Research site. (Photo courtesy of Ron McPeak and University of Santa Barbara Digital Collections [item 16822].)

FIGURE 1.19 The Blue Grama Meteorological Station (foreground) on the Sevilleta Long-Term Ecological Research site, looking east toward the Lost Pinos Mountain Front. (Photo courtesy of Doug Moore, the Sevilleta Photo Archives, and the National Science Foundation.)

FIGURE 1.20 Panoramic view of the landscape from the top of Owl's Peak in the Central Plains Experimental Range (the Front Range of the Rocky Mountains appears in the background) that represents typical habitat in the Shortgrass Steppe Long-Term Ecological Research site. (Photo courtesy of John Moore.)

FIGURE 1.21 Aerial photograph of the coastal barrier landscape that represents the ecosystems of the Virginia Coast Long-Term Ecological Research site. (Photo courtesy of John Porter.)

PART EIGHT
FLORIDA COASTAL EVERGLADES (FCE) LTER SITE

17 The Benefits of Long-Term Environmental Research, Friendships, and Boiled Peanuts

Evelyn E. Gaiser

IN A NUTSHELL

1. The Long-Term Ecological Research (LTER) program has enabled me to conduct more broadly relevant science by addressing questions within an interdisciplinary framework and to unravel the causes for surprising ecological phenomena through persistent studies and collaborations.
2. Educational opportunities within the LTER program have connected me to students from grades K–12 to graduate levels in new ways from the field to the classroom, across places from Florida to Alaska, and among disciplines in a collaborative setting.

3. The audience for my research expanded as a consequence of my experiences in the LTER program, and I have learned how to more effectively communicate integrative research to large audiences of scientists, policy-makers, and the public, often through nontraditional media.
4. The LTER program is foremost a network of people, and I have found that science evolves most successfully when ideas and information are shared voluntarily across backgrounds, disciplines, and cultures in a network of cultivated, trusting relationships.

PERSONAL OVERVIEW

The Florida Coastal Everglades (FCE) is the LTER site where I am currently the principal investigator, but the LTER program has been a part of my life for most of my career. My experiences in the LTER program began in the early 1990s when I was a graduate student at the University of Georgia, where the Coweeta (CWT) LTER site is based. Although I was not formally a part of CWT, many of my friends and professors were, so the program influenced my development as a scientist. I remember my first field trip to CWT, led by Gene Helfman and Judy Meyer, and the fun of snorkeling in mountain streams where we camped and conducted a few experiments, including examining the effects of rapid consumption of s'mores and boiled peanuts on preschool children (Judy and Gene's kids). LTER-related activities wove in and out of my graduate student experience, and the rewards of sharing of ideas, data, friendships, and boiled peanuts created in me a lifelong commitment to persistent, collaborative science. This sense of fulfillment, of being part of something larger, was reinforced at the Savannah River Ecology Laboratory (SREL), where I conducted my research. At SREL, I was investigating paleoenvironmental histories in a setting where interpretations of the past were substantiated by a rare archive of long-term field records. This experience helped me understand the importance of coordinated long-term ecological science programs in a broad sense.

Within the FCE project, I help coordinate science, education, and outreach activities across multiple institutions conducting research in the Everglades. I find that working among a large group of active collaborators toward common goals actually helps me retain focus in my own research in benthic algal ecology. For example, one of the main goals of the FCE project is to understand the origins of productivity and endemism in nutrient-poor landscapes, so studies of algal community responses to nutrients help resolve questions relevant to algae and to general ecology. Furthermore, I have tried to promote the expansion of International LTER research (Figure 17.1) because of the ways that such comparative studies have generated new discoveries and questions when coordinated with place-based research.

APPROACH TO SCIENCE

As an algal biologist, I enjoy studying the ways in which communities of producers change over long temporal and broad spatial scales, and I attempt to understand the drivers (and consequences) of such ecological change. My disciplinary research is founded in and inspired by a small community of tight-knit phycologists who are my mentors, students, and dear friends. This community has its work cut out for it; to make the most of these powerful sentinels of environmental change, we have a lot to discover about algal identities, species distributions, and drivers of community assembly. A focused effort toward a shared vision is therefore important. However, at the same time, much is to be gained by joining in more integrative efforts to "see the pond through the algae." Being

FIGURE 17.1 Scientists from the Florida Coastal Everglades Long-Term Ecological Research (LTER) program (J. Trexler, E. Gaiser, W. Loftus, and J. La Hée) collect periphyton and fish during the dry season in the Sian Ka'an Biosphere Reserve to understand the configuration of food webs in karst wetlands as part of International LTER program. (Photo courtesy of Clifton Ruehl.)

a part of the LTER program has enabled my students and me to do that. For example, our research on benthic algae in the Everglades shows both unexpectedly high levels of productivity for such a nutrient-poor habitat and high endemicity for such a geologically young ecosystem (Gaiser, Trexler, and Wetzel 2012). Collaborative studies in the LTER program have enabled us to show how this high level of algal production is partly due to hidden groundwater supplies of nutrients (Gaiser et al. 2013). Also, coordinated studies of aquatic consumers have shown how ineffective grazing is at suppressing production due to the particularly unpalatable matrix of filamentous cyanobacteria and calcium carbonate that dominate these benthic communities (Sargeant, Gaiser, and Trexler 2011). Furthermore, through comparative international studies in the LTER network, especially in Mexico and the Caribbean, we have been able to show that many of the unusual species and functions found in the Everglades are actually common in other karst freshwater wetlands (La Hée and Gaiser 2012). These discoveries have inspired a new concept of the Everglades as an icon for conservation of broadly similar yet still distinct karst wetlands found throughout the Caribbean (Gaiser, Trexler, and Wetzel 2012).

Being a part of the LTER program has enabled me to both address fundamental ecological questions by linking my research to that of other biophysical scientists and to expand the relevance of that research by conducting it within a socioecological framework. The science of the FCE program revolves around understanding the origins of estuarine productivity and the ways in which that productivity is modified by changing water availability and quality. The influences of regional water management and climate change are studied as key drivers of change. We anticipated that efforts to restore the Everglades would perform an experiment for us, rehydrating one of our study estuaries but not the other, so that we could document long-term change in a landscape-scale experiment. However, the restoration process was put on hold, mainly because of stakeholder conflicts over restoration goals. Expanding into the human dimensions of our research has been critical because it enables us to conduct research on the shared dependency of all South Florida inhabitants on the Everglades. Anthropological and hydro-political studies on human activities and uses of the Everglades shed new light on restoration conflicts, where stakeholder needs, when well understood, can lead to resolution through the shared solutions. This integration of social sciences into the FCE program was conducted in conjunction with a broader LTER network initiative to integrate science for society and the environment (Collins et al. 2011). The products of such integration underscore the importance of understanding humans as interacting agents of long-term global change (Ogden et al. 2013).

ATTITUDES TOWARD TIME AND SPACE

For me, one of the joys of long-term collaborative studies is the discovery of "surprises" that would not have been revealed without persistent records of ecological patterns and processes and the forces that drive them (Dodds et al. 2013). One of the challenges for the LTER program is to design studies that maximize this potential. For instance, all disciplines need to be at the table (in equal measure) during all phases of project planning and execution to ensure that hypotheses are being tested synoptically. Collaborators are then asked to adhere to a long-term plan that is broadly designed to allow the testing of alternative hypotheses. One example of a surprise from our long-term collections was the discovery of a prolonged spike in phosphorus export *out* of the Everglades estuary. Our prior studies had identified these estuaries as paradoxically "upside down," receiving nutrients from marine waters rather than from upstream, so this spike in export was confusing. However, studies in the FCE program of mangrove soils showed how a surge of marine muds driven to the interior by the passage of Hurricane Wilma (in 2005) were laden in phosphorus (Castenada et al. 2010), which then leached into the surrounding watershed during subsequent high tides and periods of high rainfall. The estuary was enriching itself with marine-sourced nutrients. We can now more fully understand the origins of estuarine productivity as a result of this "surprise" stemming from long-term collaborations across disciplines.

New questions easily arise from long-term studies, including those motivating mechanistic experiments to discover reasons for "surprises." In the FCE program, the discovery of such long-lasting and unexpected stimulatory influences of tropical storms on ecotone productivity in the Everglades inspired questions about whether gradual sea level rise would elicit the same types of changes through the same mechanisms. Simultaneously, we were observing the collapse of freshwater peat soils in areas exposed more frequently to saltwater during tides and storms as a result of sea level rise and depleted freshwater flows. Together these discoveries drove the need for an aggressive approach to creating a dynamic carbon budget—one that could be used to forecast the future of the massive carbon stores of the Everglades. New questions and approaches for our program were set into place, including a set of long-term experiments to determine the interaction of press (sea level rise) and pulse (storm surge) events on carbon dynamics along this changing subsidy and stress gradient.

COLLABORATION

Collaborative long-term science can reveal causes for patterns or phenomena that would otherwise be perplexing if approached in a disciplinary vacuum. However, nothing would be gained if the disciplinary rigor and focus were not there to begin with or if researchers were not in the habit of looking to other disciplines for answers. Because both activities require time and energy, researchers need to consciously maintain a comfortable balance.

Project leaders are especially vulnerable to a loss of disciplinary focus because of the time required to engage with, learn about, and make connections among disciplines. There have been times when I have felt more disconnected from developments in my own discipline than I would like, as a consequence of coordinating larger, integrative activities. As lifelong learners, academic researchers face this conundrum fairly frequently because of the fun of engaging in new activities, making connections, and learning new things. I find that it is helpful to remember that most of my collaborators are similarly motivated and are glad to share in the process of integration and synthesis. If all researchers commit to balancing disciplinary rigor with dedication to exploring broader relevance through the work of others, collaborations become much more productive and there is no real trade-off between disciplinary focus and interdisciplinary engagement.

APPLIED RESEARCH

One hallmark of the FCE program is the direct involvement of agency scientists and decision-makers as collaborators to facilitate both science and environmental decisions. We do not consider applied research as part of our outreach program. Instead, we see applications as an integral part of our core mission to design and execute science across academic and regulatory institutions. Before the FCE site was established, there was already significant scientific attention being paid to the Everglades, but a persistent framework of coordinated and thematic science was needed to unravel the complex responses of ecosystems to multiple drivers. The FCE program, established in 2000, brought new focus to research in the Everglades that underscored how the remote but expansive coastal zone responds to changing water management and climate. The FCE program also created a hub through which sound science could transform decisions about the future of the Everglades. We are far from solving the eco-political problems embroiled in the restoration process but are closer to providing alternatives for a sustainable future for South Floridians as a result of transdisciplinary studies.

COMMUNICATION

Being involved in the LTER program has expanded the ways in which I communicate science, both in formal courses as well as through other creative outlets. At my university, I established a graduate level LTER course in our curriculum that immerses students in site and network LTER science. Graduate students have also been participating as near-peer mentors of teachers and high school students through our summer research experience programs. They work with our education coordinator and local K–12 teachers to conduct novel studies and move data from the LTER network, and experiences from FCE site and other sites, into classroom curricula. Teachers then send high school students into our laboratories at the university, which, for me, has been one of the more surprisingly rewarding aspects of our education program. My laboratory has mentored two students from first-generation minority families. The first student published his project as a high school senior after winning his division in the Intel International Science and Engineering Fair, and the second student just switched her undergraduate major track from premedical to ecology. These kids are an astounding part of our laboratory and "LTER family," blending in at LTER network meetings where their confidence, knowledge, and enthusiasm seem well beyond their years. They give us all a welcome sense of hope for the future.

A final note on recent expansion of LTER communications through the classical arts: we just completed a project with a group of South Florida artists focused on diatoms as art forms. Their paintings are magnificent, and I have heard diatoms being talked about in the most unlikely places. This outreach would not have happened without an energized, grassroots LTER vision to expand communications beyond conventional outlets.

MENTORING

As the LTER network has grown in size and scope, so have the opportunities for students in all aspects of the program. In my laboratory, I can provide the disciplinary opportunities necessary for my students to find focus in their research. By enabling their engagement in our FCE program, I can help them place their research in a broader context. By participating in network science, students have had the opportunity to (1) address questions across disciplines and places and (2) learn how to participate and lead large collaborations. Recent student-led, cross-site, and cross-disciplinary research activities have been among the most insightful I have seen.

Multiauthored publications are now an anticipated part of dissertations by students in the LTER program; it has been interesting to observe (and guide) graduate programs to a new and favorable interpretation of the importance of collaborative products to a well-rounded dissertation. In this way, students in the LTER program are creating a new standard for success in ecological science. In reviewing papers and even academic job applications, I find myself looking for evidence of integrative, collaborative approaches in addition to question-driven science that has been motivated by prior knowledge. Given the dedication and enthusiasm of this younger cohort of LTER scientists, I am hopeful that the future of the LTER program is bright, despite growing fiscal and environmental challenges.

In addition, coming up with unconventional ways for reaching the public via everything from blogging to leading citizen science programs is becoming a core component of life as an LTER graduate student. The LTER program has a larger and more creative footprint than ever due to the energy, vision, and talent of these students. These collaborative skills gained by being part of a network infuse into their academic and research environment at home, as well as provide them with a community of friends with whom to share their experiences.

SKILL SET

The LTER program has recognized the importance of sharing to successful collaboration by putting a great deal of effort into developing the tools for information preservation and exchange for the field of ecology. I think that for most investigators, implementing good information management practices is akin to rolling around on tacks! Therefore, we are extremely fortunate to be surrounded by an incredibly patient and skilled team of LTER information managers. These people encourage, teach, and take time to listen to our questions and help us understand how and why to share data. In addition, many researchers in the LTER program are engaged in deploying and using information collected at higher frequency using sensors. For me, involvement in sensor-based studies has been like taking a series of short courses in electrical engineering, mechanics, and digital communications. Furthermore, interpreting data from multiple sensors and multiple sites among networks has required skills well beyond the scope of my own background, but my students and I have been able to pick these up from training workshops. Through developing tools, training, and trust, the LTER program has helped create a global environment where the process of providing, sharing, and interpreting information across scales is a routine and expected part of scientific practice.

In addition to developing such practical skills, the LTER program has taught me about how to be a better collaborator and leader. As a principal investigator, I have had

(1) to learn how to take lots of information from different disciplines and distill it down, (2) to listen deeply to other collaborators and value their opinions as much as my own, and (3) to provide opportunity for engagement across all aspects of project science and management.

PERSONAL CONSEQUENCES

For me, the LTER program is first and foremost a community of people: people who conduct science, education, and outreach together in a common, long-term framework. Long-term, collaborative science equates to long-term friendships, ones that keep growing and expanding with the footprint of the program. Within the FCE program, shared experiences in the field and on campus create a bond that then extends to lasting friendship among families. The same is true for the broader LTER network. I continue to share life experiences with my LTER friends from graduate school, and national meetings become like a family reunion. I am thrilled when my students connect with my long-term ecological friends, especially when they go to work with them. In addition, travel to interesting venues for research and meetings has been an unexpected perk of being a scientist in the LTER program. Geographical expansion of the program has driven different ways of thinking about and studying coastal ecosystems as a result of engaging scientists with different disciplinary backgrounds, cultures, and political challenges. I have found that international collaborations can add deeper discovery, creative questions, and broader societal benefits to place-based projects; observing and finding opportunities for these connections is one of my favorite aspects of the principal investigator role.

CHALLENGES AND RECOMMENDATIONS

The LTER program has a bright future ahead, mainly because of the attention that has been paid to students (at all levels). Students in the LTER program are now better trained in creating theoretically motivated research in a collaborative, persistent context; communicating their work through innovative outlets; and in the skills of collaborative science leadership than before. The greatest challenges are the obvious impediments of currently shrinking support for scientific research in the United States, the limited job market in science and education, and the ever-growing administrative responsibilities of university faculty. One way to fortify a stronger future for the LTER program is to recognize the true scope of long-term ecological research both within and outside the LTER network. The LTER program has provided an enduring model for transformative, long-term science for other ecological networks that have grown with and from this community. Although the tangible scientific products resulting from coordinated, persistent, transdisciplinary engagement in research are obvious, the intangible products derived from the LTER culture of open invitational collaboration, flat organizational structure, and informality in the scientific process cannot be underestimated. If the LTER program can consciously protect these core values of trust and sharing while embracing a broader concept of itself, there should be no obstacle to catalyzing a future that has even greater impact than that enjoyed in the past.

CONCLUSION

Experience in the LTER program has provided me many opportunities to expand the context of my research and has taught me how to be a more effective scientific leader and educator. This experience probably naturally selects for researchers who take pleasure in broad collaborations, so I have also greatly enjoyed working with friends who share my enthusiasm for project goals. For these reasons, I encourage students affiliated with the LTER program to be as engaged as possible with all aspects of science, education, and outreach, not only because of the impact it will have on science and their careers, but because it is fun. The future of ecosystem science is evolving quickly through persistent integrative, transdisciplinary, and policy-relevant research, so by embracing the concept of community established in the LTER program, young scientists can lead ecologists toward a more effective and brighter future.

ACKNOWLEDGMENTS

I am grateful for the opportunity to participate in the LTER network through support from the National Science Foundation, especially the Florida Coastal Everglades LTER program under Cooperative Agreements DEB-1237517, DBI-0620409, and DEB-9910514. I thank the many mentors and colleagues in the LTER network who fostered my career development, especially Dr. D. Childers and the co–principal investigators of the FCE LTER program, as well as all of the collaborators and students of the program who make it so enjoyable to lead. I thank J. Kominoski for sharing experiences in LTER and FCE and for reading my contribution, as well as the reviewers and the coeditors of this book for their inspiration and review of prior drafts.

REFERENCES

Castaneda-Moya, E., Twilley, R., Rivera-Monroy, V., Zhang, K., Davis, S., and Ross, M. (2010). Sediment and nutrient deposition associated with Hurricane Wilma in mangroves of the Florida Coastal Everglades. *Estuaries and Coasts,* 33, 45–58.

Collins, S., Carpenter, S., Swinton, S., Orenstein, D., Childers, D., Gragson, T., Grimm, N., Grove, J., Harlan, S., Kaye, J., Knapp, A., Kofinas, G., Magnuson, J., McDowell, W., Melack, J., Ogden, L., Robertson, G., Smith, M., and Whitmer, A. (2011). An integrated conceptual framework for long-term social–ecological research. *Frontiers in Ecology and the Environment,* 9, 351–357.

Dodds, W., Robinson, C., Gaiser, E., Hansen, G., Powell, H., Smith, J., Morse, N., Gregory, S., Bell, T., Kratz, T., and McDowell, W. (2012). Surprises and insights from long-term aquatic data sets. *BioScience,* 62, 709–721.

Gaiser, E., Sullivan, P., Tobias, F., Bramburger, A., and Trexler, J. (2013). Boundary effects on benthic microbial phosphorus concentrations and diatom beta diversity in a hydrologically-modified, nutrient-limited wetland. *Wetlands,* 34, Issue 1 Supplement, 55–64.

Gaiser, E., Trexler, J., and Wetzel, P. (2012). The Everglades. In D. Batzer, D. and A. Baldwin, eds. *Wetland Habitats of North America: Ecology and Conservation Concerns,* pp. 231–252. University of California Press, Berkeley.

La Hée, J., and Gaiser, E. (2012). Benthic diatom assemblages as indicators of water quality in the Everglades and three tropical karstic wetlands. *Freshwater Science,* 31, 205–221.

Ogden, L., Heynen, N. Oslender, U. West, P. Kassam, K., and Robbins, P. (2013). Global assemblages, resilience, and Earth stewardship in the Anthropocene. *Frontiers of Ecology and the Environment,* 11, 341–347.

Sargeant, B., Gaiser, E., and Trexler, J. (2011). Indirect and direct controls of macroinvertebrates and small fish by abiotic factors and trophic interactions in the Florida Everglades. *Freshwater Biology,* 56, 2334–2346.

18 Collaboration and Broadening Our Scope

Relevance of Long-Term Ecological Research to the Global Community

Tiffany G. Troxler

IN A NUTSHELL

1. As part of my long-term ecological research experience, I have come to recognize that individual success is not necessarily the hallmark of an effective or successful scientist. To achieve problem-oriented solutions to the grand challenges of society, service and collaboration can have more impact on ecology and society than singular scientific achievements.
2. Because of my experiences with the Long-Term Ecological Research (LTER) program, I promote the idea that collaborative research is essential training for ecosystem scientists.
3. The LTER program promoted the increasing importance of effective science communication at a time when it was not widely appreciated.
4. The LTER program demonstrated to me that engendering a spirit of collaboration was the key to building a network of scientists that could address grand challenge questions.

PERSONAL OVERVIEW

After studying anthropology as an undergraduate in Louisiana, I moved to Florida to work with Daniel Childers in the field of ecology. During my PhD work with him at Florida International University (FIU), I became involved in the Florida Coastal Everglades site (FCE) of the LTER program. During graduate school, I participated as cochair of the LTER graduate student committee. Currently, I am a research assistant professor with appointments at the Southeast Environmental Research Center and the Department of Biological Sciences at FIU. My research focuses on long-term ecosystem responses to hydrologic restoration, carbon cycling, and plant–soil interactions along environmental gradients in the Everglades. I collaborate with researchers at FCE and other LTER sites, as well as with colleagues in the International LTER (ILTER) to broaden the scope and integration of site-based, long-term research.

APPROACH TO SCIENCE

Ecosystem approaches are a hallmark of science in the LTER program, and long-term manipulations at the ecosystem-scale are numerous within the LTER network. Simple ecosystem modeling allows for the integration of responses into a few synthetic variables (e.g., soil nutrient concentrations or carbon accumulation rates). My research in the LTER program strives to identify data gaps posed by such modeling and looks for creative and robust ways to develop data sets that contain a comprehensive suite of input parameters. Using such an approach, I have developed nitrogen and phosphorus budgets for tree islands in the Everglades, contributed to ecosystem carbon budgets in mangrove sites within the FCE, and developed global scale datasets and methodologies to quantify the effects of land-use change on carbon budgets in coastal ecosystems.

Over time, my research experience has developed from the perspective of understanding human and climate perturbations at the ecosystem level, to studies at subregional to global scales. I actively work with colleagues to explicitly integrate socioecological elements into synthetic studies in which I participate, at the national and international levels, within and beyond the LTER program. In essence, I use the approaches learned and developed within the LTER program and related research to address similar questions outside of the US and International LTER networks, especially at broader geographic scales. Thus, the LTER program has influenced my ideas about how some environmental research should be performed. Although there is still a need for single-investigator, short-term studies that lead to the development of bigger, more refined questions, today's big environmental- and ecosystem-level questions are often best addressed through long-term, collaborative research. Looking beyond the doorstep of FCE encouraged the development of many of my research questions. For example, how specific do our ecosystem models need to be to capture the processes and dynamics by which ecosystems function? By viewing ecosystems from the top down, and considering the unifying processes that occur across them, we can gain insights from novel techniques, approaches, and models as well as apply them in our own place-based context. This approach, in turn, informs not only our own place-based research, but also general ecological theory and its application.

ATTITUDES TOWARD TIME AND SPACE

The program at FCE has enabled my colleagues and me to build on leveraged funding to direct core data collection to new experiments and emerging areas of transformative science. Leveraging research projects with common but clearly differentiated goals was also a strategy I came to understand as part of the LTER program. Although water resource management and restoration are and continue to be key cornerstones on which research at FCE is based, the long-term research funded by the National Park Service and South Florida Water Management District was a springboard for establishing FCE. FCE is an example of how the LTER program optimizes research and monitoring investment. This process is, in large part, a function of standardized protocols and a sustainable funding source. Management agencies want permanence and stability in research partnerships to maximize the impact of their investments. The positive feedback forged through partnerships with aligned objectives enhances the applicability and relevance of LTER science and research investment. Coordination aimed at engaging partnerships that address place-based environmental issues is the backbone of maintaining sustained research funding at FCE. This feedback serves as a win-win for research investment across management and funding agencies.

The management of long-term research and data collection requires robust methods and approaches that help to factor out inter- and intra-annual variability from ecosystem responses to the manipulation of interest. In a monitoring context, this can be particularly challenging when political will, management priorities, or management plans change, or when the magnitude of the effect is detectable only through changes in long-term trends. However, the multifaceted data sets I have leveraged with colleagues as part of the LTER program have helped to ride these storms, including maintaining long-term monitoring and research during the financial crisis.

COLLABORATION

Collaboration is not an intuitive undertaking for the critical and independent thinkers we know as scientists. Yet, the LTER program was defined for me, when I became actively involved as a graduate student of the LTER network, as endeavoring to succeed in long-term, collaborative research. I was encouraged specifically to work with my fellow students to develop something greater than the sum of our individual thesis projects and to reach out to the US LTER network leadership for support. Today, the capacity to succeed in science depends on one's ability to develop strong peer networks. These networks are necessary to develop transformative ideas and secure the support to realize those ideas as funded research products. I owe this debt of gratitude and positive experience to a small group of forward thinking (and generous) scientists in the LTER program. For those who seek it, collaboration within the community of scholars in the LTER network is an open invitation.

However, collaboration was not a word that was familiar to me before being nominated as a cochair to the graduate student committee of what was then the Coordinating Committee of the LTER network. A discussion over drinks during the annual meeting of the Ecological Society of America developed into what I viewed as a transformative idea: engendering a spirit of collaboration among graduate students as key to building a network of scientists that could address various grand challenge questions (i.e., how do altered ecosystem dynamics affect ecosystem services? [U.S. LTER 2007]). In my mind, this network was necessarily interdisciplinary and bridged the generation gap in experience, research methods, technological capabilities, and science culture, all necessary parts to build transformative ideas and solutions. Ultimately, my role as cochair led to the first LTER Graduate Student Collaborative Research Symposium (Troxler 2005) with enthusiastic cochair Chelsea Crenshaw, key supporter Dan Liptzin, and inspiration and encouragement from outgoing cochairs Rob Daoust and Jennifer Edmonds (Figure 18.1).

The Everglades is unusual among US wetlands and unique among all LTER sites because it is a coastal, subtropical wetland that has largely been engineered by humans (Noe, Childers, and Jones. 2001). As a graduate student in the US LTER network, I was constantly exposed to cross-site or network-wide efforts that endeavored to increase the

FIGURE 18.1 Graduate students from the Long-Term Ecological Research (LTER) network (from left to right, H. Dagliesh, R. Michaels, T. Heartsill-Scalley, E. Kane), and faculty mentor Nancy Grimm (in far back), collaborating to develop the proposal for the first LTER Graduate Student Collaborative Research Symposium. (Photo courtesy of Tiffany Troxler.)

geographic scope and thus the representativeness of the research that I conducted. For example, in the context of FCE, I had the opportunity to participate in initiatives to pursue international collaborative research and develop partnerships across the ILTER network. These initiatives were motivated by seeking a better understanding of the ecology of the Everglades and what features distinguish it from or made it similar to other wetland systems (Rivera-Monroy et al. 2004). This comparative aspect of FCE-led research inspired me to study the relationships between structure and ecological function in neotropical wetlands beyond the Everglades (Troxler 2007; Troxler et al. 2012).

By leveraging these relationships, I was also able to contribute to research opportunities that broadened the geographic scope of FCE science to regional and global topics. For example, a US–Mexico LTER workshop on ecohydrological research included strong representation from both countries. Other key examples where FCE science contributed to broadened research scope include global assessments of carbon stocks in sea grass ecosystems (Fourqurean et al. 2012) and global patterns of dissolved organic carbon and black carbon (Jaffe et al. 2013). These examples illustrate the power of leveraging the LTER program to develop stronger international networks and to achieve high-impact, collaborative research with large-scale applications.

For example, I see, as do others, that there is tremendous capacity and context for how long-term, collaborative research among sites in the ILTER network contributes to scientific products that inform regional to global environmental policies. Global assessments conducted across large, spatial, and long-term temporal scales reduce uncertainties and provide robust capability for predicting change at a global scale. Thus, my experience working in the LTER program helped me to develop this broader context for my research. Developing global assessments and syntheses from site-based research among the US and ILTER networks, I discovered that my research contributes to critical national and global policy issues, and advances research and education beyond what could otherwise be accomplished. This research framework sustains research productivity through a continuous stream of leveraged opportunities driven by scientific discovery, international collaboration, and environmental policy needs.

APPLIED RESEARCH

Science that can inform management and conservation of ecosystems has been the driving motivation behind my research. The LTER program did not influence my appreciation for applied science, but it may have enhanced my ability to contribute to it. Applied science encompasses my primary research interests, and I have had the opportunity to develop and pursue those interests as part of FCE, with a focus on south Florida wetland ecosystem management. However, as a student working in a highly managed system, I believed that the context for my work was somewhat limited when compared to the types of research being conducted across the LTER network. Although there are notable exceptions, the types of manipulative experiments conducted as part of research in the LTER program seemed to me to ask questions about the limits of productivity and diversity, for example, as opposed to how management of lands might be informed by long-term research. This focus clearly shifted 10 to 15 years ago with an emphasis on the ecological domain as a socioecological system that included humans as part of that system (U.S. LTER 2007). This change shifted my perspective about the direction and motivating force behind developing LTER research.

The enthusiasm for creative and collaborative work, and the long-term perspective provided by the LTER program, influenced my own motivation to gain a better perspective on how I could broaden the applicability of research to address environmental issues at national and global scales. For the past 2 years, I worked as a Program Officer for the Intergovernmental Panel on Climate Change (IPCC) Task Force for National Greenhouse Gas Inventories (TFI). The TFI develops national-scale greenhouse gas (GHG) methodological guidance in documents requested by governments through the United Nations Framework Convention on Climate Change. The TFI hosts a database or "evolving library" of data (Emission Factor Database), including country-specific data, that is used to improve the quality of GHG inventories. The potential for synergies created by linkages between efforts such as that being conducted by the IPCC and that of the US and ILTER networks is phenomenal to me, yet remains untapped. For example, the US LTER program has within its network 10 coastal, long-term research sites, and houses within its individual, site-based data repositories a smorgasbord of long-term data on components of ecosystem carbon stock and flux data. Coastal ecosystems have been reported to be among the most carbon-rich ecosystems (Donato et al. 2011; Fourqurean et al. 2012). These systems have become globally recognized as priorities for conservation, sustainable management, and restoration given their GHG mitigation potential and are now included among these strategies in global policy (IPCC 2014). A critical global policy need is a populated database that provides coastal ecosystem carbon stocks and fluxes so that national-level GHG inventories can be produced that "contain neither over- nor under-estimates, so far as can be judged, and in which uncertainties are reduced as far as practicable" (IPCC 2000). Thus, the bottleneck is neither the paucity of data, nor an established clearinghouse for these data, but instead the mechanism to synthesize data into policy-digestible forms that can be used for national GHG inventory compilation. When extended beyond the geographic scope of the United States, the unrealized potential for contributing to global environmental policy solutions is tremendous. However, sufficient understanding of the significance of this work and that of the IPCC TFI is often limited to GHG inventory and climate policy communities. Intuitively though, the value of long-term ecological research, especially of the ILTER network, is obvious: a unique capacity to contribute science that informs global climate and environmental policy to address grand challenge questions. My experiences as part of the FCE, the ILTER network, and IPCC contributed to that basis of understanding.

COMMUNICATION

Science communication, as something tangible that should be honed just like any laboratory analytical technique, was introduced to me as a theme of the first LTER Graduate Student Collaborative Research Symposium. I believe that the limit to our success as scientists is in large part linked to effective communication of results, their implications, and how they can be used in a synthetic manner to address policy-relevant science questions of significant societal importance. More effort should be allocated toward bringing together the science, policy, and education communities to not only improve opportunities for communicating but to learn how to communicate with each other. These are the collaborations that will take the lead in addressing the most pressing socioecological questions we face (U.S. LTER 2007), not only at the subregional scale but also from a global perspective.

MENTORING

To succeed in a traditional academic environment and participate or lead collaborative research requires time and effort. Multiauthored papers are common products of science in the LTER program, with author numbers often exceeding 10. Controlling for journal ranking, it seems that these kinds of multiauthored papers are not always considered as valuable as are single or lead authored publications. Long-term collaboration requires taking a "back seat" from time to time, prioritizing the contribution to a collective product rather than lead authorship. This can take time from individual research publications, which can often have more impact among peers and university administrators. The traditional academic model does not favor collaborative research, and authors of collaborative research may be at a disadvantage during promotion or tenure evaluations. Although this issue has not affected the way I review and evaluate scientific articles or proposals, it can lead to inequities in the way junior colleagues and students are evaluated for jobs, tenure, and promotion.

Regardless of the potential trade-offs brought about by collaborative or multidisciplinary research, I believe that collaborative research is essential training for ecosystem scientists. I encourage lead technical staff to participate as team players to serve the common good of those in the scholarly community at FCE. I also have tried to develop awareness in staff members of potential pitfalls of collaboration, so that these can be communicated in a timely fashion to avoid inequitable costs to our laboratory's operations.

SKILL SET

My experiences in the LTER program have directly affected my scientific tool kit. The US LTER network office provided travel support for a visit to the Virginia Institute of Marine Sciences to learn analytical laboratory techniques to accomplish my PhD research. This short interaction was significant because these colleagues donated their time to provide support and assistance, and invited my participation in field and laboratory work. It is rewarding to visit other scientists within the LTER network, not only to share techniques and approaches, but to advance the use of standard protocols that enhance comparability of data sets across sites. The more directly comparable our experiments, data collection protocols, and analytical procedures, the more robust are our cross-site and collaborative synthetic data sets and ability to increase the signal-to-noise ratio when addressing research questions. There is currently little opportunity for such domestic graduate student and laboratory exchanges. In fact, there have historically been more opportunities for international exchange through some of the National Science Foundation (NSF)'s funding programs.

PERSONAL CONSEQUENCES

Positive consequences are the innumerable opportunities that have been presented to me as a member of the LTER network. The LTER program has certainly expanded my circle of colleagues and friends. As a cochair of the US LTER network graduate student committee, I had the great fortune of visiting several LTER sites for LTER Coordinating Committee meetings. This expanded my understanding of the research being conducted across the network, and no doubt brought new insights to my own work. In my role as cochair, I co-developed and co-led the organization of the first LTER Graduate Student Collaborative Research Symposium. This brought the tremendous experience of organizing a research symposium—identifying collaborative research themes, managing a small budget, working closely with graduate student colleagues to develop the program, and working with H. J. Andrews Experimental Forest LTER hosts to bring a quality experience for all participants and contributors. The Office of International Science and Engineering at NSF also supported the international component of this symposium and funded the travel of eight International LTER graduate students. This exposed US students to the kinds of graduate student research being conducted outside the United States, stimulated novel research ideas, and contributed to an international research synthesis (Kane et al. 2008). Collectively, these opportunities fostered through the LTER community were unparalleled at the time, and they no doubt influenced my research program and professional endeavors, as well as those of other participants. One negative consequence is the unrealistic expectation that scientists in general are motivated by and wish to seek collaborative research. For instance, it is not so easy collaborating outside the LTER network as it is within the LTER network.

CHALLENGES AND RECOMMENDATIONS

The very nature of this synthesis volume epitomizes what we strive to achieve in the LTER program: innovative ideas about how to achieve collaborative, synthetic research that provides a beneficial outcome to collaborators and yields new insights not possible without it. Collaborative research can become a tragedy of the commons when some individuals provide support for these efforts that is not reciprocated. But it is this altruistic ideal that has led the LTER program to where it is today and what we need to continue to share and promote within the international community and across generations. Sustainability of a collaborative spirit and endeavor relies on the recognition that all collaborations do not necessarily yield the results we hope to achieve. However, the impact of those successful collaborations produces a whole that is greater than the sum of all efforts that led to it. This is a vision that the LTER program should communicate so as to catalyze development of participants.

CONCLUSION

Since my work within the LTER program began as a graduate student, it is difficult to reflect on how my professional identity has changed. Yet I see that without the LTER program and the excellent mentors with whom I have interacted, my vision as a scientist, both professionally and personally, would be more limited. I share my mentors' vision of how comparative, synthetic, and collaborative research brings great advantages in furthering ecological understanding, builds strong peer groups, and broadens their potential applications. A long-term approach disarticulates signal from noise, and provides the perspective to understand the difference. Moreover, working at the FCE led me to develop a passion for looking beyond FCE to gain a better understanding of the system in which I worked.

Another apparent feature of long-term, collaborative efforts is a sustained fortitude for developing critically analyzed ideas, objectives, and methods in cooperative and "organic" ways. The most pressing environmental issues of today demand interdisciplinary approaches and scientists who are trained to communicate, not only with one another and across disciplines but also across generations and to the general public. Within the LTER program, communication with the public and across generations is now more than ever considered a critical component of successful research and education programs, but I believe that more work is urgently needed to do so effectively.

ACKNOWLEDGMENTS

I thank S. Davis and editors M. Willig and L. Walker for providing valuable input on an earlier draft. I am also grateful to all friends and colleagues in the LTER program. Material is based in part on work supported by NSF grants DEB-1440485, DEB-0823293, and DEB-0218001. Any opinions, findings, conclusions, or recommendations expressed in the material are those of the author and do not necessarily reflect the views of the NSF.

REFERENCES

Donato, D.C., Kauffman, J.B., Murdiyarso, D., Kurnianto, S., Stidham, M., Kanninen, M. (2011). Mangroves among the most carbon-rich forests in the tropics. *Nature Geoscience*, **4**, 293–297.

Fourqurean, J.W., Duarte, C.M., Marba, N., Holmer, M., Mateo, M.A., Apostolaki, E.T, Kendrick, G.A, Krause-Jensen, D., McGlathery, K.J., and Serrano, O. (2012). Seagrass ecosystems as a globally significant carbon stock. *Nature Geoscience*, **5**, 505–509.

IPCC (2000). *Good Practice Guidance and Uncertainty Management in National Greenhouse Gas Inventories.* Penman, J., Kruger, D., Galbally, I., Hiraishi, T., Nyenzi, B., Enmanuel, S., Buendia, L., Hoppaus, R., Martinsen, T., Meijer, J., Miwa, K., and Tanabe, K., eds. Intergovernmental Panel on Climate Change (IPCC), IPCC/OECD/IEA/IGES, Hayama, Japan.

IPCC (2014). 2013 Supplement to the 2006 IPCC Guidelines for National Greenhouse Gas Inventories: Wetlands (Wetlands Supplement). Hiraishi, T., Krug, T., Tanabe, K., Srivastava, N., Eggleston, S., Fukuda, M., Jamsranjav, B., and Troxler, T., eds. IGES, Hayama, Japan.

Jaffe, R., Ding, Y., Niggemann, J., Vahatalo, A.V., Stubbins, A., Spencer, R.G.M., Campbell, J., and Dittmar, T. (2013). Global charcoal mobilization from soils via dissolution and transport to oceans. *Science*, 340, 345–347.

Kane, E.S., Betts, E.F., Burgin, A.J., Clilverd, H.M., Crenshaw, C.L., Fellman, J.B., Myers-Smith, I.H., O'Donnell, J.A., Sobota, D.J., Van Verseveld, W.J., and Jones, J.B. (2008). Precipitation control over inorganic nitrogen import–export budgets across watersheds: A synthesis of long-term ecological research. *Ecohydrology*, **1**, 105–117.

Noe, G.B., Childers, D.L., and Jones, R.D. (2001). Phosphorus biogeochemistry and the impact of phosphorus enrichment: Why is the Everglades so unique? *Ecosystems*, **4**, 603–624.

Rivera-Monroy, V.H., Twilley, R., Bone, D., Childers, D. L., Coronado-Molina, C., Feller, I.C., Herrera-Silveira, J., Jaffe, R., Mancera, E., Rejmankova, E., Salisbury, J.E., and Weil, E. (2004). A conceptual framework to develop long-term ecological research and management objectives in the wider Caribbean region. *BioScience,* **54**, 843–856.

Troxler, T.G. (2005). A successful year for LTER graduate students in 2004. LTER Network News, Vol. 18, Issue 1. (http://news.lternet.edu/article12.html)

Troxler, T.G. (2007). Patterns of phosphorus, nitrogen and ^{15}N along a peat development gradient in a coastal mire, Panama. *Journal of Tropical Ecology,* **23**, 683–691.

Troxler, T.G., Ikenaga, M., Scinto, L., Boyer, J., Condit, R., Perez, R., Gann, G., and Childers, D. (2012). Patterns of soil bacteria and canopy community structure related to tropical peatland development. *Wetlands* **32**, 769–782.

U.S. Long-Term Ecological Research Network (LTER). (2007). The Decadal Plan for LTER: Integrative Science for Society and the Environment. LTER Network Office Publication Series No. 24, Albuquerque, New Mexico. 154 pages.

PART NINE
JORNADA BASIN
(JRN) LTER SITE

19 A Dryland Ecologist's Mid-Career Retrospective on Long-Term Ecological Research and the Science–Management Interface

Brandon Bestelmeyer

IN A NUTSHELL

1. My association with the Long-Term Ecological Research (LTER) program has encouraged a multidisciplinary scientific approach emphasizing broad spatial scales and site-based knowledge. It also provides a solid basis from which to link science and management.
2. In my position as a federal research scientist, I do not teach university classes. When I teach in other venues and advise graduate students, my LTER experiences facilitate my ability to draw connections among disciplines that bear on particular ecological problems.
3. Multidisciplinary breadth alongside site-specific depth afforded by the LTER program is especially useful for communicating to the public. It is important to know a lot about one area (place-based knowledge), in addition to something broader.
4. Collaboration is especially important for scientists working together at an LTER site and is also important for cross-site LTER efforts addressing regional to global problems. Within-group collaboration comes rather easily when there are healthy interpersonal relationships. Cross-site collaboration requires greater effort and network-level leadership.

PERSONAL OVERVIEW

I have been a co-principal investigator of the Jornada Basin site (JRN) of the LTER program since 2006 and a research ecologist with the US Agricultural Research Service, Jornada Experimental Range (JER), since 2003. In both capacities, my research addresses land change in drylands (arid, semiarid, and dry subhumid deserts, grasslands, shrublands, woodlands). Specifically, I work on ecosystem state changes or regime shifts, including subjects such as land degradation and desertification; these may include how land managers perceive and react to state change via mental models, information, and restoration approaches (e.g., Bestelmeyer et al. 2009).

My work has been centered at the JRN in the Chihuahuan Desert grasslands of southern New Mexico and also in grasslands and woodlands of Mongolia and Argentina. My activities include those generally associated with academia (research, publishing, grants, and supervising graduate students and postdoctoral fellows) in addition to work that is applied, such as outreach through workshops, trainings, field reviews, and writing to support management or government policy. The trade-off is not teaching university courses, although leading agency workshops and trainings partially fills this niche in my scientific career.

I have grown up as a scientist associated with the LTER program, having done my PhD research on ant community responses to land use at the Shortgrass Steppe, Sevilleta, and JRN LTER sites in the 1990s, first while at Colorado State University, then as a postdoctoral fellow with the JER and JRN. Prior to that, I obtained an MS degree working on the effects of land use on ant community structure in the Argentine Chaco. The primary objective of both my MS and PhD work was to better link land management decisions about grazing, as a dominant land use in drylands, to biodiversity considerations. By the end of my PhD work, I came to recognize how little I knew about land-use decisions and their social and ecological determinants.

The LTER program provides an invaluable source of scientific information for making ecosystem management decisions because it emphasizes long-term changes in multiple ecosystem components and a holistic perspective linking these changes to a wide range of processes (including socioeconomic ones). This holistic approach has had a large impact on how I do science and my ability to link science to management.

APPROACH TO SCIENCE

In most ways, my approach to science reflects key themes in the subdiscipline of landscape ecology. I favor a landscape perspective featuring information at a variety of spatial scales, including patterns at broad (landscape to regional) scales that are captured in maps. I credit foremost my advisor, John Wiens, and his laboratory at Colorado State University for promoting a multiscaled perception of ecological pattern and process, but the LTER program encourages it as well. The large land areas of the LTER sites that I have been involved with, and the information and variety of expertise associated with them, especially soil scientists and remote sensing and geographic information system specialists, highlight the consequences of spatial heterogeneity at several spatial scales. Much of my thinking is geared toward comparisons at broad spatial scales rather than intensive manipulations at fine scales, but I have learned to value fine-scaled experiments as well, especially when they are designed to account for spatial heterogeneity and context.

I tend to favor theory integration within specific ecosystem types rather than using carefully selected cases or "big data" to examine a general theory within a particular thematic domain (see constitutive theory of Scheiner and Willig, 2011). In one project, for example, we are asking whether regime shifts in plant cover and ranch impermanence are related, and about their consequences for land change in the Chihuahuan Desert. The combination of disciplines at our LTER site inspired the question and provides a means by which to investigate it. The hope is that such multidisciplinary approaches within LTER program and similar groups can yield general approaches for investigating the function of regional socioecological systems.

ATTITUDES TOWARD TIME AND SPACE

An LTER site serves a vital and unique purpose in the global scientific enterprise. It expands our perception of time and space–time interactions. Information about historical ecosystem states, historical events or legacies, and spatial patterns of change is critical for anticipating future change as well as for understanding ecosystem structures and services. Historical information guides and conditions our interpretation of studies on contemporary ecological processes. Additionally, and distinctly, LTER sites provide a nexus for numerous kinds of data, information, and lines of inquiry, whether they are long-term or not. The studies may also occur outside the formal boundaries of the LTER site. For example, it is useful to couple broad-scale survey data (e.g., at the extent of the Chihuahuan Desert in New Mexico) to the intensive, process-based studies within LTER sites. This coupling can be used to extend inferences to management and policy-relevant scales, as well as to understand the nature of the geographic and environmental inference space of site-based studies. The linkage of deep (site-based, long-term) and broad (observations across a region or several similar regions) perspectives characterizes my general research approach.

The allocation of limited scientific resources to create long-term, landscape-level, and multifaceted perspectives of an LTER site, however, is a significant challenge. For example, LTER sites are often faced with the trade-off between allocating resources to new studies in new places or to continue long-term ones in the same place. Long-term studies have immense value, but sometimes the marginal value of a long-term study becomes lower than that of a new study for a variety of reasons, such as when the design of a long-term study limits its ability to address significant new questions. Decisions to end a long-term study should be carefully documented with respect to the trade-offs. If the long-term study is associated with adequate metadata, it can always be resumed later on.

COLLABORATION

Collaboration is the heart of a site-based LTER program. The multidisciplinarity inherent in research at LTER sites enables the integrative approaches needed for a landscape perspective. As part of a research team at an LTER site, a scientist is more likely to work with scientists from outside his or her core discipline than from within it. Such multidisciplinarity might have the disadvantage of diluting focus on particular topics, but it enables opportunities for developing novel questions and insights. A small mammal ecologist, range scientist, agricultural economist, wind erosion modeler, hydrologist, pedologist, biogeochemist, and plant ecologist walked into a bar and the bartender looks up and says, "Is this some kind of joke?" Such a multidisciplinary group does evoke the "priest-rabbi-minister" tension, but going to the bar truly is important. Long-term personal relationships among scientists and other professionals, especially those that might not work together under other circumstances, results in a level of understanding and familiarity that makes it easier to work across disciplinary boundaries.

Network-wide collaboration can also lead to new insights. One example is the cross-site LTER synthesis on abrupt transitions led by Aaron Ellison and me (Bestelmeyer et al. 2011). The group compared long-term dynamics of sea cucumbers, krill, penguins, and a desert grass using a common conceptual and analytical framework. Developing concepts to encompass such a diverse array of systems and organisms drove us to simplify and get to the core issues. But working with a group of ecologists who know their systems and organisms well allows one to explore many kinds of relationships. One of the most important revelations for me was that threshold dynamics in Adélie penguin (*Pygoscelis adeliae*) populations and black grama (*Bouteloua eriopoda*) grass stands have similar underlying causes, pointing to general classes of mechanisms with relevance to general theory. Without the LTER program, it is unlikely that I would have connected to marine biologists. Of course, it requires a lot of work to see across thematic boundaries to the bigger issues, but when you have a broad enough concept at hand (e.g., alternative states) such efforts make great strides toward consilience. Network-level leadership should play a key role in identifying these themes and leading or incentivizing cross-site collaboration.

APPLIED RESEARCH

Applied and basic sciences are typically dichotomized, but the distinction is not constructive. When the research focus of an LTER site is a landscape (or some other system such as a coast, city, ocean area, or lake), human actions are a part of the system's fabric. Perhaps "basic science" is conducted at spatial scales fine enough that the impact of human decisions is not important, on processes that are several steps removed from human actions, or in the few remaining landscapes where human activities are scarce. For most other scales and landscapes, the basic–applied dichotomy isolates knowledge about biophysical and societal processes and their powerful, interactive effects on ecological systems, prompting the recognition of "socioecological systems" that might otherwise be regarded more simply as "ecosystems." Furthermore, the basic–applied dichotomy reinforces the notion among managers that rigorous measurement, careful sampling design, and hypothesis testing are not useful to management, captured in the oft-repeated comment "we are not doing science, this is just management." This attitude represents for me a serious distortion of priorities, so I have been keen to develop research projects that directly involve land managers and policy-makers at the outset. For example, I am examining the consequences of restoration actions in the Chihuahuan Desert with the Bureau of Land Management and the prevalence of desertification with a consortium of agencies in Mongolia.

Similar efforts are underway in many parts of the ecological science community, including several LTER sites (e.g., North Temperate Lakes, Harvard Forest) that can inspire innovative approaches that reach across ecosystems. These kinds of projects should continue to be encouraged, including via network-level activities.

COMMUNICATION

In full disclosure, my spouse runs the Jornada Schoolyard LTER Program. It is easy to connect to K–12 education when an LTER site is associated with a nonprofit such as the Asombro Institute for Science Education (www.asombro.org) that is closely engaged with science in the LTER program. Furthermore, I am employed by the US Agricultural Research Service (ARS), so I speak often in nonacademic or mixed settings, with representatives of agencies, interest groups, and international development organizations. My ability to communicate about various ecological relationships in those settings has benefited from the multifaceted education provided via interactions supported by the LTER program, but I owe the lion's share of credit for my development to my ARS colleagues. In presenting to K–12 students as well as adult groups, the LTER Schoolyard Program has led me to recognize how valuable it is to reference or review the fundamentals of the scientific process in all public engagements. The successful presentation of science-based knowledge to adult, nonscientist professionals, and third graders requires similar attention to the basics of the science, in particular the idea that scientists are driven by a desire to reduce their ignorance rather than to prove something to be true.

MENTORING

The broad range of disciplines represented by principal investigators in the LTER program allows for graduate students and postdoctoral fellows to be exposed to thinking from several disciplines. I take advantage of it for my students and contribute my perspective via student committees, even though narrow disciplinary graduate training continues to be the norm in my experience (I recognize that several broad ecology training programs have been developed). Nonetheless, the often unplanned interactions among students and principal investigators within a group at an LTER site can produce multidisciplinary perspectives in spite of narrowly focused graduate programs.

In mentoring and peer review or editorial interactions with other scientists, one consequence of my LTER perspective is that I tend to be interested in learning more about the context of a particular study or set of sampling points, perhaps to an unusual degree. What I consider to be excellent research does not necessarily need to be "place-based," but it should be "place-aware." If I read a macroecological study using "big data," for instance, I will be interested in both the patterns in the residuals and the central tendencies or constraint lines that are the basis for the main conclusions. I will ask what soil or land use did a particular sample represent? Does the glossed-over heterogeneity significantly weaken the generality? Do the generalities really matter, and if they do, how should we present them to circumscribe the effects of multiple, interacting factors?

SKILL SET

The tool kit to which I have access through collaborations in the LTER program is very broad. Perhaps because it is easy to collaborate, I have become familiar with the uses and interpretations of various tools, if not proficient in using them myself. I think it is easier and more productive to collaborate with technical specialists rather than limiting my work to one technique or trying, slowly, to do it all myself. Some recent examples include the combination of vegetation monitoring data, which I work with, with high-resolution remote sensing and ecophysiological measurements, which collaborators do.

PERSONAL CONSEQUENCES

My connection with an LTER site, particularly one that features several great scientists and that is sought after as a resource for dryland management issues worldwide, has allowed me to forge connections with a diverse group of colleagues and friends, including people from other cultures and countries. For example, in a given year, I might work not only with colleagues in the LTER program, but with colleagues in Argentina, Asia, and Australia who use simulation models or who raise cattle, sheep, or goats. My colleagues therefore include scientists, government bureaucrats, ranchers, and transhuman pastoralists. Such experiences have broadened my thinking in ways that affect most aspects of my life and the lives of my family. For instance, I am fond of máte tea, boiled mutton, and Vegemite.

CHALLENGES AND RECOMMENDATIONS

Funding levels represent the biggest challenge for the LTER network. Expanding the scope of activities in the LTER program has always been limited by financial considerations. An LTER program, as I have heard it characterized, is a nest full of hungry chicks. And we know what adaptations are favored when chicks are competing in a food-limited environment (see especially "begging scrambles," Mock and Parker 1998). The problem for scientists in the LTER program is how to maximize inclusive academic fitness measured in terms of scientific output, successful site reviews, and continued funding. This optimization has to be balanced by basic decency; one generally does not decide to kick a long-term collaborator out of the nest just because he or she is suboptimal for one's academic fitness. Static budgets, the need to carefully craft a coherent and successful program, and long-term participation serve to limit direct involvement by many who would be interested in joining. There are, however, several opportunities to broaden and reinforce connections in the absence of dramatically increased funding: (1) conduct student-oriented workshops to build a common understanding of research directions and problems encompassed by the LTER program (this serves the senior researchers as well), (2) fund distantly connected researchers to attend LTER site meetings or symposia (i.e., those researchers that use the site but that do not directly interact with other scientists from the LTER program), (3) fund prospective researchers to attend LTER site meetings and symposia (i.e., sell the infrastructure without a promise of funding), (4) find mechanisms within academia and the National Science Foundation (NSF) to reward public outreach at levels commensurate with rewards for scientific production, and (5) encourage researchers to collaborate with other groups to gather or link to data "off-site" around their LTER sites.

CONCLUSION

Because JER and the JRN LTER site operate together seamlessly, their effects cannot be disentangled, and their contributions to my career are synergistic. JRN activities emphasize the generality and breadth of the ecological science and JER emphasizes the mobilization of science knowledge for land management. That combination (and tension) has framed my career. At times it can be difficult to bridge those visions and give each the attention it deserves, but attempting to do so increases the power of the science and inspires the applications. I encourage the senior scientists of LTER sites to forge direct links to those institutions responsible for the stewardship of the ecosystems within which LTER sites are embedded (where they have not already done so) and encourage young scientists to immerse themselves deeply into their messy socioecological systems and all the disciplines needed to understand them. Then take the broad conceptual models developed at LTER sites and apply them to socioecological systems in other parts of the world. How the global array of socioecological systems function, evolve, and coevolve is the new frontier of the Anthropocene.

ACKNOWLEDGMENTS

This chapter was supported by a grant from the NSF LTER program (DEB-0080412) to New Mexico State University for the Jornada Basin LTER and appropriated funds to the US ARS.

REFERENCES

Bestelmeyer, B.T., Ellison, A.M., Fraser, W.R., Gorman, K. B., Holbrook, S.J., Laney, C.M., Ohman, M.D., Peters, D.P.C., Pillsbury, F.C., Rassweiler, A., Schmitt, R.J., and Sharma, S. (2011). Analysis of abrupt transitions in ecological systems. *Ecosphere,* **2**, art 129.

Bestelmeyer, B.T., Tugel, A.J., Peacock, G.L., Robinett, D.G., Shaver, P.L., Brown, J.R., Herrick, J.E., Sanchez, H., and Havstad, K.M. (2009). State-and-transition models for heterogeneous landscapes: A strategy for development and application. *Rangeland Ecology & Management,* **62**, 1–15.

Mock, D.W., and Parker, G.A. (1998). Siblicide, family conflict and the evolutionary limits of selfishness. *Animal Behaviour,* **56**, 1–10.

Scheiner, S.M., and Willig, M.R. (2011). *The Theory of Ecology.* University of Chicago Press.

20 Tales from a "Lifer" in the Long-Term Ecological Research Program

Debra P. C. Peters

IN A NUTSHELL

1. As a long-time member of the Long-Term Ecological Research (LTER) network, first as a graduate student and scientist at the Shortgrass Steppe (SGS) site (1984–1997), then as a scientist at the Sevilleta (SEV) site (1996–present) and now as principal investigator at the Jornada Basin (JRN) site (2003–present), my professional career has been shaped almost entirely by my LTER experiences.
2. My experiences in the LTER program directly contributed to my individual-based approach to ecosystem dynamics combined with the knowledge that the dominant

ecological processes can change as the spatial extent increases, and that long-term data are critical to disentangle how these pattern–process relationships change across scales.
3. The LTER program has provided me with international experience and exposure that are valuable to my career. My opportunity to travel overseas has led to bonding experiences and new insights into other ecosystems.
4. My appreciation for the value of K–12 education and the amount of work that is involved in "doing it right" has been shaped by my experiences with the Jornada Schoolyard LTER Program.
5. One of the key challenges that I face in working at an LTER site is the tension between continuing to collect long-term observations with the need and desire to test new ideas that often result from the long-term data but then compete for resources with the collection of those data. Another challenge is in mentoring young scientists to become principal investigators, and in cultivating new relationships with potential co-principal investigators.

PERSONAL OVERVIEW

Currently, I am the principal investigator at the JRN LTER program at New Mexico State University (NMSU) in Las Cruces, New Mexico. I am also a collaborating scientist at the SEV LTER program at the University of New Mexico in Albuquerque, New Mexico. I received my BS in biology at Iowa State University in 1981 and my MS in biology from San Diego State University (SDSU) in 1983. My LTER experiences began as a PhD student at Colorado State University (CSU) through the SGS LTER program in 1984, and these continued while I was a postdoctoral fellow (1988–1989). I was the project manager for the SGS from 1994 to 1995 and a co–principal investigator until 1997. I started conducting research with Jim Gosz at SEV in 1996, which led to a number of independent long-term research projects and my role as a collaborator. I became a co–principal investigator at JRN when I was hired as a federal scientist with the US Agricultural Research Service, Jornada Experimental Range in Las Cruces, New Mexico, in 1998. On the departure of the principal investigator (Laura Huenneke) to another university in 2003, I was appointed to lead the JRN project, a position I hold currently as an affiliate faculty member at NMSU. I am currently on detail to the Office of the Chief Scientist of the US Department of Agriculture in Washington, DC, where I am the Senior Advisor for Agricultural Systems and Technology.

I have been involved in the LTER network since 1989, when I participated in a cross-site activity at NMSU that was led by Walt Conley (NMSU) and Bob O'Neill (Department of Energy–Oak Ridge). I participated in my first LTER Coordinating Committee in 1993, surprisingly held also at NMSU, which included my first tour of the Jornada. I have attended all seven LTER All Scientists Meetings to date (1990, 1993, 2000, 2003, 2006, 2009, and 2012). Also, I participated in numerous early efforts to internationalize long-term research concepts through visits to other countries. These exchanges of scientists were organized by Jim Gosz (China, 1991; Mexico, 1997; South Africa/Botswana/Mozambique, 2002) and Diana Wall (Hungary, 1994–1995). I was an elected member of the LTER Executive Committee (2004–2006) and served on the Network System Information Advisory Committee from 2006 to 2008. I also co-led the EcoTrends Project with Ariel Lugo (Luquillo LTER program) from its inception in 2004 (see Collaboration). At the site level, my leadership and expertise in landscape ecology and cross-scale interactions were instrumental in the research direction undertaken by the JRN program beginning in 2000 (Peters et al. 2004), and they are important to the facilitation of research findings on cross-scale interactions and connectivity across heterogeneous landscapes (Peters et al. 2006).

These ideas were extended to the continental scale with impacts on the development of the National Ecological Observatory Network (NEON) (Peters et al. 2008).

APPROACH TO SCIENCE

All of these experiences have collectively and interactively influenced, to various degrees, the paths that I have taken, and the way that I think about and study ecological systems. Although I was first introduced to systems ecology and simulation modeling as a master's student at SDSU, it was not until I started my PhD at CSU (1984) that I was completely immersed in ecosystems ecology and modeling within a long-term context. In 1982, the SGS LTER site had recently started, and we were a small group under the direction of the principal investigator, Bill Lauenroth. As a group, we were interested in alternative ways to view the shortgrass system from historic approaches that were systems oriented but viewed grasslands as "green slime" (i.e., homogeneous at the square meter plot scale). We were influenced by research conducted at the JRN by Bill Schlesinger and others, where individual plants and associated bare interspaces were studied as separate, yet interacting, entities, and by individual-based, canopy gap dynamics models of forests being developed by Hank Shugart, Tom Smith, and Dean Urban at the University of Virginia. Combining these two influences led to a new view of semiarid grasslands that focused on individual plants and the importance of belowground resource gaps to ecosystem dynamics (Coffin and Lauenroth 1990). We expanded these ideas spatially to understand landscape dynamics and to explain the inability of dominant perennial grasses to recover after large disturbances. Working with Osvaldo Sala at the SGS led me to appreciate the importance of long-term data, both in terms of the context these data provide for shorter-term dynamics within a site, but also that space for time substitutions (and vice versa) can break down as the spatial extent increases beyond the site. Key influences on how I approach science today were that: (1) this individual-based approach to ecosystem dynamics combined with the idea that the dominant ecological process changes as the spatial extent increases and (2) that long-term data are critical to disentangling changing pattern–process relationships across spatial scales (Peters et al. 2004; Peters et al. 2006).

ATTITUDES TOWARD TIME AND SPACE

Because of my LTER experiences, I have an appreciation for the use of long-term studies in many different ways. One of my earliest exposures to the importance of long-term data was a sliding window figure of precipitation through time shown at a SGS meeting in the late 1980s. Selecting a window size of 3 to 4 years and moving the window through time can look like a drought, a wet period, or no trend in precipitation, but within the entire 50-plus years of data, none of these short-term trends may be informative for particular questions. Although this may seem like common knowledge now, at that point in time, it was an effective way to show that one needs to put short-term data within a longer-term context. My LTER experiences have allowed me to learn more about the importance of time over the past 30 years, in terms of legacies, feedbacks and feed forwards, thresholds, and other nonlinearities (Peters et al. 2012). Many of these learning experiences have come from interactions with scientists at my own site, but I have also worked with scientists from different sites and perspectives through LTER workshops and meetings. Recently, discussions with Mark Ohman (California Current Ecosystem) about regime shifts in marine systems and with Aaron Ellison (Harvard Forest) about the need for statistical rigor in identifying thresholds have led to new ways to think about desert systems at the JRN.

COLLABORATION

At all three LTER sites, the structure of LTER as a program provided me with unique opportunities to interact closely over a period of time with a group of people with different expertise working on similar overall questions and the same goal. Although this collaboration was critical to the success of the LTER programs, it takes longer to discuss ideas and obtain consensus within a diverse group of people compared to that required in individual research projects with one or two principal investigators and a few graduate students. In our case at the JRN, it took longer to get multifaceted experiments initiated, yet the trade-off is that the collaborative experiments are more interesting, relevant, and robust with more disciplinary perspectives. We recently implemented a cross-scale experiment where the scientists quickly agreed on the important hypotheses to be tested, but it took several years of planning and discussion before everyone agreed on the experimental design, the needed equipment, and the response variables and measurements to be obtained. The resulting collaborative papers also take longer to move through the internal writing and editing stages because of multiple coauthors. In my experience, the final products are much better with a broader perspective when they include coauthors with different expertise (see Peters et al. 2006 for examples).

Similarly, the LTER network provided opportunities for me to interact closely and informally with ecologists from different disciplines (i.e., atmospheric science, pedology, social science, computer science, and informatics) who study different target organisms (from microbes to penguins) from different systems (i.e., forests, alpine, Arctic, dry valleys, marine, coastal, lakes, and coral reefs). The resultant melding of ideas through working groups, follow-up workshops from All Scientists Meetings, and cross-site funding opportunities from supplements pushed ecology and environmental sciences forward in many unforeseen ways. These interactions went beyond networking to actually doing science and writing collaborative papers within the United States and abroad (e.g., Peters et al. 2011). Personally, these interactions have been some of the major highlights of my career to date.

Because the LTER community is relatively small, there are opportunities to lead workshops and develop ideas with some of the best ecologists in the country. A good example is the EcoTrends Project, where a simple idea ("let's make long-term data products easily available to allow comparison among all LTER sites") conceived by two principal investigators (Ariel Lugo and me), led to a multiyear, 50-site collaborative effort resulting in a website of data (www.ecotrends.info) and a book (Peters et al. 2013). Along the way, EcoTrends staff "cleaned" many site-based data sets, identified flaws in the site-based system of information management at the network level, and provided input on the structure and need for the LTER Network Information System that is currently under development. There were clear challenges associated with the diversity of data from different ecosystems collected by multiple investigators through time, but keeping the group focused on a science-driven product allowed advances to be made in information management, data accessibility, and synthesis across data sets within and among sites (Peters 2010).

COMMUNICATION

My appreciation for the value of K–12 education and the amount of work involved in "doing it right" has been shaped by my experiences with the JRN Schoolyard LTER Program run by a nonprofit organization (Asombro Institute for Science Education, www.asombro.org). Full-time staff members work with many volunteers to increase scientific literacy by fostering an understanding of the Chihuahuan Desert. This amazing program serves more than 17,000 K–12 students and 1,500 adults in New Mexico and West Texas with inquiry-based science education programs each year. The JRN is now working more closely with Asombro by requiring LTER-funded graduate students to devote time to Asombro. Both the graduate students and K–12 students are benefiting from these interactions.

SKILL SET

One set of skills that I learned from my participation in the LTER program is associated with leading and working with a diverse set of people. Through time and working at multiple LTER sites, I have been exposed to different leadership styles, and different approaches to building a functional team committed to site-based, long-term research. Although each site operates differently, and the mode of operation within a site often changes with a change in lead principal investigator or management team, there are commonalities that helped me when I became a principal investigator. Based on my observations, a successful LTER program needs: (1) a leader who can articulate a vision and encourage people to work together to fulfill that vision, (2) a small group of devoted scientists and staff who will make time to keep the program going, and (3) a larger collaborative and diverse team who understands the critical need for long-term research. As a "lifer" in LTER program, I have had the opportunity to hone the skills needed as a principal investigator to develop, articulate, and fulfill a vision that fits within the LTER model. Because LTER programs involve groups of scientists, I have also learned how to deal with people with diverse backgrounds and expertise and how to get people to work together toward a common goal. For example, at the JRN, beginning in 2000, we made a major conceptual shift from an organismal and ecosystem focus to a landscape ecology perspective. We held meetings over a period of several years to allow each scientist to understand how each person's research fits into a landscape perspective. As part of this process, I learned how to see the system from all of the other perspectives and how to integrate each piece into a coherent view of the system. I have transferred this skill set to other situations where large groups of experts need to come together to think about and solve complex problems, in particular across very large spatial extents, such as cochairing the Climate Committee (2007) as part of the NEON planning meetings, and in my current position as the point of contact for the group on Earth Observations Global Agricultural Monitoring.

The second skill set that the LTER program provided me is associated with both sides of the review process: (1) developing the skills needed to be a critical reviewer and (2) then using those skills to become a better communicator and writer of proposals and papers, as well as a better organizer and leader of meetings and site or program reviews. As a graduate student and postdoctoral fellow, I had the opportunity to participate in and learn from site reviews and proposal-writing activities at the SGS. Interactions with the review team and National Science Foundation (NSF) program officers, informally through posters and small group meetings as part of the reviews, were an invaluable experience that improved my ability to communicate my research to others. As an LTER co–principal investigator, the repeat nature of these reviews at the three LTER sites where I have conducted research, and the opportunity to be on review teams for other LTER sites, provided unique insights into the process that have been incredibly valuable when leading site reviews and writing proposals for the JRN. In addition, working on LTER proposals with many scientists allowed me to develop skills that are easily translated to the development, writing, and management of large, complex, multidisciplinary proposals for other NSF programs.

PERSONAL CONSEQUENCES

The LTER program has provided international experience and exposure that have been valuable to my career (Figure 20.1). In particular, the beginning of the International LTER (ILTER) program provided the opportunity for me to travel with a diverse group of LTER ecologists to China, Hungary, Mexico, and southern Africa. The bonding experience associated with traveling for days to weeks in an unknown country with a language and culture quite different from my own cannot be overstated. In China, a small group of us (Jim Gosz, James Brunt, a woman from Norway, and me) spent time on planes, trains, and automobiles crossing the vast Asian grasslands from northeast of Beijing to Urumqi in the far northwest province, a distance of more than 3000 km. We learned to drink Mongolian milk tea and to eat whatever vegetables and meat or fish were available locally. As our leader, Jim had to try all plates brought to the table (even deep-fried scorpions and many varieties and forms of fish), and I learned from watching Jim how to be polite and gracious in difficult and uncomfortable situations. Because this was 1991, only 2 years after the Tiananmen Square incident where thousands of prodemocracy protesters were arrested, our visit to the square in downtown Beijing had a large impact on all of us. We had watched on television the tanks roll toward students in the square; now we were there being watched by armed guards. All of the international trips provided a broader context for our research and lives in the United States; the poverty we saw in Mozambique and Mexico, the number of unemployed in South Africa, the HIV-AIDS victims in Botswana, and the lasting impact of Russian power in Hungary. From these collective experiences, I continue to value my freedom of movement in the United States with an appreciation for the opportunities available, as well as the ability to work, learn, and live without extreme government constraints.

CHALLENGES AND RECOMMENDATIONS

One of the challenges I experience in working at and leading an LTER site is the tension between continuing to collect long-term observations with the need and desire to test

FIGURE 20.1 Debra Peters and numerous Hungarian ecologists at the Sevilleta Long-Term Ecological Research site in central New Mexico. (Photo courtesy of Debra Peters.)

new ideas that often result from the long-term data but then compete for resources with the collection of those data. This balance between old and new, under a fixed budget for 6 years of funding, is the crux of the LTER program, where resources required to maintain existing long-term studies are constantly challenged by resources needed to start new and often cutting-edge, exciting studies. In my experience, sites have reduced sampling frequency or intensity of old experiments, including stopping them completely, or sought external funding for the new ideas. Tipping the balance too far in one direction or the other will lead to bigger issues when the site is reviewed. I find that maintaining this balance in funding between old and new is one of the larger impediments to personal development by me and others at my site. A fixed budget for 6 years means less available support as each year passes with cost-of-living increases and the accumulation of unexpected expenses. It becomes increasingly difficult to attract new scientists to the project or to maintain resources for existing scientists as these base costs increase through time. At the JRN in 2006, we switched from an individual principal investigator allocation of resources model to an allocation to small groups of principal investigators based on collaborative research questions and productivity. Although this approach promotes collaboration, it is challenging for new scientists to become involved and for established scientists with specific questions to be funded for long time periods. We have addressed these issues through supplemental funding and finding additional sources of funding, but these have their own set of challenges. A reliable supplemental funding source would be helpful in this regard.

Another challenge that I face is in finding and mentoring young scientists to become principal investigators. Because there are many facets to being a leader, there are many alternative solutions that are currently dealt with at the individual site level. A group mentoring program may be more effective in which principal investigator candidates are trained and mentored by scientists from across the network and then apply for leadership positions as they become open. There are good examples currently in the network (e.g., H.J. Andrews Experimental Forest), where expertise in the ecosystem of the site (e.g., forests, grasslands) was less important than the personality, breadth of expertise, and synthesis ability of the person chosen to be the leader.

Finally, I find it challenging to bring new investigators into the LTER program, given the delay required to develop collaborative projects. Although these time delays are not a major concern at the program level with 6 years of LTER funding per cycle, it can be a frustrating experience for principal investigators who are used to moving quickly with little input from others and who need papers published rapidly to be considered for positions for tenure and promotion.

CONCLUSION

As a long-time member of the LTER network, first as a graduate student and scientist at the SGS, then as a scientist at the SEV, and now the principal investigator at the JRN, my professional career has been shaped almost entirely by my LTER experiences. I have been a member of the LTER program for more than 30 years and plan to continue this association into the future. Obviously, I believe that working within the LTER network provides many more advantages than limitations. My key piece of advice to aspiring ecologists is to become immersed in the LTER program to take advantage of all that individual sites and the network have to offer but to maintain a critical perspective and find ways to improve the program. There is a tendency with long-term programs to become lax, and it is the people from the next generation who will provide the greatest advances.

ACKNOWLEDGMENTS

Funding was provided by the NSF to New Mexico State University (DEB-1235828, EF-1065699, DEB-1065699) and the US Agricultural Research Service (6235-11210-007).

REFERENCES

Coffin, D.P., and Lauenroth, W.K. (1990). A gap dynamics simulation model of succession in the shortgrass steppe. *Ecological Modelling*, **49**, 229–266.

Peters, D.P.C. (2010). Accessible ecology: Synthesis of the long, deep, and broad. *Trends in Ecology and Evolution*, **25**, 592–601.

Peters, D.P.C., Bestelmeyer, B.T., Herrick, J.E., Monger, H.C., Fredrickson, E., and Havstad, K.M. (2006). Disentangling complex landscapes: New insights to forecasting arid and semiarid system dynamics. *BioScience*, **56**, 491–501.

Peters, D.P.C., Groffman, P.M., Nadelhoffer, K.J., Grimm, N.B., Collins, S.L., Michener, W.K., and Huston, M.A. (2008). Living in an increasingly connected world: A framework for continental-scale environmental science. *Frontiers in Ecology and the Environment*, **5**, 229–237.

Peters, D.P.C., Laney, C.M., Lugo, A.E., Collins, S.L., Driscoll, C.T., Groffman, P.M., Grove, J.M., Knapp, A.K., Kratz, T.K., Ohman, M.D., Waide, R.B., and Yao, J. (2013). Long-term trends in ecological systems: A basis for understanding responses to global change. USDA Agricultural Research Service Publication No. 1931. Washington, DC.

Peters, D.P.C., Lugo, A.E., Chapin, F.S. III, Pickett, S.T.A., Duniway, M., Rocha, A.V., Swanson, F.J., Laney, C., and Jones, J. (2011). Disturbance complexities and generalities emerging from cross-system comparisons. *Ecosphere*, **2**, art 81.

Peters, D. P. C., Pielke, R.A. Sr, Bestelmeyer, B.T., Allen, C.D., Munson-McGee, S., and Havstad, K.M. (2004). Cross scale interactions, nonlinearities, and forecasting catastrophic events. *Proceedings National Academy Sciences U.S.A.*, **101**, 15130–15135.

Peters, D.P.C., Yao, J., Sala, O.E., and Anderson, J. (2012). Directional climate change and potential reversal of desertification in arid and semiarid ecosystems. *Global Change Biology*, **18**, 151–163.

PART TEN
KONZA PRAIRIE (KNZ) LTER SITE

21 A Forest to Prairie Transition as a Long-Term Ecological Research Scientist

John Blair

IN A NUTSHELL

1. Being involved in the Long-Term Ecological Research (LTER) program for most of my career has greatly influenced my development as an ecologist. It has broadened my understanding and appreciation of ecological processes at scales ranging from microbial processes to ecosystem fluxes.
2. Participating in the LTER program has heightened my awareness of the critical role of spatial and temporal variability in ecological dynamics, as well as the value of long-term data for identifying directional environmental changes or assessing responses to experimental manipulations.
3. Working with other investigators at an LTER site over long periods of time has revealed the importance of a place-based understanding of ecological processes as a source of insight into complex ecological phenomena.

4. Interacting and collaborating with students and scientists having diverse research interests and backgrounds has enhanced my ability to communicate more effectively with other scientists and with the public.
5. There are some trade-offs between directing a large research program and advancing one's personal research, but the rewards of long-term collaboration are substantial.

PERSONAL OVERVIEW

I have been part of the LTER program for most of my career, from graduate student at one LTER site to principal investigator at another. I began my PhD training at the University of Georgia in 1983 under the direction of D.A. (Dac) Crossley, Jr., the first leader of the Coweeta (CWT) LTER program. My early research focused on forest ecology, including plant litter decomposition and effects of clear-cutting and regrowth on decomposer communities and forest floor processes (Blair and Crossley 1988). My first postdoctoral appointment was on a National Science Foundation (NSF) grant that I wrote to study forest-floor nitrogen dynamics using stable isotope tracers. In 1992, I joined the faculty of the Division of Biology at Kansas State University as an ecosystem ecologist. This position had been held by Tim Seastedt, another Crossley graduate student who served as principal investigator of the Konza Prairie LTER (KNZ) program and later as principal investigator of the Niwot Ridge LTER program. I was hired with the expectation that I would become engaged in the KNZ program, where my research would focus on ecosystem processes in tallgrass prairie. Becoming part of the KNZ program was a major attraction of this position for me. Within the first year I participated in an NSF site-review of the KNZ program and began attending LTER Coordinating Committee and principal investigator meetings. Over time, I assumed additional responsibilities for scientific leadership and day-to-day management of the KNZ program, leading to my transition to lead principal investigator in 1999, a role that has continued for three funding cycles.

Serving as the leader of KNZ program has broadened the way I think about and conduct ecological research and provided many opportunities to collaborate with scientists and students from other institutions and research sites. Some of my research builds on my early interests in soil processes and biogeochemistry, but my research has expanded to encompass other ecological processes (e.g., primary productivity, community assembly) and a broad range of contemporary ecological issues (e.g., climate change, altered land use and land cover, restoration). Leading a large research program also creates a number of challenges, such as coordinating activities and maintaining effective communication among a diverse and geographically separated group of researchers. There are also trade-offs in balancing the demands of managing of a large, interdisciplinary research and training program with other professional activities and aspirations. On the whole, the LTER program has had a very significant, and mostly positive, effect on my professional development and identity. It increased my visibility and standing, locally and in the broader ecological community, and provided unique opportunities to interact with a diverse group of scientists and participate in numerous planning and research initiatives, both within and outside of the LTER network.

APPROACH TO SCIENCE

As a graduate student at an LTER site, I interacted with other students and scientists conducting research on a wide range of ecological topics. As my involvement in the LTER program increased, so did my understanding and appreciation of research outside of my own sphere of interest and expertise. This was important preparation for leading an

LTER program, which is inherently interdisciplinary. The breadth of my research interests and activities increased as did collaborations with scientists in other subdisciplines of ecology and other areas of science. I remain fascinated with soil biota and their role in soil processes, but my research interests have broadened to encompass terrestrial biogeochemistry more generally along with other aspects of ecosystem ecology.

My participation in the LTER program contributed to a very inclusive view of ecological research. I understand and appreciate the value of empirical and theoretical approaches, short- and long-term experiments, environmental monitoring, and ecological modeling. Although each of these has strengths and limitations, the greatest potential for major advances comes when they are combined. That is a real strength of the LTER program: the application of multiple approaches synergistically at a focal study site and the ability to complement this with comparative studies and data from other sites. Long-term, site-based measurements of environmental characteristics and key ecological processes can reveal surprises in ecological dynamics or help identify responses to global change drivers that are not apparent in short-term studies. I value studies that build on long-term observations or take advantage of long-term experiments. In fact, many of the most important responses documented by KNZ scientists to changes in environmental drivers have taken a decade or longer to become apparent (Ratajczak, Nippert, and Collins 2014), and there are critical long-term responses to experimental manipulations that would be missed in studies that lasted only a few years (Knapp, Briggs, and Smith 2012). In addition, results supported by the LTER program highlight the value of multifactor experiments that address the interactive effects of multiple drivers, such as altered climate and altered biogeochemistry.

ATTITUDES TOWARD TIME AND SPACE

A unique aspect of the LTER program is the long-term, place-based nature of the research. This aspect has heightened my awareness of the critical importance of spatial heterogeneity and temporal variability. In tallgrass prairie, most ecological processes vary tremendously in time and space as a result of inter- and intra-annual climatic variability, landscape heterogeneity, variable fire and grazing regimes, and complex biotic interactions. As a result, decadal (or longer) time frames are often required to identify directional environmental changes or to adequately assess responses to experimental manipulations or altered ecological drivers. This means that the initial responses to an experimental manipulation or change in environmental conditions may be quantitatively, or even qualitatively, different than longer-term responses. Some of the earliest studies at the KNZ site demonstrated this with respect to responses of plant productivity to fire, which drives many ecological processes in tallgrass prairie (Figure 21.1). Our long-term record clearly demonstrates that spring fires stimulate plant productivity, while results based on short-term studies would indicate that responses to fire can be positive, neutral, or even negative (Knapp, Blair, and Briggs 1998). Similarly, plant responses to fire vary in space depending on soil depth and availability of resources. Understanding ecosystem responses to fire also requires consideration of fire history (Blair 1997).

An important lesson from the LTER program is that the value of many experiments and data sets increases with the length of data collection. Some of our long-term experiments continue to produce unexpected results after a decade or more. This presents challenges, however, regarding how to balance the need to continue ongoing experiments and long-term data collection efforts and the need to identify new questions and conceptual frameworks for each renewal proposal. At the KNZ site, we have continued our core watershed-level fire and grazing experiments as well as measurements of responses

FIGURE 21.1 John Blair lighting a dormant season fire at the Konza Prairie Long-Term Ecological Research (KNZ LTER) site. Fire is integral to the ecology of mesic grasslands worldwide, and researchers at KNZ LTER program use fire at a whole-watershed scale to understand ecological responses to different fire return intervals and different seasons of fire. (Photo courtesy of John Blair.)

to fire, grazing, and climatic variability across multiple funding cycles. However, because human activities are directly (management of grazing and fire) and indirectly (changes in atmospheric chemistry and climate) altering the key drivers of ecological processes in these grasslands, our conceptual framework has shifted to a focus on critical global change issues, including the ecology of invasions, land-use and land-cover change, human activities and water quality, and ecosystem responses to climate change. We also added new experiments and measurements to address the mechanisms of response to critical global change drivers, while continuing to assess long-term changes in the characteristics that we began to measure over three decades ago. The view of the role humans as integral components of ecosystems has clearly changed over the 30 years that the LTER program has existed, and it is clear that humans have played, and continue to play, a very significant role in shaping the grasslands of North America. Although my research does not directly address socioecological questions, my appreciation of the direct and indirect effects of human activities has increased as a result of research in the LTER program.

COLLABORATION

Much of my research involves several investigators and is highly collaborative. Working with other investigators at an LTER site has revealed the importance of interdisciplinary research for understanding complex ecological phenomena. Likewise, participating in cross-site collaborations or having access to data from multiple sites, including those within and external to the LTER network, provides unique opportunities for comparative studies and synthesis. These opportunities are important for the development and testing of general ecological theory and broadening the scope of inference of site-based science. I try to address questions that have broad applicability to other grasslands and other ecosystems, or that contribute to the advancement of general ecological knowledge. Being associated with a network of LTER sites and participating in several regionalization efforts and cross-site projects has also heightened my appreciation for the importance of collaborative cross-site research and the establishment of comparable research approaches at multiple sites.

The LTER program fosters collaboration. The KNZ program uses the majority of LTER resources to support "core" long-term experiments and associated data sets that are broadly available to investigators within and outside of the LTER program. The colocation of sampling sites for a suite of ecological responses within the same experimental watersheds promotes collaboration. Long-term experiments also provide opportunities for new measurements and new collaborations. For example, a long-term fire and nutrient addition experiment at the KNZ site led to a project focused on the genetic and genomic basis of soil nematode responses to environmental stresses and involved scientists from multiple disciplines both inside and outside the LTER program (Jones et al. 2006). Many similar examples of interdisciplinary collaborations developed around projects at the KNZ site. The LTER program also provides unique opportunities for international collaborations. For example, a small LTER supplemental grant to initiate collaborations with scientists in South Africa led to an externally funded project to assess the ecological generality of responses to fire and herbivory in grasslands on different continents (Buis et al. 2009). The opportunities in these collaborations are manifold, including new insights into mechanisms underlying ecological dynamics, broader understanding of the consequences of those dynamics, and better understanding of the generality and limits of ecological knowledge gained in one ecosystem as applied to other ecosystems. The challenges associated with such collaborations include balancing time commitments among multiple projects, particularly when these projects are done as additions to ongoing research, and coordinating activities across multiple sites. Another challenge, especially for younger scientists, is ensuring that credit is given for participation in collaborative projects. Collaboration can help to develop a record of research productivity quickly and can foster understanding and appreciation of other disciplines and other research areas within ecology. However, many institutions place a higher value on independent investigator-driven research, and ensuring that graduate students, postdoctoral fellows, and junior scientists receive appropriate credit for their contributions to collaborative research can be a challenge.

APPLIED RESEARCH

My appreciation for applying ecological knowledge to address management and conservation-related goals has increased because of the LTER program. My initial attraction to ecology was based on wanting to know more about how the natural world works. Although I recognized that humans significantly alter ecological processes, my interests were focused on fundamental ecological questions and not on applied ecology. That began to change with my PhD research at the CWT LTER site, when I studied the impacts of forest management (clear-cutting) on invertebrate communities and decomposition processes, and it has continued with the KNZ program.

Research at the KNZ site focuses on ecological responses to fire and grazing, two factors that were historically natural disturbances but that now are widely used to manage grasslands. As a result, this research is highly relevant to the management and conservation of grasslands. Results from fire and grazing research at the KNZ site are published in applied scientific journals and incorporated into extension bulletins that target land managers. Research on the ecological role of fire and grasslands at the KNZ site has been presented to regional and national Environmental Protection Agency scientists, and it is being used to develop a regional smoke management plan that addresses conservation, agricultural productivity and human health concerns. Finally, research findings from the KNZ site are used by numerous groups focused on conservation or management-related issues (e.g., Natural Resources Conservation Service, The Nature Conservancy) and have been used in the development of management plans for other grassland sites, such as the Tallgrass Prairie National Preserve, as well as for several public education and outreach efforts, such as the natural history exhibits at the Flint Hills Discovery Center.

Restoration ecology has become an important component of my research as a result of the LTER program. Most of the original tallgrass prairie was plowed and converted to agricultural use, and these areas provide great opportunities to apply and test ecological theory in the context of tallgrass prairie restoration. We initiated the first restoration experiment at the KNZ site in 1998, with a focus on the role of soil resource availability and heterogeneity in the restoration of plant community structure and ecosystem processes. Restoration has since become a core part of our program, and it is being led by a former student (Sara Baer) who is now a core collaborator in the KNZ program.

COMMUNICATION

Interacting and collaborating with students and scientists having diverse research interests and backgrounds helps me communicate more effectively with other scientists and the public. I have become familiar with important issues and questions in areas of ecology and related disciplines beyond those associated with my own research. That breadth of knowledge and a more comprehensive perspective of how the pieces fit together have had a positive effect on communication with diverse audiences. I use many examples of experiments and findings from LTER sites in both my undergraduate ecology and graduate biogeochemistry classes. I helped develop a KNZ-based exercise in the Ecological Society of America's Teaching Issues and Experiments in Ecology series, which I use in my ecology class. Travel to other LTER and International LTER (ILTER) sites has provided opportunities to learn first-hand about the other biomes and ecosystems and to incorporate this knowledge into my courses. As principal investigator, I speak to a wide range of audiences about the objectives, activities and findings of the KNZ program. Those groups include scientists within and outside of ecology, graduate and undergraduate students, K–12 teachers, university administrators, state and federal government agencies personnel, and the general public.

MENTORING

I have had numerous opportunities to mentor new faculty colleagues and new staff members, local graduate and undergraduate students, and visiting students and scientists from other US and international institutions. Mentoring is important, particularly in the context of bringing new students and faculty members into the LTER program. The LTER program is different than other site-based programs and individual investigator grants, so it is important that new researchers understand the history and goals of the LTER program, as well as specific research initiatives at a given LTER site. Mentoring in association with the LTER program also includes topics such as the importance of data-sharing and adequate documentation and metadata to facilitate data use by the broader scientific community. Mentoring requires significant time and effort, but it also has significant rewards, including training students and colleagues to think broadly, to place their research into a larger context, and to value collaboration. For example, several former KNZ graduate students who are now faculty members at Kansas State University and at other institutions remain active in research at the KNZ site and are training their own graduate students there as well. Mentoring can lead to productive, long-term professional relationships. In turn, the LTER program provides opportunities to integrate new faculty members by allowing them to jump start their research programs and to bring new ideas and fresh perspectives. Mentoring also applies to training other faculty members to assume leadership roles in the LTER program, which is an important aspect of project management.

The LTER program has broadened the way that I evaluate faculty promotions, by increasing credit given for collaborative research efforts. It has also influenced my perspective on writing and reviewing proposals. I am particularly supportive of proposals that have clear short-term goals but that also are designed to address longer-term questions and to allow for new measurements or addition of new investigators. Many of the long-term experiments that have been initiated at the KNZ or other LTER sites have attracted additional researchers asking novel questions or adding measurements beyond those that originally motivated the study. Designing long-term experiments with adequate space to accommodate additional measurements or even new treatments can greatly increase the value of those projects and can lead to new synergistic collaborations. Such long-term large-scale experiments have also been valuable as a research platform for studies by graduate students and new faculty colleagues.

SKILL SET

Although the LTER program has broadened my research interests as well as the approaches and methods that I use, I cannot say that this has had a large, direct impact on my personal scientific tool kit. However, collaborating with scientists in different disciplines and with different skills and methodological approaches has broadened the kinds of research in which I participate. Research in the KNZ program has also led to the development of unique experimental approaches and associated infrastructure for conducting manipulations in the field, such as the development of modified rainout shelters to manipulate rainfall timing and amounts over intact grassland plots.

PERSONAL CONSEQUENCES

I have had the opportunity to visit many other LTER sites, as well as ILTER and other research sites in other countries (South Africa, Botswana, Namibia, China, and Uruguay). Hosting an LTER program at KNZ has also resulted in short and extended visits to the Konza site from scientists and students from many different countries and cultures. I have met a much broader and more diverse range of scientists and students than I would likely have met otherwise. Many of these colleagues are good friends, as well as important professional contacts. In that sense, LTER networking has had a significant impact on both my professional and personal contacts. Traveling to other sites and other ecoregions also has broadened my understanding of the natural history and ecology of those regions, as well as the cultures of people in those areas.

Although participating in the LTER program has benefited my professional development in many ways, there are trade-offs associated with time and effort required to manage an LTER program and advancing one's personal research career versus facilitating the research activities of others. Serving as principal investigator of an LTER program requires a significant allocation of time for preparing proposals and reports, planning and coordinating activities, responding to requests for information, managing budgets and overseeing personnel, and other nonresearch activities. Those are important activities for the KNZ program and for the LTER program as a whole, but they take time that could otherwise be spent on expanding my personal research program. However, on balance,

I think the positive aspects of contributing to the success of a major research initiative such as the LTER program outweighs the negative aspects of reduced time for other activities.

CHALLENGES AND RECOMMENDATIONS

Junior colleagues should be aware of the ways that collaborative and independent research are valued at their institutions and adjust their roles in the LTER program accordingly. Where appropriate, use of small grants as "seed funding" could jump start the research programs of new faculty members to integrate them in the LTER program. With respect to leading an LTER program, a distributed model based on shared intellectual input, distributed leadership, and group decision-making has worked well for the KNZ site. However, additional resources for day-to-day management of the program would help alleviate the dichotomy between management and personal research. In general, more network-level resources for graduate student and postdoctoral working groups and cross-site or synthesis projects would facilitate collaboration and network-level science.

CONCLUSION

The LTER program has changed the way that generations of ecologists think about and study a broad range of ecological phenomena. It has contributed to the growing recognition that many of the most important ecological questions, and ecological problems facing society, require collaborative, long-term efforts. Growing up academically in the LTER network has provided a unique perspective on the costs and benefits of collaborative research. Participating in research in the LTER program as a graduate student can be a bridge to lifelong contacts that enrich one's personal and professional lives. Depending on one's career goals and institutional standards of evaluation, it may be important to develop both independent and collaborative research programs, or to take a leadership role in some collaborative efforts. Although there are some trade-offs between directing a large research program and advancing one's personal research, the rewards of long-term collaboration are substantial.

ACKNOWLEDGMENTS

My perspectives about the value and challenges of long-term research were conceived during my time as a graduate student with the CWT LTER program and nurtured and refined with my participation in the KNZ LTER program. I am indebted to D. Crossley, Jr., for my initiation into the LTER program and for setting me on this path as a graduate student. I am also grateful for the mentoring and support I received from former and current KNZ principal investigators and co–principal investigators as my role in the KNZ LTER program grew, as well as for the many contributions of colleagues and students that have sustained this research program for more than three decades. KNZ research is currently supported by NSF grants DEB-0823341 and DEB-1440484.

REFERENCES

Blair, J.M. (1997). Fire, N availability, and plant response in grasslands: A test of the transient maxima hypothesis. *Ecology*, **78**, 2359–2368.

Blair, J.M., and Crossley, D.A., Jr. (1988). Litter decomposition, nitrogen dynamics and litter microarthropods in a southern Appalachian hardwood forest eight years following clearcutting. *Journal of Applied Ecology*, **25**, 683–698.

Buis, G.M., Blair, J.M., Burkepile, D.E., Burns, C.E., Chamberlain, A.J., Chapman, P., Collins, S.L., Fynn, R.W.S., Govender, N., Kirkman, K., Smith, M.D., and Knapp, A.K. (2009). Controls of aboveground net primary production in mesic grasslands and savannas: An interhemispheric comparison. *Ecosystems,* **12**, 982–995.

Jones, K.L., Todd, T.C., Wall-Beam, J.L., Coolon, J.D., Blair, J.M., and Herman, M.A. (2006). Molecular approach for assessing responses of microbial-feeding nematodes to burning and chronic nitrogen enrichment in a native grassland. *Molecular Ecology,* **15**, 2601–2609.

Knapp, A.K., Blair, J.M., and Briggs, J.M. (1998). Long-term ecological consequences of varying fire frequency in a humid grassland. In T.L. Pruden and L.A. Brennan, eds. *Fire in Ecosystem Management: Shifting The Paradigm From Suppression to Prescription,* pp. 173–178. Tall Timbers Research Station. Tallahassee, Florida.

Knapp, A.K., Briggs, J.M., and Smith, M.D. (2012). Community stability does not preclude ecosystem sensitivity to chronic resource alteration. *Functional Ecology,* **26**, 1231–1233.

Ratajczak, Z., Nippert, J.B., and Collins, S.L. (2014). Abrupt transition of mesic grassland to shrubland: Evidence for thresholds, alternative attractors, and regime shifts. *Ecology,* **95**, 2633–2645.

22 Growing Up with the Konza Prairie Long-Term Ecological Research Program

Alan K. Knapp

IN A NUTSHELL

1. As someone who began working at a Long-Term Ecological Research (LTER) site prior to beginning his PhD studies, there is little doubt that the LTER program has been a major influence on all aspects of my scientific career.
2. Working within the LTER program has provided me with great appreciation for the power of collaboration, large-scale and long-term experiments, and cross-disciplinary interactions.
3. Scientists within the LTER network are among the most successful and influential in the world, and thus associating with them has many positive professional and personal consequences. Among the most valuable professional benefits are opportunities for exposure to ideas well beyond what a scientist experiences in a more typical

research environment and the opportunity to collaborate and publish with scientists who are leaders in fields other than his or her own.

PERSONAL OVERVIEW

My experience with the LTER program began in January 1982 with my employment at the Konza Prairie site (KNZ) in northeastern Kansas. I had recently completed an MS (in botany with a focus on subalpine plant ecophysiology) at the University of Wyoming, and I knew nothing about the new (at the time) LTER program. But at the urging of a fellow graduate student, Don Young (who eventually took a position at Virginia Commonwealth University and has long been involved with the Virginia Coast Reserve site), I applied for a research assistant position advertised in *Science*. This position description specifically highlighted that skills and experience were needed in abiotic measurements (i.e., installing a weather station and precipitation gauge networks and taking charge of monitoring climatic variables); these were tasks with which I had familiarity as part of my graduate program. As a lifelong resident of the western third of the United States and a fan of the mountains (often openly speaking negatively about grasslands!), I was not keen to even consider a position in eastern Kansas. But Don Young was an effective advocate and stressed the importance of keeping an open mind, something I try to stress with my students today. After presenting my research at the meeting of the Ecological Society of America in 1981, Don and I and a few other graduate students stopped in Manhattan, Kansas, as we drove cross-country from Bloomington, Indiana, to Laramie, Wyoming. Richard "Dick" Marzolf, a limnologist by training, was the principal investigator of this new LTER site, and he took us on a tour of KNZ. Although I had met many of the newly formed KNZ research group at the meeting (Tim Seastedt, Ted Evans, Martin Gurtz), this visit to the prairie, which can only be described as spectacular (it was August and the flowering stalks of the grasses were unusually tall and dense that year), sold me on moving to this ecosystem type. I was also sold on the position by (1) Dick Marzolf's promise of providing time and resources to pursue independent research in addition to the "technician" duties I needed to fulfill, and (2) the general excitement of a new site, a new program, and several young scientists all converging. Finally, given my personal uncertainty about continuing on in a PhD program at the time and a dismal job market for new PhDs, this opportunity to take a break from studies and gain research experience in a new environment—and double my graduate student salary—seemed too good to pass up. I certainly had no idea that this decision would shape the rest of my career.

Dick Marzolf was true to his word with regard to providing research time and resources. In fact, although I had an MS degree and all others hired anew at KNZ (and Kansas State University) were postdoctoral fellows or assistant professors, I felt as if I were treated and regarded as a postdoctoral fellow—with the expectation that I would conduct research both independently and collaboratively. Thus, based on research conducted during this time, I published 15 papers—made possible by tremendous support from Tim Seastedt, Dick Marzolf, and Lloyd Hulbert. This certainly increased my confidence as a scientist. And after a little more than $3\frac{1}{2}$ years with the KNZ program, I was excited to return to graduate school at the University of Wyoming (and the mountains) to pursue a PhD, never expecting to live in Kansas again. But during the 3 years that I was attending graduate school, Lloyd Hulbert fell ill and passed away. Lloyd had a tremendous influence on me (and many others) and was instrumental in the success of KNZ program, securing the site and leading the development of the watershed-level experimental design using fire and grazing, which is still in place today. Kansas State University advertised for an assistant professor position as I was finishing my dissertation, and I applied. When I was invited for an interview, I was convinced that this invitation was a goodwill gesture by my former colleagues to provide me with some much needed job interview experience. Much to my surprise, I was offered the position and began my career as a university professor in 1988. The KNZ program became my focus again almost immediately and I became the principal investigator in 1991 when Tim Seastedt, who was then an associate professor at Kansas State University and the principal investigator at KNZ site, was recruited to the University of Colorado to serve as principal investigator of the Niwot Ridge LTER site. I think it is fair to say that the LTER program has completely dominated my research career ever since.

I remained at Kansas State University until 2004, serving as principal investigator of KNZ through two 6-year funding cycles. During this time, three accomplishments were noteworthy with regard to their impact on my career as a scientist. First, we published the first site-synthesis volume (Knapp et al. 1998) in the newly commissioned Oxford University Press LTER Book Series. This cemented my transition from a leaf-level ecophysiologist to a grassland ecosystem ecologist. This book also was the first step toward fulfilling the vision of Tom Callahan at the National Science Foundation (NSF) of having such a volume from each site on his bookshelf. Today, about half of the sites in the LTER network have such volumes published (with Oxford or other publishers) and, perhaps not coincidently, I serve as the chair of the Publications Committee for the LTER network. Second, in 2000, Melinda Smith and I began conducting a comparative analysis of data on the patterns and controls of aboveground net primary productivity (an LTER core area) from the 11 LTER sites with the longest data sets (Knapp and Smith 2001). Although certainly not the first multisite LTER data synthesis article, this research was successful in demonstrating how a network-level approach to ecological research could provide new insights into fundamental ecological processes. Furthermore, this analysis strongly influenced my future research approaches. Today I find ecological research to be far more compelling when conducted in a comparative manner, including multiple sites and biomes rather than being focused on a single site. Third, during this period John Blair was recruited to Kansas State University and transitioned to assume principal investigator status during the 1997 to 2003 LTER funding cycle. Transitions in leadership of research programs expected to operate for decades were recognized early as important for the success of the LTER model. Kansas State University showed great foresight in anticipating this need and making this hire. Today, few would disagree, either locally or at the network level, that John Blair has been one of the strongest leaders and most successful of LTER principal investigators to date.

In 2004, I left Kansas State University to take a position as the senior ecologist in the Graduate Degree Program in Ecology at Colorado State University. But I have maintained my research interests in grasslands as well as many projects at KNZ. I actively collaborate and publish with Kansas State University ecologists and others that work at KNZ and many of the graduate students I advise work there as well.

APPROACH TO SCIENCE

I have always been interested in basic and fundamental ecological questions. My experience in the LTER program has not changed this and indeed, the breadth of the Program has allowed me to stay the course with this interest. My early experiences in LTER program gave me a great appreciation for the power of collaboration, large-scale and long-term experiments, and cross-disciplinary interactions. My graduate training in a botany department tended to be more focused and strongly experimental (short-term), and I was certainly comfortable with this mode as well. After two decades in the LTER program, I became distinctly more synthetic, theoretical, and strongly comparative in my experimental approaches. These shifts were driven in part by the collective value that the LTER program (including the scientists and the NSF) placed on synthesis and theory, and partly by a natural interest in assessing the generality of knowledge gained in an ecosystem that I had studied intensely for 20 years. My appreciation of human impacts on and interactions with ecological systems has certainly increased over the years as well—but I think that this is true of most ecologists and is not a consequence of involvement in the LTER program.

ATTITUDES TOWARD TIME AND SPACE

My graduate training and early career interests were at the organismal level and my professional identity, albeit nascent, was as a plant ecophysiologist whose spatial scale of interest varied from the stomata to the leaf, with a temporal scale of seconds to minutes to at most a growing season. After more than 20 years of involvement in the LTER program, my professional identity is as an ecosystem ecologist with a spatial scale of interest that extends to the globe with temporal scales ranging from the growing season to decades. The LTER program is 99% responsible for that transformation, although it is difficult to know whether or not I would have remained a leaf-level ecophysiologist had I not been involved in the LTER program.

It is impossible to collect long-term data and not learn to appreciate the value of such data sets. Because monitoring is not typically considered to be a fundable research approach, despite the fact that long-term experiments and studies require consistent and usually continuous data collection, new questions and a periodic recasting of an LTER site's research within a fresh conceptual framework is required. This is really not a difficult task. Indeed, a rich and diverse array of long-term data sets make it much easier to do this than completely re-inventing new research projects every 3 to 6 years within the context of more traditional funding programs at agencies such as the NSF, the US Department of Agriculture, or the US Department of Energy.

COLLABORATION

My experiences with collaboration in the LTER program have mostly been positive. There have been exceptions, where impatience on my part or a lack of communication among colleagues has led to disagreements and some periods of unhappiness. But these were all resolved in short order. To elaborate, my impatience as a scientist (and person) has probably caused me the most frustration as a collaborator. Indeed, one has to learn to accept that when working with groups, there will always be some or several colleagues who will not be able to make progress at the rate that you would expect. Working alone alleviates this problem, of course, but the power of collaboration for me is that the collective skill set that can be brought to bear on a problem is much greater than that which I possess as an individual. Dealing with this frustration is the only real option. Over the years, I have become almost dependent on collaborations, but I have also become much more selective in who I collaborate with (personality compatibility and complementary skill sets are critical for how I select potential collaborators now).

APPLIED RESEARCH

My experience in the LTER program is completely responsible for shifting my research career from mountainous habitats to grasslands. Because grasslands are so strongly managed and valued for forage production for domestic livestock, it is relatively simple and prudent to place the results of even the most basic research in the context of this dominant agricultural land use. I seriously doubt that such involvement with ecosystem services would have happened if I had continued studying basic ecophysiological processes in herbaceous plants in subalpine environments.

COMMUNICATION

The most obvious impact of the LTER program on my teaching has been by increasing my breadth of experience. Having attended dozens of science-based LTER meetings and having visited most LTER sites and listened to countless research presentations across many disciplines, I believe that my involvement in the LTER program has offered invaluable educational opportunities for me, opportunities that I have incorporated into the classes that I teach.

MENTORING

Because most of my graduate students work at LTER sites, my LTER experiences strongly affect how I guide students through their graduate research projects. I do this in two ways. First, I encourage students to write one chapter of their dissertation based primarily on long-term data that are collected at one or more LTER sites. It is usually not difficult to use core data to address specific questions relevant to the student's research project. Second, I insist, where feasible, that short-term student projects make use of long-term data to provide context for their results. In addition, I actively promote collaboration with other graduate students in the LTER program and try to provide support for students that begin such collaborations early in their careers.

My appreciation of the collaborative approach for conducting science, gained through the LTER program, has strongly influenced the ways in which I evaluate progress of junior colleagues (tenure and promotion decisions) and applicants for faculty positions. Compared to many of my colleagues, I tend to be much less stringent in my expectations of junior faculty to be the lead author on papers or the lead principal investigator on grants. The most complex environmental problems we must tackle require large multidisciplinary teams to be successful. But in large teams, most collaborators cannot be lead authors or lead principal investigators, and we cannot discount the accomplishment of team members for this reality. If we do, then there is less incentive for such teams to form. Thus, my experiences in the LTER program have taught me to relax the traditional expectations of independence and primary leadership in evaluating colleagues and the way that we train students. A certain degree of independence and leadership will always be necessary to demonstrate, but I require less than most of my colleagues. I credit strongly young faculty that can fund their laboratories through collaborative grants (even if they are never the principal investigator).

SKILL SET

Having served as principal investigator of an LTER site, my initial perspective on moving into administrative roles has been solidified. Directing the science and providing some vision at an LTER site was a challenge that I enjoyed, even if not always entirely successful, but my experiences also confirmed to me that I am poorly trained in the skills necessary to manage people. I think my time as a principal investigator improved those skills, but probably like most principal investigators in the LTER program, it was the science that drew most of my attention.

In some sense, my experiences in the LTER program may have stunted the growth of my scientific tool kit. Being so strongly collaborative with colleagues who have skill sets that are very different from mine, has led me to be familiar with and appreciate tools from many fields inside and outside of ecology. But I have mastered few of these, preferring instead to have both access to these tools *and* the insights my colleagues provide about those tools, rather than learning and using them myself.

PERSONAL CONSEQUENCES

There is a certain camaraderie inherent in the LTER program, particularly among those that have spent more than a decade or two in it. Because many scientists in this cohort are among the most successful and influential in the world, there are professional and personal consequences—all positive—for being a part of this "club." Chief among the consequences that can be attributed to an enduring association with the LTER program is an increase in scientific self-confidence and exposure to ideas well beyond what one experiences in an academic department or at a meeting where one inevitably listens to presentations in one's own subject area. But perhaps most important professionally is the opportunity to collaborate and publish with scientists who are leaders in the field. My experiences in the LTER program have certainly provided opportunities for travel abroad through the International LTER. Some of these travels have led to long-term collaborations in South Africa and China, with a subsequent strong appreciation for their cultures.

CHALLENGES AND RECOMMENDATIONS

I believe that the LTER program today is generally and broadly considered a success, and thus is attractive to young (and old) scientists, both of whom will jump at the chance to become involved. So attracting "new blood" to the network is not a problem. The major problem today is the limited resources available to support this new blood. Although new and fresh ideas are certainly needed, so are the collective experiences and insights that can only be provided by those with a long-term association with sites and the culture of the LTER program. How to balance both of these needs, with few if any new resources, is a major impediment to intellectual growth of the network. Although the refrain "we need more money" is certainly not new, in this case some funding mechanism is needed that can either (1) target early-career scientists to ensure that their addition to the roster of scientists at LTER sites does not displace existing team members or (2) support late-career scientists in a way that values their more synthetic and advisory roles, which can be essential for catalyzing the careers of young scientists.

CONCLUSION

As someone who installed the meteorological station array of rain gauges at a new LTER site in the early 1980s, as well as attending some of the very first network and All Scientists meetings (when the LTER network was fewer than 12 sites), it is difficult to gauge the impact that the LTER program has had on my success as a scientist. I consider myself very fortunate to have had the opportunity to grow with the LTER program as well as being a result of the program. Students and young scientists joining the "LTER club" today face a very different set of circumstances and challenges than I did decades ago. But although the opportunities for success are different when joining a mature program compared to joining one that is finding its way, I believe that affiliating with today's

LTER program is a wise career move. However, as Tom Callahan noted in 1984, "neither NSF nor the LTER investigators intend to make LTER data the exclusive province of scientists associated with the LTER projects. In fact, the intent is exactly the opposite, and it is hoped that the scientific community at large will come to regard the data sets as valuable resources" (Callahan 1984). Thus, scientists, students, and educators today need not be affiliated with an LTER site to benefit from the program; this represents a broader impact that all in the LTER network appreciate. Furthermore, Callahan predicted that, "in the future ecologists who wish to perform broad comparisons of ecosystems attributes would base a major proportion of the work on existing LTER data sets." Within 20 years, Callahan's vision has certainly been realized in my own work (Knapp and Smith 2001, Knapp et al. 2008) as well as in the research of many others (e.g., Jackson et al. 2002; Huxman et al. 2004; Parton et al. 2007).

ACKNOWLEDGMENTS

My continued LTER involvement at the KNZ site is supported by NSF grants DEB-0823341 and DEB-1440484. I am grateful to the past and present KNZ scientists and the dedicated Konza Prairie Biological Station staff for their combined efforts in maintaining a world-class ecological research program.

REFERENCES

Callahan, J.T. (1984). Long-Term ecological research. *BioScience* **34**, 363–367.

Huxman, T.E., Smith, M.D., Fay, P.A., Knapp, A.K., Shaw, M.R., Loik, M.E., Smith, S.D., Tissue, D.T., Zak, J.C., Weltzin, J.F., Pockman, W.T., Sala, O.E., Haddad, B.M., Harte, J., Koch, G.W., Schwinning, S., Small, E.E., and Williams, D.G. (2004). Convergence across biomes to a common rain-use efficiency. *Nature*, **429**, 651–654.

Jackson, R.B., Banner, J.L., Jobbágy, E.G., Pockman, W.T., and Wall D.H. (2002). Ecosystem carbon loss with woody plant invasion of grasslands. *Nature*, **418**, 623–626.

Knapp, A.K., Briggs, J.M., Collins, S.L., Archer, S.R., Bret-Harte, M.S., Ewers, B.E., Peters, D.P., Young, D.R., Shaver, G.R., Pendall, E., and Cleary, M.B. (2008). Shrub encroachment in North American grasslands: Shifts in growth form dominance rapidly alters control of ecosystem carbon inputs. *Global Change Biology*, **14**, 615–623.

Knapp, A.K., Briggs, J.M., Hartnett, D.C., and Collins, S.L. (1998). *Grassland Dynamics: Long-Term Ecological Research in Tallgrass Prairie.* Oxford University Press, New York.

Knapp, A.K., and Smith, M.D. (2001). Variation among biomes in temporal dynamics of aboveground primary production. *Science*, **291**, 481–484.

Parton, W., Silver, W.L., Burke, I.C., Grassens, L., Harmon, M.E., Currie, W.S., King, J.Y., Adair, E.C., Brandt, L.A., Hart, S.C., and Fasth, B. (2007). Global-scale similarities in nitrogen release patterns during long-term decomposition. *Science*, **315**, 361–364.

23 Born and Bred in the Long-Term Ecological Research Network

Perspectives on Network Science and Global Collaborations

Melinda D. Smith

IN A NUTSHELL

1. I am a plant community and ecosystem ecologist who has conducted research within the context of the Long-Term Ecological Research (LTER) network from the beginning of my scientific career, now almost two decades ago. My research has benefited greatly from site-based research at the Konza Prairie (KNZ) LTER site, as well as

from network-level syntheses utilizing data sets and knowledge produced by the collective of LTER sites.
2. My involvement in the KNZ LTER site, in particular, has shown me the strength of conducting site-based research, yet my involvement in synthesis activities within the LTER network and beyond has illuminated the limitations of site-based research for addressing cross-site comparative research. To this end, I have been and continue to be a strong proponent of highly coordinated, multisite experiments, and much of my research is comparative in nature.
3. Being involved in the LTER network from the start of my research career has made me a scientist who is well aware of the benefits and power of collaborative, multidisciplinary research. Because of the benefits and breadth of experiences that I have received from such research endeavors, I encourage my graduate students and postdoctoral fellows to also become involved in such research, and I recognize the positive impact collaborative, multidisciplinary research can have on beginning investigators.
4. I believe that individuals outside of the LTER network (ranging from established principal investigators, to young investigators, to graduate students) are often not fully aware of the benefits of being involved in the LTER network or of the advances in ecological understanding that it has made possible. Thus, there is a need for the LTER network to be more proactive and creative in the ways that it attracts new researchers to get involved in the site-based or network-level research. Ultimately, the LTER network will only benefit from increased involvement by new investigators, who also could serve the role of leading the LTER network in the future.

PERSONAL OVERVIEW

I have been affiliated with the LTER program since beginning as a graduate student at Kansas State University. My master's research was carried out at the Konza Prairie Biological Station, which was among the first group of LTER sites funded by the National Science Foundation (NSF). The KNZ LTER site is managed by the Division of Biology at Kansas State University, and there was and continues to be strong encouragement in the department for graduate students to conduct ecological research within the context of KNZ. It was during my time as a master's student that I began to realize the benefits of conducting research at an LTER site (i.e., extensive knowledge of the site, existing research infrastructure and experimental platforms, long-term data sets, logistical support). At the time, KNZ already had been in existence for more than a decade. With this knowledge in hand, I chose to conduct my PhD research at KNZ as well, and I set out to take full advantage of the benefits of a LTER site to enhance my own research. For example, I used long-term data sets and existing long-term experiments to address questions central to my dissertation research (e.g., Smith and Knapp 1999; Smith and Knapp 2001). Also, I worked with my advisor to compile long-term data sets from multiple LTER sites to assess relationships between precipitation variability and productivity across a broad range of ecosystems (Knapp and Smith 2001) (Figure 23.1). I benefited greatly as a young investigator from being involved in KNZ, and I continue to benefit now as an associate professor from my involvement with the site—and more generally the LTER network. All of my research, which aims to understand factors and processes that shape plant communities and ecosystem structure and function, in some way relies on LTER sites or products (data sets). Thus, when I consider how my experiences in the LTER program have influenced me as a scientist, it is not necessarily a matter of how the LTER program

FIGURE 23.1 Plant species composition data collection at the Extreme Drought in Grasslands (EDGE) experiment located at the Shortgrass Steppe Long-Term Research Ecology site. (Photo courtesy of Alan Knapp.)

has changed my perspective or how I do science, but rather how would these factors have been different had I not been involved in the LTER program.

APPROACH TO SCIENCE

The LTER network has shown me the strength of conducting site-based research, yet my involvement in synthesis activities within the LTER program and beyond (e.g., National Center for Ecological Analysis and Synthesis [NCEAS]) has illuminated the limitations of site-based research for addressing cross-site comparative research. Thus, while I continue to be very interested in asking site-based questions, my involvement in the LTER program has made me intrigued to pursue whether the dynamics and mechanisms identified as important at KNZ or at any other LTER sites are important at other sites within the network or in other ecosystems. Thus, one of the main ways the LTER program has affected me is through the recognition of the need for comparative research to address questions related to the impacts of global changes on ecosystem structure and function (Smith, Knapp, and Collins 2009; Fraser et al. 2013). Although we can begin to see tantalizing differences and similarities among LTER sites (and more generally among different ecosystem types) in their responses to different global change drivers, what remains to be tested is whether LTER sites (or ecosystem types) will differ in their sensitivity to similar manipulations of global changes, such as climate change and nitrogen deposition. To this end, I am a proponent of highly coordinated, multisite experiments (e.g., the Nutrient Network; Smith, Knapp, and Collins 2009; Fraser et al. 2013), and much of my research is comparative in nature. This comparative, highly coordinated, network-level research involves conducting experiments and studies using identical protocols and making measurements using identical methodologies, with the ultimate goal of allowing seamless integration of data across sites and experiments (Fraser et al. 2013). Along with this notion is the recognition that young scientists and students will need to be involved in such research to make it vibrant, and that these young investigators have the potential to reap large rewards from comparative, networked science.

ATTITUDES TOWARD TIME AND SPACE

When considering time and space, I would say that my involvement in the LTER program has most strongly affected my attitude toward time and its role in ecological studies. It is clear from the body of research from the LTER program that has emerged over the past 30 years that a long-term perspective is often critical to fully capturing dynamics and elucidating drivers of change (Knapp et al. 2012). Indeed, much of my research has a longer-term perspective. Many experiments and studies that I have conducted (or am currently conducting) are longer term (> 5 years in duration) than typical studies (1–3 years in duration). Moreover, the conceptual frameworks that I base much of my global change research on have an explicit long-term element to them (i.e., Smith, Knapp, and Collins 2009; Smith 2011). Fortunately, my involvement with KNZ has allowed me to incorporate a long-term perspective in much of my research and provided assistance to maintain such research. Although I have sought, when relevant, to continue studies over the long-term with additional extramural funding, when such funding is not available I make every effort to continue the research by involving students and postdoctoral fellows in my laboratories in these studies. By doing this, I expose them to the benefits of long-term research, as well as provide them with potential research platforms (i.e., long-term experiments that can be used to address additional questions) to carry out their own independent research in the future.

COLLABORATION

As a scientist in the LTER program, I have had the opportunity to become involved in multiple types of collaborations. These have included externally funded collaborative experiments involving scientists in my own discipline, as well as those outside my area of expertise. An example of the former is a collaborative experiment funded by the NSF Ecosystems Program that we (Alan Knapp, Scott Collins, John Blair, and I—all investigators in the LTER program) conducted at KNZ and at Kruger National Park in South Africa to assess the impacts of alterations in fire regimes and loss of large herbivores on plant community and ecosystem function and structure. An example of a collaborative study with investigators representing multiple disciplines is a multi-investigator study funded by the Department of Energy's Terrestrial Ecosystem Science Program, which aimed to link genomic to ecosystem level responses to climate change at KNZ.

Other collaborative experiences have included working groups aimed at synthesizing data or scientific knowledge from the LTER network (e.g., Knapp et al. 2012; Smith et al. 2015), as well as working groups at the NSF-funded NCEAS (e.g., Knowledge Network for Biocomplexity working group) and National Evolutionary Synthesis Center (NESCent; e.g., Origins of C4 Grasslands working group). What these experiences have taught me is that much can be gained by collaborating with other researchers and that my science is almost always improved by the different perspectives that collaborators can bring to a project. Despite these positive aspects, there are challenges associated with any collaboration. For example, not all partners in collaborations participate equally, and there can be difficulties with communication across disparate disciplines (ecological genomics vs. ecosystem ecology and even ecology vs. evolution!). What I have learned from my experiences over time is that collaborations and collaborative partners must be carefully selected and that these endeavors can be more challenging but even more rewarding than individual-investigator research.

APPLIED RESEARCH

Overall, I would say that my experiences in the LTER program have definitely affected my appreciation of applied science. Although much of my research addresses fundamental ecological questions (e.g., factors that shape community structure and drive dynamic relationships between community and ecosystem structure and function), this quest for theoretical understanding is often couched within the context of global environmental change (Smith, Knapp, and Collins 2009). For example, some of the recent research in my laboratory has focused on the impacts of climate changes and climate extremes (e.g., drought, heat waves on plant community structure and ecosystem function), as well as on how eutrophication (increased deposition on nitrogen and phosphorus) may have an impact on plant community change and ecosystem processes. This incorporation of ecological theory into applied research focused on the impacts on global environmental change was strongly influenced by my involvement in KNZ, which has since its inception taken this same perspective.

COMMUNICATION

My experiences in the LTER program have provided me with a broader perspective ecologically than if I had not been involved in it. For example, I often look beyond the site where I conduct much of my research (KNZ—the tallgrass prairie ecosystem) to how the responses and dynamics I observe there compare to other sites or ecosystems, both within the LTER network and beyond. This comparative perspective comes out in my research (two of the major projects in my laboratory are explicitly comparative), as well as in my teaching, where I often use comparative examples as teaching tools. My involvement in the LTER program also has allowed me to expand the broader impacts of my research through the NSF-funded LTER Schoolyard and Research Education for Undergraduates activities, whereby I have the opportunity to have my research conveyed to a much broader audience than I could ever reach on my own and where opportunities for training young scientists is expanded.

MENTORING

To date, all of my graduate students and postdoctoral fellows have been involved in research at LTER sites, particularly at KNZ. This pattern occurs, in part, because I conduct much of my research at KNZ, as well as at several other LTER sites. Therefore, my students are introduced early on to the LTER network. But I also believe that they conduct research at LTER sites because of the many benefits of working at these sites: long-term data, extensive site knowledge, existing experimental platforms, and logistical support. This support allows the students to build on previous research and exploit existing collaborations and ongoing studies. Because of the benefits and breadth of experiences that I have received from collaborative and multidisciplinary research, I encourage my graduate students and postdoctoral fellows to also get involved in such research. As a result, most of my students and postdoctoral scholars have augmented their own research with collaborations with other LTER-affiliated graduate students or investigators. Furthermore, nearly all have been involved in or have led LTER working groups and other collaborative efforts, such as symposia or organized oral sessions at annual meetings of the Ecological Society of America. In all cases, I believe that the students and postdoctoral fellows have benefited from these experiences, have experienced the challenges of such efforts, and will carry these perspectives forward as they become established scientists.

My extensive exposure to collaborative and multidisciplinary research also has made me more open to appreciating this kind of research from other investigators. It has also made me aware of how important this type of research can be for addressing novel questions. Thus, I think that I am more receptive to multiauthor papers, to young investigators being involved in such research, and for new research that involves multiple types of investigators.

SKILL SET

This is a difficult issue to address, given that I have been involved in the LTER program since the start of my research career. However, my involvement in the LTER program has likely affected my scientific tool kit by exposing me to data synthesis opportunities that utilized multiple data sets from KNZ, as well as data sets from multiple LTER sites. It is through my involvement in these activities that I honed my data management and documentation abilities and statistical skills. However, it is not easy to say, given my general interest in statistics, whether or not I would have worked to build these skills over time in the absence of involvement with the LTER program.

PERSONAL CONSEQUENCES

It is difficult to gauge the full impacts that the LTER program has had on me, given my involvement from the start of my research career. But if I were to step back and ask myself whether I would likely be doing things differently or be collaborating with different people had I not been involved in the LTER program, I would say the answer is "yes." I conduct a majority of my research at LTER sites, most of the people I collaborate with are LTER investigators, and many of my friends are either products of the LTER program or are current LTER investigators. It is likely that I would be friends and collaborate with a different circle of people had I not been involved with the LTER program, yet my circle is larger and more diverse and dynamic as a consequence of being involved. Furthermore, my success as a scientist can be attributed to a large degree to my involvement in the LTER program over the past two decades. In addition, my involvement in the LTER program has facilitated my pursuit of different research avenues, in part through the fact that KNZ (and more generally the LTER network) provides an excellent platform for conducting novel research, as well as through the support provided by the LTER network for working groups and other data synthesis activities.

CHALLENGES AND RECOMMENDATIONS

Individuals outside of the LTER network (ranging from established principal investigators, to young investigators, to graduate students) are often not fully aware of the benefits of being involved in the LTER program or of the advances in ecological understanding that it has made possible. There is clearly a need for increased communication by scientists in the LTER network with the broader scientific community. However, I also think that the LTER program may be perceived by many scientists as a closed community, with benefits reaped only by those investigators affiliated with LTER sites. Such ideas arise, in part, from what I have heard from investigators not involved in LTER research. Some of these perceptions could be overcome by increasing access to LTER data sets and publicizing these datasets for use by the broader scientific community. In addition, mechanisms for investigators not affiliated with the LTER program to start conducting research at LTER sites could be made available, such as funding specifically targeted toward developing new research projects within the LTER network or new data synthesis activities utilizing LTER data sets. Importantly, these opportunities would not only be made available to the LTER community (i.e., via supplemental funding requests) but also could be open for competition by the broader scientific community. These initiatives could potentially facilitate new collaborations among investigators currently affiliated with LTER sites and those investigators that have no experience with the LTER program, thereby providing a new avenue for recruiting "new blood" into the LTER network.

CONCLUSION

Being involved in the LTER program from the start of my research career has made me well aware of the benefits and power of collaborative, multidisciplinary research. LTER sites embody these traits, involving investigators from multiple disciplines working together toward common research goals. All of my current research is collaborative in nature, often involving multiple disciplines, and I believe that it benefits greatly from this. Furthermore, some of my early experiences with conducting comparative and synthetic research were only made possible by the LTER network (e.g., Knapp and Smith 2001), either through the wealth of data readily available via the LTER network or through my involvement in LTER-funded working groups focused on data synthesis activities. I continue to infuse these approaches in my research and have evolved from participating in such efforts to leading these research endeavors within the LTER network. I encourage young investigators to take advantage of the LTER network and other existing (e.g., the Nutrient Network) or emerging networks (e.g., National Ecological Observatory Network, Drought-Net); the benefits for individual research programs and for developing lifelong collaborations can be quite significant. However, to continue to enlist the next generations of researchers in such networks will require communication by the networks as well as recognition by the scientific community and academic administrations that participation in such networks is an important aspect of any successful scientist's research portfolio.

ACKNOWLEDGMENTS

Thanks to the Konza Prairie Biological Station and staff for their invaluable support of numerous research projects over the years. Thanks also for support from the NSF-funded KNZ LTER program.

REFERENCES

Fraser, L.H., Henry, H.A.L., Carlyle, C.N., White, S.R., Beierkuhnlein, C., Cahill, J.E., Jr., Casper, B.B., Cleland, E., Collins, S.L., Dukes, J.S., Knapp, A.K., Lind, E., Long, R., Luo, Y., Smith, M.D., Sternberg, M., and Turkington, R. (2013). Coordinated distributed experiments: An emerging tool for testing global hypotheses in ecology and environmental science. *Frontiers in Ecology and the Environment*, **11**, 147–155.

Knapp, A.K., and Smith, M.D. (2001). Variation among biomes in temporal dynamics of aboveground primary production. *Science*, **291**, 481–484.

Knapp, A.K., Smith, M.D., Hobbie, S.E., Collins, S.L., Fahey, T.J., Hansen, G.J.A., Landis, D.A., La Pierre, K.J., Melillo, J.M., Seastedt, T.R., Shaver, G.R., and Webster, J.R.. (2012). Past, present and future roles of long-term experiments in the LTER Network. *BioScience*, **62**, 377–389.

Smith, M.D. (2011). An ecological perspective on extreme climatic events: A synthetic definition and framework to guide future research. *Journal of Ecology*, **99**, 656–663.

Smith, M.D., and Knapp, A.K. (1999). Exotic plant species in a C_4-dominated grassland: Invasibility, disturbance, and community structure. *Oecologia*, **120,** 605–612.

Smith, M.D., and Knapp, A.K. (2001). Size of the local species pool determines invasibility of a C_4-dominated grassland. *Oikos*, **92,** 55–61.

Smith, M.D., Knapp, A.K., and Collins, S.L. (2009). A framework for assessing ecosystem dynamics in response to chronic resource alterations induced by global change. *Ecology* **9,** 3279–3289.

Smith, M.D., La Pierre, K., Collins, S.L., Knapp, A.K., Gross, K.L., Barrett, J.E., Frey, S.D., Gough, L., Miller, R.J., Morris, J.T., Rustad, L.E., and Yarie, J. (2015). Global environmental change and the nature of aboveground net primary productivity responses: Insights from long-term experiments. *Oecologia* **177,** 935–947.

PART ELEVEN
LUQUILLO (LUQ) LTER SITE

24 Confessions of a Fungal Systematist

D. Jean Lodge

IN A NUTSHELL

1. The Long-Term Ecological Research (LTER) program has not influenced my basic approach to science.
2. The LTER program has reinforced my approach to mentoring, and it has increased my opportunities to mentor students through the LTER-associated Research Experiences for Undergraduates Program.
3. LTER program has greatly enriched my collaborative network and expanded my research in directions that I would not have otherwise pursued; similarly, I have expanded the research and perspectives of my collaborators.
4. My involvement in the LTER program has changed my perspective in reviewing grant proposals and manuscripts.

PERSONAL OVERVIEW

I have been a co-principal investigator or senior personnel at the Luquillo site (LUQ) of the LTER since its inception in 1988. My MS was on fungal population genetics and epidemiology of a plant pathogen, and my PhD work involved a study of the ecology of arbuscular and ectomycorrhizal fungi associated with cottonwood and willow, with a minor in entomology. I was employed as an ecosystem ecologist for the first 9 years of my professional career as a research scientist with the University of Puerto Rico, Center for Energy and Environment Research, which later became the Terrestrial Ecology Division. My early research in the LTER program focused on the role of arbuscular mycorrhizal fungi in plant colonization of landslides in collaboration with plant ecologists and physiologists in the "disturbed plant group." Hurricane Gilbert struck Jamaica in 1988, shortly after I had measured vegetation there, so I returned to Jamaica with a group that was studying migrant bird habitat and helped to remeasure plants. I used this opportunity to design the tree damage protocol that was used in 1989, when Hurricane Hugo struck the Luquillo Experimental Forest in Puerto Rico (the location of LUQ) (Zimmerman et al. 1994). Consequently, I was nicknamed "Hurricane Hattie" by my collaborators at the Coweeta LTER site.

Throughout my career, I have used my graduate training in ecology and soil microbial ecology to make important estimates of fungal and bacterial biomass and nutrient immobilization, and to determine what factors control spatial and temporal patterns in fungal distributions, abundance, and diversity (Lodge and Cantrell 1995; Lodge 1997). I received additional training to run a radioactive phosphorus tracer experiment to show fungal translocation of phosphorus by leaf decomposer fungi in microcosms at the Institute for Terrestrial Ecosystem Studies in England.

After arriving in Puerto Rico in 1982, I began describing new species of fungi with additional training in taxonomy and systematics at the Field Museum of Natural History in Chicago and the Royal Botanical Garden, Kew, in England. I took a position in tropical fungal systematics with the Center for Forest Mycology Research in the US Forest Service in 1992, which is when the main focus of my research changed to fungal classification. Many years ago, at the second LTER All Scientists Meeting, more than 30 LTER researchers who considered themselves to be taxonomists or systematists gathered in a circle after lunch and introduced themselves. As one of the first, I introduced myself as a fungal systematist masquerading as an ecosystem ecologist. Subsequently, everyone else in the circle confessed to being a "closeted" systematist or taxonomist, as if it were a gathering of "systematists anonymous." Based on my experiences and of others in the group, it appears that the "big tent" approach to ecology at most LTER sites has provided a niche where systematists and taxonomists can survive in a research climate that has generally lost sight of the importance of species.

APPROACH TO SCIENCE

If I had not been involved with LUQ, I probably would not have become involved in one of my main ecological research foci: the roles of mushrooms in leaf litter decomposition, nutrient cycling, and erosion control (Lodge et al. 2008). I am probably better known now as an ecologist than I am as a systematist, largely as a consequence of my research in the LTER program. I, and most other systematists in the LTER program, am concentrated in disciplines with highly diverse groups such as fungi and invertebrates, and these organisms play critical roles that influence ecosystem processes such as herbivory, plant survivorship, seed dispersal and pathology, decomposition, and nutrient cycling. My work and that of other taxonomists and systematists in the LTER program largely focus on the effects of keystone or dominant species and functional groups. The groups of organisms we work with, such as decomposers, have often been treated as belonging to "black box" compartments in ecosystem processes. Our research shows, however, that the species or functional groups inside the black boxes influence rates of ecosystem processes and the fate of carbon and nutrients. For example, we showed that the species of microfungi that are dominant early decomposers of particular leaf species decompose their preferred hosts faster than do dominants from other leaf species (Santana, Lodge, and Lebow 2005). Also, we showed that the presence of a different functional group of decomposers (mushrooms that degrade lignin) greatly accelerated decomposition beyond that caused by microfungi (Santana, Lodge, and Lebow 2005; Lodge et al. 2008). In addition, we confirmed that lignin-degrading mushrooms were inhibited by nitrogen loading (Lodge et al. 2008) and canopy opening from hurricane wind damage (Lodge and Cantrell 1995). Predicting responses to disturbance, nitrogen loading, or climate change can be difficult without knowledge of the species or functional groups that mediate ecosystem processes and how they respond to stress.

Like many of the taxonomists and systematists in the LTER program, I was crosstrained in ecology. I sought ecology training in graduate school, partly because there were (and still are) more positions available in research and academia for ecologists than there are for systematists. The cross-disciplinary training that many taxonomists and systematists received is preadaptive to the interdisciplinary research in the LTER program. Not only does ecological training allow systematists and ecologists to occupy niches in research in the LTER program, but it also hones skills in explaining ideas and principles to those from different disciplines.

ATTITUDES TOWARD TIME AND SPACE

My research has always been oriented toward changes in time and space, and that has not changed with my involvement with the LTER program. Although I have a greater appreciation of socioecological interactions through my involvement, I have not yet incorporated it into my research program.

COLLABORATION

If I had not been involved in the LTER program, I would not have become involved in collaborative research on primary succession in landslides, secondary succession following hurricanes, or the effects of hurricanes on vegetation composition and structure (Zimmerman et al. 1994), nutrient immobilization, and litter deposition and decomposition (Lodge et al. 2008). Similarly, I doubt that most of my coauthors in the "disturbed plant group," who were stream chemists and forest ecologists, would have become involved in the autecology of basidiomycete leaf decomposers or their roles in nutrient cycling and erosion control (Lodge et al. 2008). Although I can easily say that my involvement in research at LUQ has led me to be more multidisciplinary, collaborative, synthetic, and insightful in my research, I cannot say that my research is more theoretical or comparative because of that involvement.

One of the downsides of my early involvement in the LTER program was having been "plugged into" gaps in research proposals because I had useful skills rather than the desire to carry out those particular aspects of research. Although research on arbuscular mycorrhizae is important and intellectually challenging, I had no desire to continue mycorrhizal research after having examined a mind-numbing number of samples under the microscope for my doctoral dissertation. Instead, I trained graduate students interested in that type of research. I have learned to be more selective and to say "no" much more to invitations for collaborative research, so that I have enough time to pursue my main interests and what I consider to be important. My main criteria for saying "yes" to a collaborative research request is whether it intrigues me, and whether it is what I do best. I say "no" when others could better take on a particular aspect or when it is mind-numbing.

APPLIED RESEARCH

Although I appreciate applied aspects of research, I cannot say that my experience in the LTER program has altered my views. Although some of my research publications from the LTER program are in part or mostly applied (e.g., Miller and Lodge 2007; Lundquist et al. 2011), I have always had dual basic and applied aspects in my research and student training. My research and publications from the LTER program have fostered collaborations with foresters and forest pathologists (e.g., Lundquist et al. 2011) as well as with national forest ecosystem managers.

COMMUNICATION

As a full-time government researcher, I do not regularly teach classes, although I do train graduate students and give guest lectures and workshops. Most of my teaching is through outreach or mentoring activities.

MENTORING

LUQ has a strong Research Experiences for Undergraduate (REU) program that has provided valuable training for undergraduate students and mentoring opportunities for me. In addition, I have also mentored high school students in their science fair projects. The REU students are trained by a dedicated staff person in designing research, statistical analysis, and presentation; they are part of a social cohort; and I encourage them to help each other with their projects. The REU application process is very competitive, and I can select highly motivated students with interest in areas of research similar to mine. Based on my experiences in collaborative research in the LTER program, I am careful to not push students into projects. Once a student is selected, I like to see what topic lights up their imagination and engages their thought processes. Otherwise, the LTER program has not influenced the way that I mentor students or junior faculty members; rather, it has provided me with more opportunities and an ideal environment in which to do so.

My research experiences in the LTER have altered how I evaluate research proposals and manuscripts. I am now quick to look for underpinning paradigms that are being proposed or tested and whether the proposed research or results are able to support or refute the paradigms. Also, I place more value on proposals that can leverage data by being colocated or coordinated with other studies. In addition, I look for the applicability of the results to solving problems and understanding responses in complex ecosystem processes.

SKILL SET

Research in the LTER program required an expansion of my skills in ecology to include extracting labile nutrients from soil, quantifying fungal and microbial biomass, working with radioactive isotopes to trace phosphorus translocation between litter cohorts, and making mass balance calculations. I have also stretched my skills in statistical analyses, mostly through collaborations with others both inside and outside the LTER network. I have also learned how to analyze microbial communities using molecular methods through my collaborations with other researchers at LUQ.

One of the most valuable skills I learned through my collaborative research in the LTER program was how to effectively work with a large interdisciplinary team. I learned from Lawrence Walker's leadership of the "disturbed plant group" to elicit prospective titles for manuscripts or sections of manuscripts at the beginning of a collaborative project, get task and author commitments for each title, and then make adjustments to authorship as needed until the research is published. Those skills served me well in leading a self-assembled group of 34 mycologists in a 15-year project to revise the higher-level systematics of a fungal family (Lodge et al. 2013).

PERSONAL CONSEQUENCES

When living on a small island, contacts with researchers who are visiting to work in the LTER program are a critical part of our social network. The off-island researchers bring fresh perspectives, knowledge, and ideas, and represent various cultures. Parts of my holidays are often spent with our collaborators when they come to Puerto Rico, and I take time when we can to visit them in their homes and home institutions. My life would be much poorer without the social network provided by association with the LTER program.

CHALLENGES AND RECOMMENDATIONS

The most difficult aspect of research in the LTER program is the review process for renewal proposals. The funding for each site is generally only sufficient to maintain infrastructure, including some large-scale manipulations, long-term measurements, and a few critical key people such as data managers and site managers. Consequently, there is only a meager amount of funds for scientists and their students to conduct research or for costly cutting-edge research. In essence, funding for research in the LTER program goes more to support sites as a platform for other research proposals. This platform is quite valuable, but the evaluation of renewal proposals is influenced by reviewers who are biased toward short-term results and testing of cutting-edge hypotheses. It is a struggle each time to reinvent a site's program to meet the long-term goals of the program while simultaneously addressing the short-term bias of the reviewers.

The LTER network was designed to be used for cross-site analyses and comparisons, but there is little financial incentive to accomplish that goal. The funding of cross-site workshops that evolve from the All Scientists Meetings are helpful in fostering cross-site comparisons and publications, and there are some funds for students to work at multiple sites, but funding levels and incentives are insufficient to motivate and support cross-site experiments. Unless more resources are devoted to cross-site comparisons, I do not think that the LTER program will live up to its potential as a network.

CONCLUSION

My involvement in research in the LTER has not greatly influenced my approach to science or mentoring, but it has greatly expanded my skills and my opportunities for mentoring and for collaborative research. My early experiences in collaborative research at LUQ reinforced something I learned from watching my graduate student classmates: that if someone is not inspired by their research project, and they do not experience "the fire in the belly," they will not bring their project or thesis to its successful completion. I think the LTER program provides a critical foundation for collaborative research. The most important skill I learned from my research collaborations in the LTER program was to negotiate agreements on commitments and expectations at the beginning of a collaborative project, and then revisit those plans on a regular basis. Cross-disciplinary training is critical to launching a successful career, not only in relation to the LTER program, but also in light of the complex ecological and social problems that we face with global climate change. I do not think, however, that the LTER program will live up to its full potential as a network unless the evaluation system and the funding structure are changed. I recommend that students and junior colleagues negotiate authorship agreements up front in collaborative research projects and seek opportunities for training in other disciplines whenever possible.

ACKNOWLEDGMENTS

The author thanks the National Science Foundation (NSF) for support from grants BSR-8811902, DEB-9411973, DEB-0080538, DEB-0218039, DEB-0620910, and DEB-1239764 to the Institute for Tropical Ecosystem Studies, University of Puerto Rico; to the International Institute of Tropical Forestry US Forest Service, as part of the Luquillo LTER; and for NSF Biotic Surveys & Inventories grant DEB-9525902 to the Research Foundation of SUNY at Cortland in a joint venture with the US Forest Service Forest Products Laboratory. The US Forest Service and the University of Puerto Rico provided additional support.

REFERENCES

Lodge, D.J. (1997). Factors related to diversity of decomposer fungi in tropical forests. *Biological Conservation*, **6**, 681–688.

Lodge, D.J., and Cantrell, S. (1995). Fungal communities in wet tropical forests: Variation in time and space. *Canadian Journal of Botany*, (suppl. 1), S1391–S1398.

Lodge, D.J., McDowell, W.H., Macy, J., Ward, S.K., Leisso, R., Claudio Campos, K., and Kühnert, K. (2008). Distribution and role of mat-forming saprobic basidiomycetes in a tropical forest. In L. Boddy and J.C. Frankland, eds., *Ecology of Saprobic Basidiomycetes*, pp. 195–208. Academic Press, Elsevier, Amsterdam.

Lodge D.J., Padamsee, M., Matheny, P.B., Aime, M.C., Cantrell, S.A., Boertmann, D., Kovalenko, A., Vizzini, A., Dentinger, B.T.M., Kirk, P.M., Ainsworth, A.M., Moncalvo, J.-M., Vilgalys, R., Larsson, E., Lücking, R., Griffith, G.W., Smith, M.E., Norvell, L.L., Desjardin, D.E., Redhead, S.A., Ovrebo, C.L., Lickey, E.B., Ercole, E., Hughes, K.W., Courtecuisse, R., Young, A., Binder, M., Minnis, A.M., Lindner, D.L., Ortiz-Santana, B., Haight, J., Læssøe, T., Baroni, T.J., Geml, J., and Hattori, T. (2013). Molecular phylogeny, morphology, pigment chemistry and ecology in Hygrophoraceae (Agaricales). *Fungal Diversity* **64**, 1–99.

Lundquist, J.E., Camp, A.E., Tyrrell, M.L., Seybold, S.J., Cannon, P., and Lodge, D.J. (2011). Earth, wind, and fire: Abiotic factors and the impacts of global environmental change on forest health. In: J.D. Castello and S.A. Teale, eds. *Forest Health: An Integrated Perspective*, pp. 195–244. Cambridge University Press, Cambridge.

Miller, R.M., and Lodge, D.J. (2007). Fungal responses to disturbance—Agriculture and Forestry. In K. Esser, P. Kubicek and I.S. Druzhinina, eds, *The Mycota, 2nd Ed., IV, Environmental and Microbial Relationships*, pp. 44–67. Springer-Verlag, Berlin.

Santana, M., Lodge, D.J., and Lebow, P. (2005). Relationship of host recurrence in fungi to rates of tropical leaf decomposition. *Pedobiologia*, **49**, 549–564.

Zimmerman, J.K., Everham, E.M. III, Waide, R.B., Lodge, D.J., Taylor, C.M. and Brokaw, N.V.L. (1994). Responses of tree species to hurricane winds in subtropical wet forest in Puerto Rico: Implications for tropical tree life histories. *Journal of Tropical Ecology*, **82**, 911–922.

25 A Glimpse of the Tropics Through Odum's Macroscope

Ariel E. Lugo

IN A NUTSHELL

1. The philosophy of research in the Long-Term Ecological Research (LTER) program expanded what I learned in graduate school from H. T. Odum by providing an approach for a holistic understanding of ecological processes in the tropics.
2. Participation in the LTER program enabled collaborations with many talented people from many parts of the world and enabled the mentoring and education of a new cadre of tropical natural and social sciences students.
3. By expanding the opportunities for research and analysis at larger scales, the LTER program allowed me to address tropical ecosystem responses to such phenomena as

hurricanes, floods, landslides, and past land uses and to do so at the appropriate scales of time and space.
4. Paradigms of tropical forest resilience and adaptability in the Anthropocene emerged from research at the Luquillo (LUQ) LTER site.

PERSONAL OVERVIEW

I first became aware of the LTER program in 1978 as I walked by the White House in Washington, DC, with Sandra Brown, then an intern on the President's Council on Environmental Quality (CEQ), and Wayne Swank, a US Forest Service employee on detail with the National Science Foundation (NSF). I was a staff member at CEQ, and W. Swank explained to us a new long-term ecological research program that he was helping develop at the NSF. Although the first cadre of sites appeared to have been selected, I was immediately captured by the concept and expressed my interest in developing a proposal for a tropical site in Puerto Rico. Little did I know at the time that my whole scientific career was about to change, in part because of the LTER program, but also because I was to become a US Forest Service scientist. The first 30 years of my US Forest Service career would be heavily influenced by the LTER program and the people I worked with while developing a new way of thinking about tropical forest ecosystems.

I am an ecologist trained at the Universities of Puerto Rico and North Carolina at Chapel Hill. My experience before becoming part of the LTER program involved (1) teaching at the University of Florida at Gainesville and (2) government work at the Commonwealth (Puerto Rico Department of Natural Resources) and federal (President's Council on Environmental Quality) levels. Since 1979, I have worked for the US Forest Service at the International Institute of Tropical Forestry.

APPROACH TO SCIENCE

The LTER approach to ecological research fostered by the LTER program affects all the professional activities that one engages in, bar none. However, this approach, a powerful tool for understanding ecosystem dynamics, has to be projected with care in the review of articles and proposals, mentoring of students, and the evaluation of the careers of colleagues because not all aspects of scientific endeavor need to be long term to be effective. It made sense for the LTER program to aggressively push its fundamental research notions of observation of long-term phenomena at a time when most ecological research was short-term in nature. Today, 30 years after Callahan (1984) introduced the LTER program to the scientific community, its approach is engrained in the social and ecological sciences and the prospects for the future for that approach are not in question. However, the key to ecological understanding of the world is to recognize that both short-term and long-term research is required to advance social and ecological knowledge. Also, long-term ecological research requires the long-term commitment that only government agencies such as the NSF and US Forest Service can provide, usually involving particular sites with the research infrastructure and research history to support long-term research. Moreover, not all scientific careers or science products require long-term research approaches. Thus, we would be shortsighted to evaluate proposals, papers, and professional careers exclusively from a long-term perspective. Instead, we have to be cognizant of the long-term nature of ecosystem functioning and make sure that short-term research is conducted in ways consistent with the principles, paradigms, and discoveries resulting from research in the LTER program. Fundamentally, research supported by the LTER program allows scientists to be more balanced and holistic when evaluating papers, proposals, students, and professional careers.

ATTITUDES TOWARD TIME AND SPACE

The LTER program has expanded both the level of my synthesis work and the scale at which I visualize ecological phenomena. Like most ecologists between the 1960s and 1980s, my research focus was on local ecosystems (mostly mangroves and tropical wet forests and plantations) and internal ecosystem functioning. I was trying to understand how these ecosystems functioned in the context of Holocene conditions. At the time, landscapes and historical effects of humans on ecosystems were matters to which I paid little attention, and when I did, it was usually in a limited academic context. However, the LUQ-based research of Thompson et al. (2002) was a major event in my career because it uncovered almost overnight the importance of the past actions of people on tropical forest regeneration and set me on a course of expansion of research scope to the whole landscape level. At the Institute, this epiphany led to the establishment of our geographic information system laboratory that now functions with more than 15 people and is the center of a significant fraction of the research that we conduct pantropically. This expansion in the scale of analysis has also led to increased levels of synthesis because of the broad scope of the subject, the need to dispel myths about tropical lands, and the changes that have taken place in the context under which we now conduct research. There is a greater social pressure for scientists to explain what they do and its relevancy to people. Synthesis of the state of knowledge is the most powerful tool available for scientists to explain what they do and the significance of research for improving social wellbeing.

Time and space issues are front and center in the research mandate of the LTER program. Of these, it is easier to justify the temporal dimension because continuing to collect information at a site or an experiment always yields surprises and, more importantly, an understanding of patterns. Space is more difficult to deal with because it requires attention to more sites, which increases the cost of research and adds the difficulty of identifying and studying other sites. For managing purposes, continuity of temporal data collection is essential and so is the understanding of spatial variability. Nevertheless, in academic circles continuous data collection at a site is termed *monitoring*, and in some places it is viewed as a somewhat inferior activity lacking scientific basis. This attitude is also built into the LTER program, where sites must somehow revamp research questions every 6 years if they are to secure funding continuity. In the US Forest Service, we have learned that continuous data collection leads to improved understanding of natural phenomena, improved hypothesis development and testing, and more effective ecosystem management (Lugo et al. 2006). We have long-term plots that have allowed collection of the same data for more than 70 years and through several turnovers of principal scientists; these data are still yielding important insights to critical questions from basic and applied perspectives. Thus, a plot established initially to understand silvicultural phenomena now also yields information about carbon storage and turnover as well as subtle responses to environmental change. Therefore, the secret of success in long-term monitoring of ecological phenomena is to constantly examine the data with an eye to detecting answers to new and emerging questions about ecosystems, questions that were not considered at the time the monitoring began.

COLLABORATION

My ecosystem thinking, multidisciplinary approach to research, and collaborative approach to studying tropical forests were acquired from H. T. Odum during the radiation experiment in Puerto Rico in the early 1960s (Odum and Pigeon 1970). The LTER program introduced me to a group of superb colleagues from different parts of the world with whom I developed partnerships to work on novel ways of thinking about the ecosystems of the Luquillo Mountains. For example, we had a strong aquatic ecology group in the LUQ LTER site, whose initial interest was in aquatic populations such as shrimp, but evolved into the study of riparian zones and eventually terrestrial processes that supported the aquatic fauna. Because of those interactions combined with climatic and geochemical studies, we began to visualize the Luquillo Mountains as a gigantic wetland that during periods of heavy rainfall could be completely draped by a water film through which aquatic organisms could move unimpeded by the aquatic–terrestrial discontinuities prevalent during dry seasons. The incredible diversity of aquatic species in tank bromeliads growing on elfin forests on the tops of the mountains began to make sense under this wetland paradigm that would not have emerged through traditional discipline-bound research.

We were successful in pursuing new ways of thinking about tropical forests because of the friendship and respect that we developed as a research team at LUQ. Team cohesion started while writing the proposals to establish LUQ, later when faced with the awesome effects of Hurricane Hugo, and finally while developing special issues of journals, and a synthesis book (Brokaw et al. 2012) on the state of knowledge about the tropical forests of the Luquillo Experimental Forest (LEF). Those relationships allowed us to jointly bridge the gap between population and ecosystem approaches to ecology. These two contrasting views of how ecological systems function had defined my graduate and later my early professional career. I credit the LUQ as the place where we developed a communication scheme that allowed us all to realize how artificial those academic silos can be and how the understanding of complex ecological systems benefit from all conceivable approaches, including those outside the natural sciences (i.e., the social sciences and indigenous knowledge). For example, when we conducted a synthesis of the various paradigms that explained the Luquillo Mountain ecosystems (Lugo et al. 2012b), one of the gratifying surprises was the similarity of concerns with hurricane effects among the Taínos who coined the term *juracán*, and present populations on the island that must adapt to the storms.

The LTER program is a conduit for research at multiple levels, spanning local to international engagement, including collaborations within and between institutions, as well as within and between disciplines or career types. I have experienced all this throughout my involvement in the LTER program and have been lucky to incorporate this diversity of talent into the research program of my own institution. The most dramatic example is the development of a social ecological program at the Institute, a program that received seed funding from the NSF and the US Forest Service, and which now involves two staff social scientists at the Institute and a very productive and large group of scientists and students from the social and natural sciences working together on an urban watershed (see sanjuanultra.org). This program has its roots in the LUQ. Ironically, the LUQ desisted in its social ecological efforts, perceiving an NSF policy against such an approach, but we in the US Forest Service are benefiting enormously from the integration of the social and natural sciences. The Institute has also benefited from international collaborations from a number of tropical and temperate countries, but such types of interactions have been an important part of our organization since its inception in 1939.

APPLIED RESEARCH

The Institute is focused on tropical forest management (synonymous to conservation) and functions in close collaboration with tropical forest managers from commonwealth, state, private, and federal jurisdictions. We have pursued that US Forest Service mission for 75 years, which means that the mission of the Institute was determined well before the inception of the LTER program at NSF. However, the LTER program has enriched and improved the delivery of the mission with its focus on the functioning of the ecosystems of the Luquillo Mountains. The reason for this is that LUQ is not only an Experimental Forest, it is also a national forest in the US Forest Service National Forest System, and when the LTER program began, most of the US Forest Service's research focus within the Luquillo Mountains was management oriented as opposed to ecologically oriented. Beginning with the radiation experiment of H. T. Odum, the scope of ecological research in the LEF expanded considerably, thus greatly increasing the range of ecological issues addressed by the research (Lugo et al. 2012b). Odum's work led to the Luquillo LTER program, and forest management benefited by the increase in ecological knowledge and the applications of that knowledge to forest conservation (Lugo et al. 2012a). As an example, forestry research documented the timber value of low elevation forests, whereas ecological research documented the importance of upper elevation forests for clean water as well as the importance of free-flowing rivers to the migration of aquatic fauna. The designation of wilderness and wild and scenic rivers in the Luquillo Mountains is in part a result of ecological research, whereas forestry research led to the restoration of degraded low elevation lands.

COMMUNICATION

The Institute has sponsored long-term ecological research since 1941 and has had involvement with K–12 schools since 1986. The benefits from involvements of the LTER program in these activities have been in the rise in the level and intensity of the research and education programs. The growth has been significant, and we have reached a level of involvement with the LTER Schoolyard Program that would not have been possible with our usual program funds. As an example of how important the LTER program has been for K–12 education in Puerto Rico, prior to establishment of LUQ, the Institute had formal agreements with five high schools involving some 100 students and teachers, annually. In fiscal year 2013, we reached approximately 20 schools and 6,000 K–12 students and teachers, all Hispanic minorities. This leads me to conclude that the LTER program has greatly expanded our capacity to systematically and on a long-term basis reach a larger number of K–12 students and teachers, thus increasing the pool of minority candidates that can be encouraged to pursue scientific careers (Figure 25.1).

MENTORING

Involvement with the LTER program increases the opportunities for mentoring of students and young scientists. As a LUQ scientist, I have come in contact with a large cadre of students and faculty members that I would not have had the opportunity to meet, much less interact with, in the absence of our LTER site. Moreover, I have observed that the quality of these students and faculty members is above the norm, which makes the interactions more challenging and fun. In some cases, those that I have mentored reach levels of development well beyond my level of understanding of what they do, and they reach those levels in spite of my limitations. These experiences have changed my approach to mentoring in that I am now less intrusive into a student's or beginning scientist's research plans because I realize that the level of change and innovation that they bring to our site is best not contaminated with potential biases that one develops over a lengthy scientific career. However, promoting collaboration and transdisciplinary research among upcoming students and staff members remains a challenge because academia still has a strong inertia against these trends. Regardless of how I project these issues, I find that the response to transdisciplinary approaches depends on the individual, with some being more inclined than others to collaborate within and outside disciplines.

FIGURE 25.1 Ariel Lugo (back right) faced the toughest audience of his career, with students of the Wyoming Park Elementary School at Ocala Florida. Valda Niznik, school principal (back left), and second and fifth graders, Lindsey Lugo and Gracie Lugo, respectively, stand by their grandfather after his presentation. (Photo courtesy of Ariel Lugo.)

PERSONAL CONSEQUENCES

It is ironic that the success of the LTER program in changing my views about the interactions between people and ecosystems has driven me away from participation in the LTER program, and into the new world of the effects of the Anthropocene on the urban environment with its novel ecosystems. Humans have altered more than 70% of the biosphere and in Puerto Rico, for example, only 25% of its forests are native forests typical of the Holocene (Martinuzzi et al. 2013). In contrast, the LUQ is the best example of a mature and little disturbed native forest on the island (Harris et al. 2012), and colleagues in the LTER program intend to continue to study this site for as long as possible (a good thing). However, what about the rest of the island's forests, which comprise 75% of the forest cover? Moreover, Puerto Rico now has about 23% of its geographic area in urban cover with 27,311 ha of urban forests, and 99% of its population concentrated in urban areas. Most forests in Puerto Rico are novel forests with novel species composition that respond to past and present human activities (Lugo and Helmer 2004). It would appear that to effectively conserve the forest resources of the island and provide a healthy environment to islanders, one would have to establish a strong research presence in urban locations and in novel forests. Yet, these are areas that ecologists least prefer to study, thus creating a knowledge vacuum and a research opportunity that I have elected to pursue at the expense of my participation in the LTER program. Thus, research at LUQ led me to socioecological research, which led me to abandon participation in the LTER program!

The greatest personal reward from my participation in the LTER program has been the people that I have met and with whom I have worked. I have always believed that one of the benefits of being a scientist is that you meet and work with smart people. But I learned from C. S. Holling that "smarts" is not the only criteria for successful collaborations—you have to have fun as well. You have to like the people you work with, and you must all have a good sense of humor to ensure a superb scientific collaboration. From our first Luquillo meeting at Lucy's Inn (at El Verde, Puerto Rico) in the early 1980s to the times that the seven coeditors of the LUQ synthesis book met in my office to conceptualize our book (Brokaw et al. 2012) and argue endlessly about how we visualized the functioning of the ecosystems of the Luquillo Mountains, I have had fun in spades. The core collaborators that founded the LUQ program were successful because our functioning was based on friendship. This is why even if we were empiricists, theoreticians, population biologists, or ecosystem scientists, it did not matter. We were friends and friends learn to understand each other. So, friendship is the one personal gain that I derived from collaboration at LUQ and the LTER program.

CONCLUSION

From the very beginning my experience in the LTER program was transformational. The first two All Scientist Meetings at Estes Park were unbelievable experiences for me. They expanded my intellectual horizons and the number and types of people with which I interacted. Our research on hurricane effects and other disturbances of tropical forests attracted national and international attention, and opened numerous opportunities for travel to meetings and universities to report our results and debate emerging paradigms of resiliency, when most of the field had the notion that tropical forests were fragile. More recently, the socioecological approach has once again expanded the scope of interaction and intellectual inquiry. This later transformation, which is mostly outside the scope of LUQ, is even more significant and intellectually rewarding than the one we experienced when studying the ecosystems of the Luquillo Mountains. Socioecological research uncovers a new cadre of urbanites, nongovernmental organizations, communities, government agencies, and academic disciplines of unprecedented magnitude and scope. The study of San Juan brings my research more in alignment with two mainland LTER sites that focus on the cities of Baltimore and Phoenix.

The LTER program has been extremely successful and it must guard against complacency, which could lead to lowering the intellectual status of the program. I am particularly troubled about the divide between the social and natural sciences in the network. The network is overwhelmingly oriented toward the natural sciences and does not appear to know how to deal effectively with the social sciences. The merging of perspectives and traditions into a socioecological approach requires enormous intellectual investment and significant levels of change by participants from both biophysical and social sciences. Maintaining a competitive and exclusionary environment will not lead to a successful implementation of social-ecological research. The LTER network also has to break its inertia with the Holocene and burst into the Anthropocene. Based on my experience with LUQ, where friendship prevails, the jump to social ecology and the Anthropocene is not easy, and it is unlikely to occur without strong leadership from the NSF. However, if the NSF is organized in silos, then the chances of NSF providing leadership of such significance is unlikely. Traditional academic disciplines are likely to continue to maintain the NSF and LTER programs in a traditional trajectory, while the rest of the world moves in a different direction. The solution might have to come from society itself demanding answers to the new conservation issues of the Anthropocene.

ACKNOWLEDGMENTS

This work was done in collaboration with the University of Puerto Rico. M. Alayón edited and improved the manuscript. I also received valuable insights for improving the manuscript from G. González, M. R. Willig, and anonymous reviewers.

REFERENCES

Brokaw, N., Crowl, T.A., Lugo, A.E., McDowell, W.H., Scatena, F.N., Waide, R.B., and Willig, M.R., eds. (2012). *A Caribbean Forest Tapestry: The Multidimensional Nature of Disturbance and Response*. Oxford University Press, New York.

Callahan, J.T. (1984). Long-term ecological research. *BioScience*, **34**, 363–367.

Harris, N.L., Lugo, A.E., Brown, S., and Heartsill-Scalley, T., eds. (2012). *Luquillo Experimental Forest: Research History and Opportunities*. U.S. Department of Agriculture, Washington, DC.

Lugo, A.E., and Helmer, E. (2004). Emerging forests on abandoned land: Puerto Rico's new forests. *Forest Ecology and Management*, **190**, 145–161.

Lugo, A.E., Swanson, F.J., Ramos-González, O., Adams, M.B., Palik, B., Thill, R.E., Brockway, D.G., Kern, C., Woodsmith, R., and Musselman, R. (2006). Long-term research at USDA's Forest Service's Experimental Forests and Ranges. *BioScience*, **56**, 39–48.

Lugo, A.E., Scatena, F.N., Waide, R.B., Greathouse, E.A., Pringle, C.M., Willig, M.R., Vogt, K.A., Walker, L.R., González, G., McDowell, W.H., and Thompson, J. (2012a). Management implications and applications of long-term ecological research. In N. Brokaw, T.A. Crowl, A.E. Lugo, W.H. McDowell, F.N. Scatena, R.B. Waide, and M.R. Willig, eds. *A Caribbean Forest Tapestry: The Multidimensional Nature of Disturbance and Response*, pp. 305–360. Oxford University Press, New York.

Lugo, A.E., Waide, R.B., Willig, M.R., Crowl, T.A., Scatena, F.N., Thompson, J., Silver, W.L., McDowell, W.H., and Brokaw, N. (2012b). Ecological paradigms for the tropics: Old questions and continuing challenges. In N. Brokaw, T.A. Crowl, A.E. Lugo, W.H. McDowell, F.N. Scatena, R.B. Waide, and M.R. Willig, eds. *A Caribbean Forest Tapestry: The Multidimensional Nature of Disturbance and Response*, pp. 3–41. Oxford University Press, New York.

Martinuzzi, S., Lugo, A.E., Brandeis, T.J., and Helmer, E.H. (2013). Geographic distribution and level of novelty of Puerto Rican forests. In R.J. Hobbs, E.S. Higgs, and C. Hall, eds. *Novel Ecosystems: Intervening in the New Ecological World Order*, pp. 81–87. John Wiley & Sons, Sussex, UK.

Odum, H.T., and Pigeon, R.F., eds. (1970). *A Tropical Rain Forest*. National Technical Information Service. Springfield, Virginia.

Thompson, J., Brokaw, N., Zimmerman, J.K., Waide, R.B., Everham, III, E.M., Lodge, D.J., Taylor, C.M., García Montiel, D., and Fluet, M. (2002). Land use history, environment, and tree composition in a tropical forest. *Ecological Applications*, **12**, 1344–1363.

26 Taking the Long View
Growing Up in the Long-Term Ecological Research Program

Whendee L. Silver

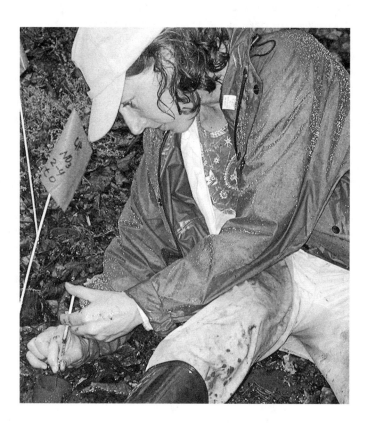

IN A NUTSHELL

1. The Long-Term Ecological Research (LTER) program has shaped me as a scientist by providing a collaborative environment and the opportunity to take a long-term, large-scale perspective in my research. I share this perspective with students by incorporating the principles, questions, and data from such research into my teaching.
2. Working at an LTER site, and one that is based in Puerto Rico, has allowed me to increase the diversity of my laboratory and our graduate program by facilitating the recruitment of women and minority students.

3. Personal experiences with science and data management in the LTER program, particularly the bad experiences, have helped me to improve as a communicator in the broadest sense.
4. Although being a scientist in the LTER program has contributed to my career in many positive ways, it has also presented challenges to my work–life balance.
5. To maintain its leadership role, the LTER program needs to remain an open network welcoming new scientists, new ideas, and thus new potential for discovery.

PERSONAL OVERVIEW

I grew up, professionally speaking, in the LTER program. In 1989 as a new PhD student, I was strongly encouraged (i.e., told in no uncertain terms!) to explore research opportunities in the Luquillo Experimental Forest in Puerto Rico. My mentors had developed a graduate field course in Puerto Rico that I participated in and later helped teach. Puerto Rico was their first venture into the tropics, one that was made easier by the fact that Puerto Rico is part of the United States and provides almost all of the conveniences of home. As one of my professors, Tom Siccama, liked to remark, Puerto Rico was "just like Connecticut, only different!" Puerto Rico was not, however, my first venture into the tropics. I had traveled, studied, and worked in Central and South America and the Pacific since my sophomore year of college and considered myself to be a tropical veteran. I felt at home in tropical rain forests, and had envisioned my PhD research taking place at some remote field site, in a foreign country, far from civilization: just me, my tent, the jungle, and the animals.

This vision was not to be. When my PhD committee heard that a consortium of scientists had been awarded an LTER grant, I was sent to Puerto Rico to explore research ideas that might complement the newly established Luquillo (LUQ) LTER program. My mentors themselves "grew up" academically during the era of the International Biological Program and the large Atomic Energy Commission effort. They had a deep appreciation for multidisciplinary, collaborative, "big science." I, of course, had no experience with large-group science, knew none of the scientists in the LTER program, and was justly intimidated.

Now, in hindsight, I realize how fortunate it was that I, for once, listened to my committee. I am currently professor of ecosystem ecology at the University of California (UC) Berkeley and a co–principal investigator at the LUQ site. My research at the LUQ site focuses on the causes and consequences of global change in tropical ecosystems. I study the (1) biogeochemistry of greenhouse gas production and consumption and (2) the effects of climate and land use on carbon and nutrient dynamics. I have served several roles at the LUQ site, from student to executive committee member. In addition to site-based research, I have helped lead cross-site synthesis actives such as an analysis of the Long-Term Intersite Decomposition Experiment. Being part of the LTER program from the early stages of my career helped shape my perspectives and approach, and provided unparalleled opportunities. It has also presented some challenges.

APPROACH TO SCIENCE

The pursuit of science is similar in many ways to constructing a large building. The foundations (i.e., paradigms) need to be broad and strong to provide support. Wings and stories (i.e., specific hypotheses) should adhere to a blueprint of the building's purpose and functionality. The edges and the top of the building can provide opportunities for new construction, alter the scope or purpose of the building (i.e., transformative discoveries), or enhance the utility or elegance of the existing structure. As scientists, we continually build on information from the past, making adjustments, replacing bits and pieces, pushing knowledge in new directions. An example from my own research is our work on soil redox dynamics. Our work built on the foundation established by innovators such as the wetland biogeochemist William Patrick, who designed a field-deployable chamber for in situ oxygen measurements, and tropical ecologist Ariel Lugo, who recognized that upland tropical forests behaved as "slope" wetlands.

Being part of an LTER program has influenced the topics that I have studied and the techniques that I have used. However, it has done so in subtle ways. The collaborative nature of research in the LTER program has sometimes forced me to collect data that does not stand alone but contributes to a larger story. I say "forced" because collecting data that may not, by themselves, tell a story increases the risk of data being orphaned and unpublished. At the same time, collaborating with colleagues in different disciplines has taught me about new approaches to data analysis, and identified new avenues of research as collaborative activities unveiled new directions for our evolving scientific edifice at the LUQ site. Being part of an LTER program has also allowed me to collect data that eventually led to separately funded primary research. These projects generally arose from ancillary findings, interesting or unexpected patterns, and results that were not easily explained by the primary hypotheses being tested in the core project. The three best examples of this from my research were two novel pathways in the nitrogen cycle (Silver, Herman, and Firestone 2001; Yang, Weber, and Silver 2012), and one interesting development in the carbon cycle (Hall and Silver 2013). Research on these topics emerged as we measured patterns in redox dynamics and associated biogeochemistry at different spatial and temporal scales in the forest. Working at an LTER site, which was also a US Forest Service long-term research site, meant that I had access to vast stores of background data on ecosystem physical and chemical characteristics for a wide range of sites in the forest and long-term records on climate, soil oxygen, and litterfall production, which guided and complemented our research.

ATTITUDES TOWARD SPACE AND TIME

Apart from LTER and related programs (i.e., the National Science Foundation [NSF]–funded Long-Term Research in Environmental Biology), most research funding in environmental biology comes in 3- to 4-year increments. However, many ecological phenomena change slowly. With short funding cycles, we ecologists are forced either to exchange space for time (often scientifically unsatisfying and with a suite of analytical and empirical problems) or focus on short-term phenomena and questions that tell us what is happening at the moment but give little insight into the temporal and spatial robustness of our conclusions. Using the previous analogy, this approach increases the probability that we will continue to add bricks to the building while losing sight of the strength of the foundation or the overall functionality or beauty of the structure.

Even small amounts of guaranteed funding over many years allow us to broaden our scientific perspective. Low-intensity, long-term projects sometimes lead to surprises. For example, it was generally assumed that redox dynamics, among the primary drivers of greenhouse gas emissions in soils, were limited to processes operating at the microsite scale. We collected soil oxygen data every 2 weeks for 8 years from 60 chambers along an elevation gradient in the Luquillo Experimental Forest to determine the scale of variability in redox dynamics. The temporal and spatial richness of this data set showed us that soil redox dynamics were synchronized at large spatial scales, allowing us to begin to link a suite of metabolic processes (such as methane production and consumption) across ecosystems and over time. Similarly, we found that soil respiration followed distinct seasonal cycles that varied in magnitude across years, improving our understanding of the drivers of soil CO_2 emissions in space and time. This research showed that climate and solar cycles could be used to model soil redox dynamics, even in humid tropical forests that are characterized by low seasonality relative to more temperate ecosystems.

COLLABORATION

Ecosystem ecology and biogeochemistry are interdisciplinary fields and thus are generally collaborative. The growth of ecosystem ecology over the past several decades has helped foster collaboration as new questions and discoveries require new skill sets. For example, extending the carbon cycle from soils, plants, and animals to the atmosphere required collaborations with atmospheric scientists. Similarly, following carbon and nutrients across the land–water interface required collaborations between terrestrial and aquatic scientists. The explosion of models as a tool in ecosystem ecology led to new collaborations with engineers, physicists, and computer scientists. This ongoing need for new skill sets is one of the primary drivers of collaboration. The LTER program has helped to foster collaboration by creating a platform (site-based science) for researchers to come together to address broad, complex, interdisciplinary questions.

The value of site-based science cannot be underestimated when it comes to building collaborations. Sometimes collaborations emerge from unexpected sources. On one of my early trips to Puerto Rico, I found myself staying at El Verde Field Station with most of the researchers on the newly funded grant to the LUQ site. Field station life facilitates interaction; actually, it necessitates interaction as people eat, drink, sleep, and work in close quarters. It was an exciting time: the researchers were meeting to plan experiments and discuss how they would coordinate this broadly multidisciplinary research effort. The PhD students quickly bonded. We commiserated about challenges, set up schedules to help each other in the field and laboratory, and generally enjoyed teaching and learning from each other. This experience allowed us to remove the boundaries we normally set up for ourselves as disciplinary scientists. As I began to explore collaborations with colleagues in other disciplines, it became clear that the only real limits were those imposed by my willingness to venture outside of my comfort zone. Perhaps the most memorable collaboration-spawning event of my career was a day spent sliding down steep muddy slopes in the rain with the late geomorphologist Fred Scatena, discussing the potential role of topographic position in ecosystem nutrient dynamics. We subsequently collaborated on 12 publications and undoubtedly would have continued if Fred's life had not been cut tragically short. By focusing efforts at a common field site, often with communal housing, dining, and laboratory facilities, the LTER program has fostered collaborations across a wide range of scientists.

Collaborations are not without challenges, and they are not "free." Good and lasting collaborations require significant time commitment, mutual respect, and the willingness to learn enough about your collaborator's expertise to be able to meld data collection, analyses, and writing phases of research. Most of us learn to collaborate by trial and error or by watching how mentors and colleagues manage their professional relationships. The LTER program provides a mechanism for this sort of learning. To enhance successful collaborations, especially across disparate disciplines, the community should develop more formal training activities for faculty, graduate students, and postdoctoral fellows focused on developing collaboration skills. These could include finding the balance between contributing to a group while maintaining one's own identity, how to be a good leader, and how and when to be a good follower. Developing these skills may help researchers avoid wasted effort or minimize unsuccessful collaborations, which can sap energy and resources and endanger the professional future of young scientists.

APPLIED RESEARCH

My LTER-related research would generally be characterized as "basic science," although one hopes that it would have an application down the road. Most of the research

I conduct outside of the purview of the LTER program is much more applied. There are two reasons why I focus on basic science for my LTER-related work. First, the nature of the peer-review process for LTER research requires that participants produce high quality basic science. Whether this is derived from the goals of the NSF or from the attitudes of the larger peer-review community is unclear, but strong, theory-based, hypothesis-driven science is a requirement for successful LTER proposals. Such science can be conducted in an applied setting. Although the ecological community has been slow to embrace it, applied research is becoming more common and programs at LTER sites such as the Kellogg Biological Station, the Central Arizona–Phoenix site, and the Baltimore Ecosystem Study site are helping. Second, it is more difficult to fund basic than applied scientific research in today's funding climate; thus, I choose to focus on basic science for my LTER-related research. Procedural changes at NSF have slowed the rate of scientific funding. This, together with an overall poor economic climate for academic pursuits, makes it more difficult to fund basic research in the ecological sciences. There is considerable evidence that basic research is one of the foundations of US competitiveness in the global marketplace (http://www.commerce.gov/sites/default/files/documents/2012/january/competes_010511_0.pdf). The LTER program plays an absolutely critical role in funding high-quality basic science.

There are other ways in which research in the LTER program can have application. The LUQ program is housed in a national forest. The research arm of the US Forest Service is distinct from the forest management branch, although each group informs the other. Being based in a national forest helps scientists in the LTER program stay informed about management issues and keeps managers aware of potentially relevant local science. Although the potential is great for increased interaction between managers and scientists, this has primarily taken place on an ad hoc or individual basis and not at a programmatic level. More dialogue between scientists in the LTER program and managers would undoubtedly increase application of the science produced and provide more opportunities for scientists to conduct applied science at LTER field sites.

COMMUNICATION

I have become an active science communicator, stemming from an LTER-related experience. From 2004 to 2006, I was involved in a network-wide activity to analyze data from a large-scale, long-term decomposition experiment, and I was a lead author on a paper that was published in *Science* (Parton and Silver et al. 2007). The media office at Berkeley wanted to do a press release (the first time I had been involved in one), and came up with a catchy headline "Rotting leaf litter study could lead to more accurate climate models." The day they posted this, I received a barrage of phone calls from journalists. Who would have thought that rotting leaves had so much allure! The outcome of my initial foray into science communication was disappointing. I was misquoted several times, and the story was thoroughly mangled by a few media outlets. At the same time, I was engaged in other research with significant policy implications and general appeal and knew that I was likely to have more interaction with the media and the public. Thus, I pursued training through the Leopold Leadership Program and a Google Science Communication Fellowship. These activities helped me to minimize the media disasters (but not eliminate them) and to better incorporate science communication and outreach into my career.

NSF mandates that the data collected with NSF funds be made available to the public. When I first started working at the LUQ site, I viewed this requirement with a fair amount of dread. As a young scientist, I had been warned about having my ideas scooped, or worse yet, having my own data published before I had a chance to write it up myself. I was overwhelmed by the effort that it took to organize and upload data and metadata. There never seemed to be enough time to collect the data, analyze it, and publish the results. How was I going to find the time to organize the data for public distribution? Although the LUQ site had a knowledgeable data manager, the infrastructure was not in place at that time to make this process easy or efficient. Over time, however, the system improved and became an integrated component of my research program. I am still far from expert in providing the metadata and data formats to make my data as useful as possible to others, but I think I have improved. I have never been scooped and no longer fear data sharing. Data management is now discussed at the inception of projects in my laboratory, not just at the end. My students are expected to be aware of and follow data management policies. Furthermore, I have made use of LTER data in courses, which shows students the value of data sharing. I still wish it were easier to enter data and search for it, but I suppose that this, too, will continue to improve over time.

I expect that most scientists feel that their research has benefited society in some way, either directly or indirectly. Benefiting society is an explicit goal of all research in the LTER program. Most of my research at the LUQ site has focused on the causes (greenhouse gas emissions) and consequences (drought, changing storm frequency) of global change. Understanding how tropical forests respond to environmental change and how this feeds back to climate change drivers can help decision-makers legislate or manage for human and environmental health. My department at Berkeley is explicitly designed to address key environmental issues of relevance to society. Thus, our research and the training provided to my students and postdoctoral fellows help support this component of the LTER mission. Teaching students and training the next generation of scientists is perhaps the greatest contribution to society associated with our research.

MENTORING

NSF requires that funded research advances discovery and understanding while promoting teaching, training, and learning; broadens participation of underrepresented groups; enhances infrastructure for research and education; broadens dissemination to enhance scientific and technological understanding; and generally benefits society. These broader impacts are required in addition to the transformative science researchers must conduct in a highly competitive environment with less funding (per scientist) per year. A tall order! However, it is important to remember that for society, the broader impacts are the most immediate and obvious result of most scientific research.

As a professor, teaching, training, and mentoring are a major part of my job description. I have involved K–12, high school, undergraduate, and graduate students, as well as postdoctoral fellows, in LTER-related activities. This is clearly one of the most satisfying aspects of my career. I have fostered and mentored young people as they discover science, get "hooked," exceed their expectations, overcome limits and difficulties, apply and get into graduate schools, receive competitive postdoctoral fellowships, find full-time employment, and get tenure. It does not get much better than that! The LTER program encourages and supports educational outreach and training through the Research Experience for Undergraduates (REU) program and through sharing information among interested researchers and sites. I have taken advantage of REU opportunities, avidly read newsletters, and attend lectures at national meetings on education and training. I have spoken to graduate students and postdoctoral fellows at an LTER-sponsored national forum about career choices. I explained why I think that being involved in an LTER site at any level is a great idea for all the reasons outlined above.

Working in Puerto Rico has made it easy to increase participation of underrepresented groups. Two of our Puerto Rican technicians have gone on to PhD programs, and

I have recruited several outstanding Hispanic students to the UC Berkeley PhD program. Two of my former Hispanic PhD students worked on LTER-related research, went on to receive competitive postdoctoral fellowships, and are currently employed as tenure track faculty at top research universities. As a female faculty member, I have trained and mentored a large number of women students and postdoctoral fellows, many of whom are involved in research in the LTER network. I have also mentored other women faculty and greatly benefited from being mentored by women. My experience is that the LTER network is very friendly and welcoming for women and minority scientists.

SKILL SET

The LTER program can facilitate the improvement of infrastructure for science when funding is available. In my experience, this occurred through supplemental grant competitions. We applied for and received funding to replace and upgrade meteorological equipment for long-term climate monitoring and purchase analytical instrumentation for research. In this way, the LTER program has added to my scientific toolbox. This type of funding is critically important for the success of long-term ecological research. Equipment wears out and needs to be replaced. New technologies become available that can improve the efficiency, precision, or accuracy of measurements. Network-scale and cross-site research has allowed me to broaden the scope of my research and provide a mechanism for infrastructural improvements.

PERSONAL CONSEQUENCES

Being a scientist in the LTER program has affected my personal and professional life in many ways. I live far from the LUQ site—a full day's travel at least. The time I am in Puerto Rico is time away from my family. It is too far to go for short trips, although I occasionally do just that, spending more time on airplanes than on the ground. Spending 1 to 3 weeks in the field has meant missing important events such as class plays, parent–teacher conferences, birthdays, and anniversaries. As a scientist I do not like to share this information; I worry that it makes me sound less dedicated and committed than my colleagues. However, the reality is that this is painful and an unfortunate consequence of my professional choices. At the same time, when I go to Puerto Rico I can fully engage in science with no other distractions or responsibilities. I become "100% scientist," something that is impossible when I am at home or in my office. I value these trips tremendously, because it professionally renews and refreshes me. LUQ may be different from some of the other LTER sites in that most of the researchers must travel to get to the field. In my experience, this has made us a tightly knit, highly collaborative group. We live and work together at the field site, with few distractions, and it seems that all of us are on a constant "research high." It is wonderful, and has created a productive, supportive, enjoyable environment.

CHALLENGES AND RECOMMENDATIONS

In summary, the LTER program has clearly influenced many aspects of my career and the ecological sciences in general. My almost two decades of support (financial and infrastructural) from the LTER program has facilitated the development of a large body of science, trained 12 PhD students, 10 postdoctoral fellows, countless undergraduates, and yielded approximately 100 publications. This summer we will sample my thesis sites 25 years after they were established. I owe this opportunity to the vision of the LTER program and the commitment to long-term science of the International Institute of Tropical Forestry of the US Forest Service, an LTER partner.

What can the LTER program do better? As LTER scientists, and as a network, we need to improve our communication skills. We need to inform the public and policy makers of the value of long-term ecological research. We need to advocate more strongly for continued and enhanced funding to support a long-term perspective in science, both within and beyond the LTER program. We need to practice and improve communication with each other to enhance and facilitate collaboration. We need to train our students and postdoctoral fellows how to be successful communicators and collaborators. We also need to help them develop their leadership skills, as we will be passing the torch to them as the next leaders of the LTER program. The LTER network needs to remain an open network, maintaining space for new contributors and collaborators. Special or dedicated funding lines would greatly facilitate this. Finally, we need to train our institutions to respect, value, and reward large-scale, long-term collaborations such as those in the LTER program. New, collaborative science networks are being developed that have learned from the LTER model (e.g., the Critical Zone Observatory, the National Ecological Observatory Network, Urban Long-Term Research Area). Institutional support at all levels of science is critical to maintain the success of these valuable ventures.

CONCLUSION

In summary, "growing up" academically and professionally in the LTER program has afforded me many opportunities and experiences that I value greatly. I have learned the worth of strong collaborations, and how to be an effective collaborator. Moreover, I have learned the value of taking a long-term perspective in my research, not simply viewing my career as a compilation of 3 or 4-year projects, but of a lifetime of inquiry. This long-term perspective is at the heart of LTER and what will continue to distinguish this program and LTER scientists into the future.

ACKNOWLEDGMENTS

Funding was provided by NSF grant DEB-0620910 to the Institute for Tropical Ecosystem Studies, University of Puerto Rico, and to the International Institute of Tropical Forestry US Forest Service, as part of the LUQ LTER program. W. L. Silver received additional support from the US National Institute of Food and Agriculture, McIntire Stennis project CA-B-ECO-7673-MS.

REFERENCES

Hall, S.J., and Silver, W.L. (2013). Minerals masquerading as enzymes: Iron oxidation stimulates soil carbon decomposition. *Global Change Biology,* **19**, 2804–2819.

Parton, W., Silver, W.L., Burke, I.C., Grassens, L., Harmon, M.E., Currie, W.S., King, J.Y., Adair, E.C., Brandt, L.A., Hart, S.C., and Fasth, B. (2007). Global-scale similarities in nitrogen release patterns during long-term decomposition. *Science,* **315**, 362–364.

Silver, W.L., Herman, D.J., and Firestone, M.K. (2001). Dissimilatory nitrate reduction to ammonium in tropical forest soils. *Ecology,* **82**, 2410–2416.

Yang, W.H., Weber, K.A., and Silver, W.L. (2012). Nitrogen loss from soil via anaerobic ammonium oxidation coupled to iron reduction. *Nature GeoScience*, **5**, 538–541.

PART TWELVE
MOOREA CORAL REEF (MCR) LTER SITE

27 Kelp Forests, Coral Reefs, and the Long-Term Ecological Research Program

Synergies and Impacts on a Scientific Career

Sally J. Holbrook

IN A NUTSHELL

1. Involvement with the Long-Term Ecological Research (LTER) program has enabled me to ask novel and exciting science questions at larger spatial and longer temporal scales than I could have otherwise. It has enhanced my ability to engage in interdisciplinary collaborative research.

2. The LTER program has afforded my graduate students a variety of opportunities that have enhanced their training and experiences as early career scientists. My undergraduate students learn about LTER research findings in my classes and have the opportunity to work as research assistants in the field and the laboratory.
3. My experiences with LTER-funded research have made me aware of the importance of community and K–12 outreach, and it has provided me opportunities to plan such activities.
4. Engaging in the LTER program has provided me with a myriad of opportunities to collaborate with other sites and groups to address network-level science questions. My collaborators include investigators from within the LTER network, as well as international scientists.

PERSONAL OVERVIEW

My experience in the LTER network began in 2000, when the Santa Barbara Coastal (SBC) LTER project was established, and expanded in 2004 with the founding of the Moorea Coral Reef (MCR) LTER site. I have been a co–principal investigator at both of these sites since their inception. Because I am a marine community ecologist, my research interests and those of my graduate students are closely aligned with the goals and activities of both sites. My LTER network-level experiences include a 3-year term on the LTER Executive Board, participation in several LTER All Scientists Meetings, and a network-sponsored working group on abrupt state shifts.

Currently, I am a professor of ecology in the Department of Ecology, Evolution, and Marine Biology at the University of California (UC) Santa Barbara. My disciplinary background is population and community ecology, and prior to my involvement with the LTER program, my research and that of my students focused mainly on questions related to population dynamics and species interactions. Involvement in LTER projects has enabled me to greatly expand the spatial and temporal perspectives of my research, including a cross-site synthesis of abrupt transitions in ecological communities (Bestelmeyer et al. 2011), as well as landscape-scale analyses of responses to disturbances in the kelp forests of the SBC site (Byrnes et al. 2011) and on the coral reefs of the MCR site (Figure 27.1) (Adam et al. 2011; Adam et al. 2014). It has provided many opportunities to interact with scientists, both in the United States and internationally, that otherwise would not have been possible.

APPROACH TO SCIENCE

My involvement with two LTER sites has been a transformative experience. Both the SBC and MCR sites provide a platform for collaborative interdisciplinary research on spatial and temporal scales that simply cannot be achieved by individuals or small teams of investigators, and this has affected the type of research questions I can ask as well as the potential to work collaboratively. The LTER sites also afford exceptional opportunities for training students at all levels, as well as outreach and communication of scientific findings to the public. Participating in the LTER network has allowed me to interact with a broad range of ecologists who work in different ecosystems, from terrestrial to marine, and who address a range of ecological issues with a strong temporal perspective. Finally, I am principal investigator of a recently funded National Science Foundation (NSF) grant through their program on Science, Engineering, and Education for Sustainability, which is examining ecological resilience in a coral reef socioecological system. Our integrated

FIGURE 27.1 Sally Holbrook conducting field research at the Moorea Coastal Reserve Long-Term Ecological Research site in French Polynesia. (Photo courtesy of Melissa Schmitt.)

social and natural science approach is addressing questions about resilience, sustainability, and adaptive capacity of Moorea's coastal system. This project, and my involvement in it, is a direct result of my LTER experiences.

ATTITUDES TOWARD TIME AND SPACE

Prior to my involvement with the LTER program, the major focus of my research was on population and community dynamics of marine reef organisms, primarily fishes. I explored such questions as the effect of biotic interactions (interspecific and intraspecific competition, predation), habitat availability, and environmental forcing on patterns of distribution, abundance and dynamics of focal species. Although these issues continue to be of great interest to me, in recent years I have collaborated more extensively to address a variety of new science questions, many of which cover larger spatial and longer temporal scales than I would have previously thought possible. This is clearly a result of the fact that LTER sites are intended to examine ecological processes over multi-decadal time scales, and they receive continuous funding to support consistent collection of data. As a result, time series data on critical physical and biological variables are freely available. Experiments are carried out over decades (or longer). At both the SBC and MCR sites, field experiments are being conducted on spatial and temporal scales that as a single investigator I simply could not do. For example, physical disturbances (mainly from winter storms) periodically result in loss of the foundation species (giant kelp) on near shore reefs in California, and these disturbances may increase with climate change. We designed a long-term field experiment to explore effects of variation in frequency of winter storm disturbance to loss of giant kelp on the structure and function of the kelp forest community. It enabled one of my graduate students to assess impacts of disturbance on primary productivity and a postdoctoral fellow to study changes in the food web structure of the kelp forest produced by disturbance. In addition to long-term experiments, the time series data are a rich research resource for both investigators and students at LTER sites. Recently, one of my graduate students who has expertise in fisheries modeling used SBC time series data collected by Russ Schmitt and me to explore the mechanisms driving the population dynamics of black surfperch, an abundant kelp forest inhabitant. We used our three-decade-long data set on age-structured fish abundance and their food resources and demonstrated the importance of explicitly incorporating temporal variation in food supply in commonly used fisheries models. These examples illustrate just a few ways in which the spatial and temporal scales of the science questions that I can ask have been affected by LTER experiences.

COLLABORATION

Engaging in the LTER program has provided me with a myriad of opportunities to collaborate with other sites and groups to address network-level science questions. The LTER All Scientists Meetings are a platform for interactions across the broad ecological community, where attendees can engage in workshops on a huge range of topics. I joined a workshop on ecological state shifts at the 2009 All Scientists Meeting. The group was excited about the possibility of pursuing the topic further, using time series data from LTER sites in analyses. We obtained funding from the LTER network office for two additional meetings. Multidecadal (28–58-year) data sets from four LTER sites (California Current Ecosystem, Jornada Basin, Palmer Antarctica, and SBC) were utilized to explore whether and how often community state shifts occurred, delineate underlying mechanisms, and assess whether state shifts can be forecasted in time series data. The workshops convened scientists at different career stages (including graduate students) who work in different systems on a variety of organisms (sea cucumbers, grasses, penguins, zooplankton). We were challenged to find common ground both conceptually and analytically to make a scientific advance and produce a publishable product (Bestelmeyer et al. 2011). From the onset, our efforts were made possible by the LTER program, starting with the initial interactions at an All Scientists Meeting, to funding for working groups from the network office, to the time series data that had been gathered at the LTER sites. Examples of my more recent collaborations include a SBC–MCR cross-site comparison of trophic relationships in reef ecosystems (Page et al. 2013) and an interdisciplinary study of how biological and physical processes affect cross-shelf transport on reefs of Moorea (Leichter et al. 2013).

Operating within the LTER network has increased my awareness of cross-site interactions and predisposed me to move outside the network and engage other sites and groups. One good example of this is the Coral Reef Environmental Observatory Network (CREON). This is a grassroots effort that I cofounded with scientists from Taiwan and Australia. Its goal is to use our study sites as test beds for the development of underwater observing systems, as well as to develop middleware for real-time data availability. The latter is critically important in the marine environment, where physical oceanographic instruments are often deployed for months or longer and data are only available on recovery. As a result of CREON, over the years MCR has engaged in technology transfer with scientists who work on the Great Barrier Reef, Australia, and in Taiwan at the Kenting International LTER (ILTER) site. Our site has been a test bed for developing instruments and real-time data transmission. I have been working with several MCR colleagues and researchers in Thailand to deploy physical oceanographic instrumentation and initiate biological sampling for time series at their study site on Racha Island, with the possibility that that site might become part of the ILTER network in the future. None of these activities would have taken place if the MCR site did not exist.

APPLIED RESEARCH

My LTER experience has not really changed my appreciation of applied science. I have had a long-standing interest in natural resource management and conservation, and a small portion of my research activity has consistently been devoted to those issues. What my LTER involvement has done is cause the focus of those activities to change somewhat. Prior to 2000 and the initiation of the SBC LTER, I had worked, for example, on several projects relating to impacts of oil and gas development in the nearshore marine environment, as well as one that focused on development of restoration techniques for surfgrass. As my time constraints increased with the inception of the MCR LTER, my focus shifted to issues more centrally related to the SBC and MCR LTER projects and to my core interests in fish ecology. The more applied research that I do now focuses on fish conservation and management issues. Because the MCR LTER is a recognized entity within its local community on Moorea, it has afforded me enhanced opportunities to interact with stakeholders, particularly local fishers.

COMMUNICATION

My undergraduate teaching has been positively affected by my LTER experiences. In virtually every course I teach, findings from LTER research are presented and identified as such. I use examples to illustrate effects of climate change, the value of long-term, landscape scale studies in contributing to understanding ecological processes, and the role of interdisciplinary approaches in science, to name a few. Beyond that, I acquaint students with the LTER mission, and more specifically, with the broad range of activities carried out at the SBC and MCR sites. Involvement in the LTER program has definitely involved me in the planning of a variety of community and K–12 outreach activities, but my direct involvement in them is somewhat limited.

MENTORING

LTER sites afford several types of experiences that are unique and in my opinion contribute greatly to student training. First, because SBC and MCR both involve relatively large research groups, there are ample opportunities for tiered mentoring, a process that I strongly believe is crucial to professional training. Undergraduate students, graduate students, postdoctoral fellows, and investigators work in teams in the laboratory and in the field. For example, a number of my graduate students have taken advantage of summer grants awarded by my department that fund graduate-undergraduate student mentorship teams for summer research. The undergraduate student helps the graduate student with his or her research, and in turn the undergraduate initiates research for a Senior Honors Thesis under the mentorship of his or her partner. In the case of MCR, the site subsidizes the travel, laboratory fees, and research costs for the undergraduate mentee, enabling them to spend 3 months working in the field at MCR. Graduate students in turn are mentored not only by the postdoctoral fellows, but also by other LTER personnel, including the information manager and site technicians. Not only do they learn valuable field and laboratory techniques, tools for data analysis, and so on, but they all become aware of the importance of data management and data accessibility.

A second benefit that students derive from association with the LTER program is that they become members of an academic research community. My students (at all levels) attend our annual LTER site meetings. The MCR has 18 investigators from six academic institutions (UC Santa Barbara, California State University Northridge, UC San Diego, UC Santa Cruz, Duke University, University of Hawaii), and we place a high value on training a diverse group of students. Students from the various campuses become acquainted with everyone associated with the project. They are exposed to interdisciplinary science and a diversity of research approaches. Increasingly, we are having students move among the MCR institutions for different stages of their training. Several undergraduates who have worked in my laboratory have moved to California State University Northridge, for example, to do MS degrees, and then on to work with another MCR investigator at yet a different campus to pursue a PhD. An MCR postdoctoral fellow from UC Santa Barbara is now a faculty member at the University of Hawaii. This phenomenon is a direct outcome of the mentorship and community- building that occur within the MCR program.

A third benefit to students is the provision of unusual opportunities that likely would not occur in the absence of their participation at an LTER site. An example of this is MCR's contribution to the LTER Children's Book Series, *Kupe and the Coral*. Out of curiosity, an MCR graduate student attended a workshop a few years ago on the Book Series at an LTER All Scientists Meeting in Estes Park. She became sufficiently interested and motivated that she proposed to author a children's book for the series. The project started when she was a graduate student (at the University of Hawaii) and was completed during her postdoctoral fellowship at UC Santa Barbara. Support for the production of the book was provided by NSF as part of its effort to disseminate findings of LTER science to the broader community. The story involves the discovery of the coral life cycle by a young Tahitian boy. A unique aspect of the book is that it has several bilingual versions of five different languages (English, French, Tahitian, Spanish, Paumotu), because it is designed to reach classrooms in French Polynesia, Hawaii, and California. The graduate student author did the Spanish translation herself. The project also drew in other MCR participants: one investigator aided in translation and another (in partnership with an MCR undergraduate) did the photography on which the illustrations were based. Clearly, these types of experiences are enriching for our students and in this case, were only possible because of the student's affiliation with the LTER program.

SKILL SET

All LTER sites have websites that, among other tasks, make available data from the time series and other studies. Information management and dissemination of data are central goals of LTER sites, and MCR and SBC perform this function far more efficiently and better than an individual investigator could. In my own research I have always valued long-term data. However, being involved with LTER sites and the LTER network has sharpened my appreciation of the importance of making such data easily accessible, and the technical challenges involved in data management. In my experience with the LTER program, I have learned a great deal about the importance of data management and accessibility, and techniques for accessing and using archived data sets.

PERSONAL CONSEQUENCES

There have been a variety of personal consequences arising from my participation in the LTER program. Through our regular meetings (e.g., Science Council) and other events (e.g., workshops, All Scientists Meetings), I have had the opportunity to interact with many different types of ecologists. This has led to new friendships, which I value highly. I especially enjoy the chance to interact with people outside my own discipline (marine ecology), and to visit other sites in the network and become acquainted with their science. The LTER program has also directly or indirectly resulted in extra travel, which has been mostly a positive experience.

CHALLENGES AND RECOMMENDATIONS

The existence of LTER sites encourages larger scale, longer-term ecological research, and often projects are done by groups of investigators and students. Data may arise from site time series or long-term experiments that may have been conducted by others. I view this as a very positive phenomenon that enables ecologists to do "big science," but it can pose challenges for how we evaluate early career scientists for academic appointments, grants, and tenure. Group science often obscures individual contributions, which can be great or large to any specific endeavor. In addition, early career scientists can be overshadowed by the senior investigators (in terms of perceived intellectual contributions). These are challenges our field must cope with as we move forward into an era of greatly increased collaboration and larger scale projects.

Many ecologists have not been trained in an environment where data sharing is valued and accepted, and the LTER program has done much to change those views. My graduate students make use of LTER data and increasingly view data sharing as the norm rather than the exception. This will only increase in the future as more and more ecological data become widely accessible. However, increasing data sharing via truly sound mechanisms of data management remains a challenge, technically, financially, and culturally.

Another challenge that faces all LTER sites is to bring the next generation of scientists into positions of leadership. This will require an open policy for people wishing to engage at LTER sites and active mentorship to develop their scientific and leadership skills.

CONCLUSION

I view the effect of my LTER experiences on my career to be overwhelmingly positive. Involvement with the LTER program has enabled me to ask different science questions at larger spatial and longer temporal scales than I could have otherwise. It has afforded my students a variety of opportunities that have enhanced their training and experiences as early career scientists. I have interacted with many wonderful people, both as LTER colleagues and scientific collaborators. My unequivocal advice to junior scientists would be to get involved.

ACKNOWLEDGMENTS

I would like to acknowledge my colleagues at the SBC and MCR LTER sites, including the many investigators, students, and technicians who contribute so much to the success of these projects. Financial support from the NSF (OCE 1236905, OCE 1232779 and earlier awards) and the Gordon and Betty Moore Foundation is deeply appreciated. Thanks to N. Davies, F. Murphy, and H. Murphy, as well as the staff of UC Berkeley Richard B. Gump South Pacific Research Station and UC Santa Barbara for logistic support.

REFERENCES

Adam, T.C., Brooks, A.J., Holbrook, S.J., Schmitt, R.J., Washburn, L., and Bernardi, G. (2014). How will coral reef fish communities respond to climate-driven disturbances? Insight from landscape-scale perturbations. *Oecologia*, **176**, 285–296.

Adam, T.C., Schmitt, R.J., Holbrook, S.J., Brooks, A.J., Edmunds, P.J., Carpenter, R.C., and Bernardi, G. (2011). Herbivory, connectivity and ecosystem resilience: Response of a coral reef to a large-scale perturbation. *PLoS ONE*, **6**, e23717.

Bestelmeyer, B.T., Ellison, A.E., Fraser, W., Gorman, K., Holbrook, S.J., Laney, C.M., Ohman, M., Peters, D.P.C., Pillsbury, F.C., Rassweiler, A., Schmitt, R.J., and Sharma, S. (2011). Abrupt transitions in ecological systems: What have we learned? *Ecosphere*, **2**, 129.

Byrnes, J.E., Reed, D.C., Cardinale, B.J., Cavanaugh, K.C., Holbrook, S.J., and Schmitt, R.J. (2011). Climate driven increases in storm frequency simplify kelp forest food webs. *Global Change Biology*, **17**, 2513–2524.

Leichter, J.L., Alldredge, A.L., Bernardi, G., Carlson, C.A., Carpenter, R.C., Edmunds, P.J., Fewings, M.R., Hanson, K.M., Holbrook, S.J., Hench, J.L., Nelson, C.E., Schmitt, R.J., Toonen, R.J., Washburn, L., and Wyatt, A.S.J. (2013). Biological and physical interactions on a tropical island coral reef: Transport and retention processes on Moorea, French Polynesia. *Oceanography*, **26**, 52–63.

Page, H.M., Brooks, A.J., Kulbicki, M., Galzin, R., Miller, R.J., Reed, D.C., Schmitt, R.J., and Holbrook, S.J. (2013). Stable isotopes reveal trophic relationships and diet of consumers in temperate kelp forest and coral reef ecosystems. *Oceanography*, **26**, 180–189.

28 The Long-Term Ecological Research Construct for Understanding Dynamics of Coral Reef Ecosystems and Its Influence on My Science

Russell J. Schmitt

IN A NUTSHELL

1. The Long-Term Ecological Research (LTER) program facilitated my scientific growth in terms of the questions I can address; the tools, approaches, techniques and data to which I have access; and the diversity of intellectual and disciplinary expertise that I can tap. As a consequence, I am asking questions that cut across much larger spatial and especially temporal scales, and my research projects are more interdisciplinary, complex, and integrated.
2. My ability to mentor students at all levels has been transformed by the variety of resources and opportunities afforded by the LTER program. One consequence is that these students are better prepared to become engaged globally.
3. My role in the LTER program has required me to communicate scientific issues and findings to a broad audience. I have become more interested in the translation of science findings to public policy and practices to help conserve key functions of threatened ecosystems.
4. My involvement with the LTER program has enabled me to forge a much larger circle of national and international collaborators to address questions that require a network of similar sites. The LTER construct has enabled me to broaden the scope of my research by expanding the interdisciplinary nature of my collaborations and the diversity of tools at my disposal.

PERSONAL OVERVIEW

My involvement with the LTER program began in 2000 when I joined the Santa Barbara Coastal (SBC) site as an associate investigator, and it expanded in 2004 to include being the principal investigator of the newly established Moorea Coral Reef (MCR) site. I am privileged to continue to serve as principal investigator of MCR and as an associate investigator at SBC.

My research interests center on ecological processes and feedbacks that drive the dynamics of populations and communities. Prior to my involvement in the LTER program, I conducted my research projects either alone or with a small group of like-minded collaborators to address such issues as regulation of (marine) animals with open populations or the effect of indirect interactions on coexistence of species (Figure 28.1). These projects taught me some of the limitations of "small science," particularly when exacerbated by a lack of relevant long-term data.

APPROACH TO SCIENCE

My experience with the LTER program has greatly altered my approach to science. Ecology is a discipline where questions traditionally have been addressed by individuals or small groups who typically have a focus that is constrained in space, time, and scope. In addition, it currently is a discipline where theory, models, and empiricism are not consistently coupled tightly, reflecting in part the ongoing progression of ecology from its purely descriptive origin in the last century to a mature, predictive science. As a field experimental ecologist, I have been frustrated with some of the limitations associated with the standard approach to ecology. With a rapidly increasing human footprint on ecological systems globally, most of my colleagues and I recognize the acute need for environmental science in general and ecology in particular to predict the dynamics of linked socioecological systems and the consequences of those dynamics to ecosystem functions of value to society.

FIGURE 28.1 Russell Schmitt diving with an underwater camera at the Moorea Coastal Reserve Long-Term Ecological Research site. (Photo courtesy of Melissa Schmitt.)

Meeting these challenges requires a shift in approaches and tools. I have come to appreciate that the LTER program enables scientists to effectively address ecological phenomena that occur over long time and large spatial scales by providing a network of temporally stable research platforms that focus on a common set of fundamental issues across representative ecosystems. In a sense, then, the LTER program can be viewed as a "big science" element of US-funded ecology. Before joining the LTER program, my research was limited by familiar approaches and tools (e.g., Schmitt 1985). Since joining the LTER program, my questions have shifted to encompass drivers of long-term community dynamics, particularly with regard to exploring disturbance ecology in the context of slowly changing environmental drivers associated with global climate change and ocean acidification.

ATTITUDES TOWARD TIME AND SPACE

Prior to my involvement in the LTER program, I used marine reef fishes to conduct short-term field studies to explore population regulation (e.g., Holbrook and Schmitt 2003). Many of my results were counter to dogma at the time and engendered debate (Osenberg et al. 2002). Time series data on the actual dynamics of these fish populations would have provided a strong test of our model predictions and a context in which to judge the ecological importance of the regulatory mechanisms that my studies revealed. This experience reinforced to me the largely unrealized need to connect short-term process studies with the actual dynamical behavior of systems, which is the fundamental model of science in the LTER program.

This construct also has greatly broadened the scope of my research by expanding the interdisciplinary nature of my collaborations and the diversity of tools at my disposal (e.g., Leichter et al. 2013). As a consequence, I am able to address science questions on spatial and temporal scales (Schmitt and Holbrook 2007; Adam et al. 2011) and across levels of biological organization that otherwise are challenging or impossible to do without the infrastructure and interdisciplinary framework of an LTER site. For example, this enhanced capacity, coupled with time series data on key environmental and biological attributes, enables my colleagues and me to address a fundamentally important contemporary issue for coral reef ecosystems (processes that affect ecological resilience). I am exploring the possibility that observed shifts from corals to fleshy seaweeds in some tropical reef communities are difficult to reverse. If true, this process has tremendous management implications, yet it is an issue that is unresolved and highly contentious. The LTER program provides an ideal construct to address these types of issues, and, as it turns out, the MCR site is an ideal model system. The controversy regarding reversibility of coral to algal community shifts on coral reefs has not been successfully addressed previously because it requires the intellectual capital, breadth of expertise, logistical support, long-term knowledge base, and data-rich resources that only an LTER-like entity can provide. In this context, the long-term data and experiments from the MCR site are producing tremendously exciting findings that are tangible evidence of the impact of the LTER program in enhancing my ability to address problems of larger spatial and temporal scales.

COLLABORATION

The LTER network is part of a much larger International Long-Term Ecological Research (ILTER) network, and it also forges research collaborations with a number of other LTER-like sites. The MCR site is the only US LTER site that is not physically situated on US-controlled territory (the site is in an overseas territory of the Republic of France and operates under an agreement between the University of California and the Territorial Government of French Polynesia). Hence, MCR investigators inherently have an international focus. However, in addition to collaborating with local French and Tahitian scientists, involvement with the MCR program has enabled me and my colleagues to forge a much larger circle of international collaborators to address questions that require a network of similar sites. For example, we collaborate with scientists from Australia (e.g., Messmer et al. 2011) to determine whether the loss of coral diversity will have the same effect on fish communities along the geographic gradient in biodiversity across the Pacific. We also cooperate with coral biologists from Taiwan and Japan to understand how the same species of corals in different localities respond to differences in oceanographic regimes. These international research collaborations have enhanced my ability to generalize results from the MCR site, as well as help reveal reasons underlying context-dependent responses.

APPLIED RESEARCH

Although I have long been interested in the application of basic science to applied issues, the LTER program greatly altered my ability to conduct strategic science that has immediate value for resource managers and environmental policy makers. Prior to my involvement in the LTER program, my applied worked focused largely on assessments of the environmental impacts of planned development projects (e.g., Osenberg and Schmitt 1992). Although I was satisfied in the sense that this effort contributed directly to society, I was frustrated in that I basically was focusing cutting-edge techniques to measure how badly natural resources were being degraded, rather than proactive measures to conserve, preserve, or restore critical ecological resources. The MCR program provided me with the perfect opportunity to shift my applied research perspective to understanding the key ecological interactions and feedbacks that underpin ecosystem resilience, determine the vulnerabilities to human activities, and provide scientific information on which to base resource management practices. One of the things that excited me the most is the relevance that my findings about ecological resilience have for resource managers and policy-makers. For example, we know the crucial value of parrotfishes in preventing a disturbed coral reef from shifting to a seaweed-dominated system and that the nursery habitat for parrotfishes must be protected from degradation (Adam et al. 2011). Consequently, I personally have become more interested in the translation of my science findings to public policy and practices to help conserve key ecosystem functions of coral reefs. Lessons relevant to sound resilience practices and policy are clear, general, and transferable, and I think these are key contributions of studies across the LTER network.

COMMUNICATION

My undergraduate teaching has been positively affected by my experiences in the LTER program. In virtually every course that I teach, findings from research at LTER sites are presented and identified as such. I use examples to illustrate effects of climate change, the value of long-term, landscape-scale studies in contributing to understanding ecological processes, and the role of interdisciplinary approaches in science, to name a few. Beyond that, I acquaint students with the mission of the LTER program, and more specifically, with the broad range of activities carried out at the SBC and MCR sites. Involvement in the LTER program has engaged me in the planning of a variety of community and K–12 outreach activities, but my direct involvement in them is somewhat limited.

MENTORING

The nature of science throughout the LTER network, together with the international setting of the MCR, enable my colleagues and me to train graduate students and postdoctoral fellows in a much more synthetic, integrative, and global fashion. Students take full advantage of time series data to explore ecological responses to environmental drivers as well as to motivate field experiments or inform, parameterize, or test models. Students and postdoctoral fellows associated with LTER sites also have interdisciplinary approaches and techniques ingrained into their scientific thinking and experience, and they therefore tend to be more broadly trained than is typical in single-investigator laboratories. Students and postdoctoral fellows have access to the tremendous intellectual capital of their particular LTER site as well as the LTER network as a whole. The conduct of their science and scope of their research projects also benefit greatly by the physical infrastructure and logistical support of LTER sites. A substantial added-value component is mentoring in proper data documentation and information management. In addition to developing a skill set in this regard, students are exposed to a culture where data sharing is the norm, which contrasts greatly with historical practices and expectations in the environmental sciences. Finally, one of the things that the LTER program has fostered well is a sense of community; graduate students and postdoctoral fellows self-identify with the community and take great pride in participating in the network. MCR students and postdoctoral fellows have taken great advantage of the network aspect of the LTER program by, for example, participating in or organizing cross-site and other synthetic projects that take advantage of time series data and intellectual capital.

The MCR site is one where graduate students gain tremendous international experience. The need for US scientists to operate at a high level in an international research environment is paramount, particularly given the global nature of many pressing environmental issues. An effective mechanism to nurture the development of such a community is to expose scientists to meaningful international experiences early in their careers. One way undergraduate, graduate students, and postdoctoral fellows associated with the MCR gain such an experience is by conducting field work at the MCR site in French Polynesia, where they are immersed in the local Polynesian culture and interact with Tahitian and French scientists (including graduate students). This has resulted in several of our students and postdoctoral fellows forming international research collaborations. In addition to site-based international experiences, graduate students from the LTER program have participated in a number of highly enriching activities beyond the MCR site. For example, the MCR program has partnered with programs at coral reef sites in Taiwan and Okinawa (Japan) to exchange graduate students and postdoctoral fellows.

Early-career scientists at the MCR site also have the unusual opportunity to engage in international capacity building. For example, the MCR program is partnering with a research group from Walailak University, Thailand, to foster development of a coral reef ILTER site on Racha Island, Thailand. In 2013, four graduate students from the MCR and SBC LTER sites accompanied four senior personnel from MCR, including me, to an International Coral and Reef Fish Workshop on Racha Island. The purpose of the workshop was to train Thai students and scientists in techniques for gathering biological time series information on coral reefs. The US graduate students gave presentations on relevant topics and assisted with all training activities. Finally, graduate students from the MCR program actively participate in K–12 and public outreach efforts. They are involved in developing content and delivering it to K–12 students through our Schoolyard Program and partnerships with local schools in the United States and in French Polynesia. Of course, as the principal investigator of the MCR program, I have become far more engaged in public outreach and education that I ever would (or could) have been without the construct and resources of an LTER program.

SKILL SET

Ironically, perhaps the greatest increase in my personal scientific skill set from being involved in the LTER program centers on information management, including current standards for documentation, archival, and data providence, among others. The scale at which the LTER information management is done is enormous yet dynamic, so this continues to be an area where my colleagues are still learning.

Intellectual byproducts from the creative activities of the scientific community in the LTER network as a whole also have enhanced my scientific tool kit. For example, I participated with other scientists in developing analytic frameworks to explore the nature of abrupt community shifts at several LTER sites using long-term data on community dynamics and underlying environmental drivers (Rassweiler, Schmitt, and Holbrook 2010; Bestelmeyer et al. 2011). A critically important issue to scientists, resource managers, and policy-makers alike concerns the type of nonlinear dynamics that underlie abrupt shifts in community composition, which determine whether the shift is readily reversible when conditions abate. I can now address this issue because, among other things, the logical frameworks developed by scientists in the LTER program provided the key to the design of field experiments that can unambiguously distinguish between the two types of abrupt (nonlinear) community shifts; virtually all previous experimental explorations of this issue used flawed designs. Hence, my involvement in the LTER

Program has provided me with new analytical tools that enabled me to tackle this crucial issue when the opportunity arose at the MCR site.

PERSONAL CONSEQUENCES

Of course as a principal investigator of a new site, there have been personal costs and benefits. Among the costs have been a reduction in my ability to personally lead single investigator–initiated research projects from inception through to publications, but this has been more than offset by the far broader range of collaborative projects with which I have been involved. Other personal costs of leadership involve taking responsibility for failure and spending enormous time program-building, which of course are offset by successes and a genuine and palpable sense of community within the MCR program. Leadership of an LTER site requires a huge investment of time and energy, but the synergy is such that I get more out of the experience than the time and energy I devote to it. To me, the two greatest personal pluses of being involved in the LTER program are (1) getting to meet, interact, and often collaborate with a tremendous set of world-class scientists, and (2) getting to travel to many interesting national and international places in an LTER capacity. The greatest personal consequence is the great expansion in my circle of friends.

CHALLENGES AND RECOMMENDATIONS

The model of place-based science in the LTER program in which long-term observational data, multiscale process studies, and models are tightly integrated has been tremendously successful. However, by its nature, the LTER program is not highly optimized for addressing questions that require a network of sites. Certainly, cross-site comparisons have arisen from the LTER network, but there are limitations to functioning as a network of research sites of a program that comprise a series of different, representative ecosystems. Of course, scientists in the LTER network seek to understand general ecological phenomena and enhance comparability among sites by focusing on the same five core research themes. Although several insightful cross-site synthesis efforts have taken advantage of this structure, a number of challenges hinder rapid progress. Chief among these are limitations related to discovery, acquisition, and integration of relevant data among different sites, which occur despite the adherence of sites to the open-access data policy of the LTER program. To a large extent, this limitation reflects the relatively primitive state of information management of the environmental sciences in general, as well as a distributed but idiosyncratic data management structure that is used by the community of sites. Nonetheless, the LTER program has had a pioneering role in promoting a culture of data sharing in environmental sciences, as well as in facilitating advances in information management.

Another limitation to network-level science concerns the "one off" collection of sites within the community. Although this distribution enhances the diversity of ecosystems being explored, the trade-off is that it hinders comparisons across similar ecosystems. A sensible solution to this would be to expand the NSF-funded LTER network to include multiple examples of the same ecosystem. This expansion would require a substantial investment of more resources into the LTER program. Additionally, each current LTER site is grossly underfunded relative to its potential and expectations, and my engineering colleagues are aghast at the small investment made in the LTERs relative to equivalent entities in their fields. Of course, this represents the appallingly low level of resources the United States invests in basic science in general.

CONCLUSION

In summary, the LTER program has enabled tremendous growth in my science in terms of the questions that I can address; the tools, approaches, techniques and data to which I have access; and the diversity of intellectual and disciplinary expertise that I can tap. My scientific thinking has evolved and my research projects are far more interdisciplinary, complex. and integrated. I have benefited greatly from a much greater level of collaboration with scientists from other fields, locations, ecosystems, and indeed cultures. The research issues I now find of greatest interest still are motivated by ecological theory but have much more relevance to policy makers and resource managers. They also encompass scales (time, space, integration across levels of biological organization) that I otherwise could not address through a "little science" approach or without long-term data on both the biotic and abiotic domains. My ability to mentor students at all levels has been transformed, and these students are well prepared to be engaged globally. Ecologists at all career stages would benefit greatly by leveraging the intellectual, data, and logistical infrastructure of the LTER program as it offers a myriad of unparalleled research opportunities. Indeed, ecological research would benefit tremendously from mechanisms that would engage a much larger segment of the US ecological research community in the LTER network.

ACKNOWLEDGMENTS

It is with pleasure that I acknowledge my MCR LTER colleagues, notably S. J. Holbrook, P. J. Edmunds, and R. C. Carpenter, as well as the many other investigators, students, and technicians who have made the research successful. Logistic support from N. Davies, H. Murphy, F. Murphy, and the rest of the amazing staff of the University of California Berkeley Richard B. Gump South Pacific Research Station on Moorea (home of field operations of the MCR LTER) is gratefully acknowledged. I also deeply appreciate the financial support of the NSF (OCE 1236905 and earlier awards) and the Gordon and Betty Moore Foundation.

REFERENCES

Adam, T.C., Schmitt, R.J., Holbrook, S.J., Brooks, A.J., Edmunds, P.J., Carpenter, R.C., and Bernardi, G. (2011). Herbivory, connectivity and ecosystem resilience: Response of a coral reef to a large-scale perturbation. *PLoS ONE*, **6**, e23717.

Bestelmeyer, B.T., Ellison, A.E., Fraser, W., Gorman, K., Holbrook, S.J., Laney, C.M., Ohman, M., Peters, D.P.C., Pillsbury, F.C., Rassweiler, A., Schmitt, R.J., and Sharma, S. (2011). Abrupt transitions in ecological systems: What have we learned? *Ecosphere*, **2**, 129.

Holbrook, S.J. and Schmitt, R.J. (2003). Spatial and temporal variation in mortality of newly settled damselfish: Patterns, causes and co-variation with settlement. *Oecologia*, **135**, 532–541.

Leichter, J.L., Alldredge, A.L., Bernardi, G., Carlson, C.A., Carpenter, R.C., Edmunds, P.J., Fewings, M.R., Hanson, K.M., Holbrook, S.J., Hench, J.L., Nelson, C.E., Schmitt, R.J., Toonen, R.J., Washburn, L., and Wyatt, A.S.J. (2013). Biological and physical interactions on a tropical island coral reef: Transport and retention processes on Moorea, French Polynesia. *Oceanography*, **26**, 52–63.

Messmer, V., Jones, G.P., Munday, P.L., Holbrook, S.J., Schmitt, R.J., and Brooks, A.J. (2011). Habitat biodiversity as a determinant of fish community structure on coral reefs. *Ecology*, **92**, 2285–2298.

Osenberg, C.W., and Schmitt, R.J. (1992). Detecting ecological impacts caused by human activities. In R.J. Schmitt & C.W Osenberg, eds. *Detecting Ecological Impacts: Concepts and Applications in Coastal Habitats*, pp. 3–16. Academic Press, New York.

Osenberg, C.W., St. Mary, C.M., Schmitt, R.J., Holbrook, S.J., Chesson, P., and Byrne, B. (2002). Rethinking ecological inference: Density dependence in reef fishes. *Ecology Letters*, **5**, 715–721.

Rassweiler, A., Schmitt, R.J., and Holbrook, S.J. (2010). Triggers and maintenance of multiple shifts in the state of a natural community. *Oecologia*, **164**, 489–498.

Schmitt, R.J. (1985). Competitive interactions of two mobile prey species in a patchy environment. *Ecology*, **66**, 950–958.

Schmitt, R.J., and Holbrook, S.J. (2007). The scale and cause of spatial heterogeneity in strength of temporal density dependence. *Ecology*, **88**, 1241–1249.

PART THIRTEEN
NIWOT RIDGE (NWT) LTER SITE

29 Top of the World Collaborations

Lessons from above Treeline

Katharine N. Suding

IN A NUTSHELL

1. My involvement at the Niwot Ridge Long-Term Ecological Research (LTER) site began when I was an undergraduate summer research assistant, and it has extended through a postdoctoral fellowship, a tenured professorship, and now a leadership role in the program. I focus on alpine tundra plant diversity, plant–soil interactions, and how environmental changes may influence community dynamics over time and space.
2. Cross-site synthesis work has been one of the most valuable experiences of my career, enabling me to ask more general questions and produce more influential work than I could have done with a focus at a single site. Such comparative research has allowed

me to interact with a fabulous group of scientists that has strongly influenced my professional development. These scientists remain strong role models for me.
3. My experiences in the LTER program have formed my model of education and training, emphasizing experimental and observational approaches, quantitative methods, and data management and sharing. I think it is the best way to approach the difficult and complex ecological questions facing our society today.

PERSONAL OVERVIEW

My involvement in the LTER program started in college, when I decided to study for one semester at the University of Colorado. During that semester, I took a class from Marilyn Walker, who was part of the Niwot Ridge (NWT) LTER program. Marilyn's class did not go to the tundra or even focus on alpine systems. However, when time came to figure out what to do over the summer, I asked her if I could be her research assistant. She gave me the chance to work at Niwot Ridge (Figure 29.1).

I spent the summer before my senior year at 3,500-m elevation, recording point quadrat vegetation data in permanent plots. The snow was late to melt that year, so I spent much of June in the Institute of Alpine and Arctic Research's loading dock, painting thick black stripes on 2.5-m long PVC poles to be used to measure snow depth. When snow melted enough to allow access on the entrance road, I went up to Niwot Ridge for the first time. Marilyn and I both pulled sleds packed with snow depth poles and fiberglass warming chambers up snow-covered slopes. It was from Marilyn, while we were pulling those packed sleds, that I got my first bit of great advice concerning the LTER program: take a breath for every stride, and if you go slow and steady, you will never have to stop.

The diversity in the tundra was fantastic. I liked learning to identify plant species. I also liked being part of a group of researchers who hiked up to the ridge daily to study all sorts of things: snow, wind, plants, pikas, and soil. This experience was in contrast to the undergraduate thesis research that I did the previous spring where I spent many long days alone, figuring out plant identification, the best ways to sample, and the best ways to

FIGURE 29.1 Discussion on Niwot Ridge with Amy Miller (center) and Katharine Suding (right). (Photo courtesy of William Bowman.)

locate my plots. Particularly when you are starting out, there is a fine line between independence as a researcher (i.e., training yourself) and being part of a team (i.e., learning from others). Working at NWT was a great way to balance on that line.

I went to graduate school at the University of Michigan in a program that prided itself on having graduate students develop their own projects. I learned a lot from that experience, had a fantastic advisor, and was glad to develop a dissertation project myself. I found my own research site, where few if any ecologists had worked before. Graduating, I coauthored with my dissertation advisor, but no one else.

I went back to NWT to conduct postdoctoral research. It was not because I recognized the opportunities it presented (I should have), and it was not because NWT really needed a postdoctoral fellow (it did not). I begged a couple of people and was lucky that they were able to create a position where I could do some research and some teaching, with partial funding from the prime NSF grant to the site.

It was during my postdoctoral fellowship that I recognized the value of being part of the LTER network. For one, due to the depth of the site knowledge that had accumulated from the project, I was able to begin writing proposals soon after my arrival. The one semester that NWT funded me produced 3 years of funding from a Mellon Foundation grant and a research program that I continue to expand, 12 years later. At the same time, I started to work at the Kellogg Biological Station (KBS) with Kay Gross. My project with her was not related to the KBS LTER program, but it was a start in getting to know several LTER sites.

APPROACH TO SCIENCE

The best sort of collaboration is with scientists who think differently than you do. This type of collaboration is also the most difficult to start, because the learning and communication curves are often quite steep. Working with a team of multidisciplinary scientists as part of NWT has allowed me to venture into some areas that I might not have before. I started out as a plant community ecologist with a focus on plant interactions. My time at NWT has broadened my expertise to include aspects of biogeochemistry, landscape ecology, and microbial ecology. Although I will always collaborate with others in these fields, I am confident leading projects, taking measurements, and advising students across these boundaries. The most interesting questions sit on edges between disciplines, and this experience has allowed me to better understand those edges.

My experiences with the LTER program have also taught me the value of building science rather than starting from scratch. Ecological questions invariably depend on general phenomena, as well as the natural history and processes specific to a particular system. A typical LTER site has accumulated years of knowledge along many different lines of investigation. There is much value in asking general questions with this knowledge as a foundation, enabling a more nuanced and deeper understanding of dynamics.

ATTITUDES TOWARD TIME AND SPACE

I remember a transformative moment in graduate school when Earl Warner put on the board: Pattern → Process → Traits. "Ah, so that is how you do ecology!" I thought. Although there are lots of other approaches in science, this still resonates strongly with me: observe a pattern in space or time, then do an experiment to test the process responsible for the pattern, then expand to the evolutionarily constrained traits or mechanisms that might account for the process. The LTER program is well suited to this scientific approach.

In many ways, documenting an ecological pattern is the most fundamental step in ecological research. Yet, it is a step that is sometimes difficult given the emphasis on hypothesis testing and experimentation. In many proposals, the argument starts with the assumption that the pattern is well documented, and the proposed research starts with hypotheses about processes. The emphasis on monitoring in the LTER program fills this missing link. For instance, at NWT, I often include links with global change and effects on biodiversity as rationale for my work. Although that is certainly a valid argument, we are finding that there are many areas of alpine tundra that are not changing over time (Spasojevic et al. 2013). Looking more closely at our monitoring data in different areas of the site, it appears that the tundra is most responsive at its lower boundary with forest and its upper boundary with barren areas. Although this observation is not surprising, the long-term data across space provided supporting evidence and a much more nuanced view of the detailed patterns of the responses.

I also appreciate the ability to run experiments over times that are more appropriate to the ecology of the system. It is often only after many years that a response occurs, or that longer-term responses replace more transient ones (Suding et al. 2008). A 3- to 4-year study is often not adequate to understand these dynamics, and in many ways, these longer-term experiments can act as patterns to feed more detailed process studies. For instance, a key plant species took 4 years to respond in an experiment that combined nitrogen additions and species removals. After 8 years, it was clear that nitrogen was causing a drastic decline in its abundance, even when its competitor was removed. These results allowed us to start a project asking whether microbial interactions at higher nitrogen levels might be a key factor (Farrer et al. 2013). Even after 12 years, we are seeing an overall decline in diversity that follows the loss of the key species by several years.

COLLABORATION

One strong influence of the LTER program is the opportunity to conduct cross-site syntheses. Laura Gough, then a new faculty member, and I, then a postdoctoral fellow, submitted a proposal to the LTER network office in 2002 to synthesize results from nitrogen fertilization experiments across LTER grassland and tundra sites. Building off a very successful effort addressing nitrogen fertilization effects on diversity, our aim was to look at how shifts in composition related to declining diversity. The ensuing support from the LTER network allowed us to have a series of meetings, share data and insights about natural history, and discuss analytical approaches. The type of collaboration it enabled, across institutions, expertise, and sites, was invaluable and resulted in some of the papers of which I am most proud (e.g., Suding et al. 2005).

Although my research questions have evolved over the years, the synthesis work has been and continues to be very rewarding, both in terms of research productivity and its influence on my professional development. Some of my most influential work has come from these efforts: 15% of my publications and 30% of my citations have resulted from work done under the auspices of the LTER program as cross-site synthesis.

Cross-site collaboration also comes with its challenges. For one, funding is mostly for meetings and not for time. Although meetings can be productive, the productivity often depends on preparation prior to and completion after the meeting: an unfunded responsibility for a very busy group of people. For example, the nitrogen fertilization synthesis involved compiling data on the responses of more than 900 species in 28 different experiments. Before any questions could be asked, I spent quite a lot of time putting together very disparate, site-level data sets. As is often the case, to compare across sites, we wanted to add additional data on species traits, which required even more work. In a fortunate turn of events, we were able to identify a talented graduate student, Elsa Cleland, and convince her to incorporate our effort into a National Center for Ecological Analysis and

Synthesis postdoctoral fellowship. Her involvement elevated the focus on the project and the requisite data management, and allowed all participants to make substantial progress when we did meet.

Although Elsa is not formally affiliated with any LTER site, she continues to be involved in collaborations related to synthesis (e.g., Cleland et al. 2013). Although it is good that people can be involved in cross-site work this way, it remains difficult to come in from the outside, identify data sets, and collaborate. It is not easy even within the LTER network. For instance, recently I was asked by a new postdoctoral fellow at NWT whether there was a cross-site data set on soil carbon. Although this is not my area of expertise, I knew most sites made these measurements and was thinking it was something I could help find. I had very little luck until I asked another LTER scientist who was more involved with this type of data. Surprised, I tried to find soil carbon data in site databases. There was no shortage of data sets, yet without site-level knowledge it would be very difficult to select the dataset appropriate for cross-site comparative work.

APPLIED RESEARCH

I am very active in applied work, predominantly on questions related to invasion and restoration ecology (e.g., Suding 2011). Interestingly, invasive plant species are not an issue at NWT, and restoration is not a management need. Therefore, I tend to wear two hats, one relating to invasion and one relating to diversity and environmental change. It is this second hat that includes higher elevations (alpine environments) at NWT. Although I suspect that some people know my work mostly from one or the other of these hats, the underlying concepts and tackled mechanisms are much the same, regardless of the system. I find that these differences mostly surface in proposal rationale and funding sources, rather than across my research group or in our approaches to science. I am not ready to abandon either.

Although some LTER sites and ecosystems are definitely more predisposed to applied work, another important aspect in applied work is the establishment of relationships with land managers and other stakeholders. I find that these relationships often depend on personal trust and communication. Although gaps between basic and applied science are easy to bemoan, identifying research that does address these gaps is challenging. I am a firm believer that relationships and collaboration are essential to make progress. More time at the field site and with stakeholders at NWT will undoubtedly lead to more opportunities to do applied work related to mountain systems.

COMMUNICATION

My involvement in the LTER program has not strongly affected my involvement in K–12 and schoolyard programs. I find that most of my outreach is related to local work, being based in California. I look forward to being more involved with the LTER Schoolyard Program as I transition my research group to Boulder, Colorado. I also am excited about ways to use the recent EcoTrends book (Peters et al. 2013) to encourage students to ask questions and synthesize data across many sites.

I have taken the LTER model of research and synthesis to the classroom, allowing students to build knowledge and synthesize ideas and data toward a directed end. Such synthesis also allows students to think about how their perspectives can advance science and how to use preexisting work to develop robust arguments. Although these products have not focused on anything related to NWT, in several cases they have produced meaningful synthetic papers (Eitzel et al. 2012; Hallett et al. 2013).

MENTORING

I would not like to practice science in isolation, and I enjoy the working group meetings and conferences. This is one thing that attracts people to the LTER network. These personal interactions have been essential to my professional development because they are where I get lots of advice, and, more importantly, where I see how my role models work. I do not think that I would be very good at networking without opportunities such as these. Instead, I would have more likely interacted with people from my home institution, or at my professional stage, and would have missed out on these broadening experiences.

I hope that my mentorship of younger scientists reflects the lessons that I have learned from my mentors in the LTER program. I try to emphasize long-term professional growth and the importance of establishing a network of advisors. I try also to show a bit of my personal side, as my mentors did with me (e.g., that I go home to my kids, that I love to play soccer), as I realize how helpful it is to see work and all other parts combined in a life.

I am at the stage of my career where I am convincing people in my laboratory to join working groups, on top of their dissertation or postdoctoral work, because of their value. Being able to ask questions at one site, perhaps with more detailed focus on mechanism and process, and then expand to patterns over time within a site as well as across sites is a powerful model of research that the LTER program facilitates. For instance, one of my graduate students led an effort to produce a cross-site synthesis paper on compensatory dynamics and its relationship with precipitation (Hallett et al. 2014). The patterns across sites allowed her to capitalize on natural variability in precipitation and provided strong rationale for experimental mechanistic tests at a single site. However beneficial this cross-site work is, it still often suffers from the unfunded mandate that makes balancing site-level research and cross-site research difficult.

SKILL SET

The value of data and sharing data is one clear lesson that has stemmed from my experiences in the LTER program. There are so many questions that can be asked with a single data set. Sharing data poses little risk of someone taking your idea, and the value of sharing includes learning new techniques and the fun of working with new people. My involvement with the LTER program has also (1) affected my development of data management skills and (2) emphasized the importance of including statistical code with analyses and advantages of modeling based on pattern-based empirical datasets. Within my laboratory group, I have tried to instill this same perspective: having a central and accessible data repository where everyone must put their data, statistical code, and metadata. Several years ago, my laboratory group decided to invest time to continue a long-term data set at a non-LTER site, and at a recent laboratory retreat we started synthesis efforts similar to the ones done across LTER sites. Even at this level, the value of such efforts is large and collaboration within my research group has increased markedly.

PERSONAL CONSEQUENCES

More than the unique research platform, I have greatly benefited from working with the people who make up the LTER network. Many of my most influential mentors are linked with the LTER program. It is from them that I learned how to do science and how to be a scientist. This is where the LTER network is at its best, attracting fantastic scientists that are also generous mentors and collaborators.

CHALLENGES AND RECOMMENDATIONS

Getting younger people to become involved represents an important challenge for the LTER network. In my experience, people are eager for the opportunities that the LTER program can provide. However, the amount of funding that a particular LTER site can allocate to a researcher to facilitate involvement is small. An important question is how to get someone involved at the site long enough to generate some data that can facilitate further external funding. It is hard to maintain a research program on just funds from a particular LTER site, and while the long-term nature of the sites makes it easier to start a project, this initial hurdle is substantial.

On the flip side, once someone has invested time in a LTER project, it is not easy to stop. This strong continuity and commitment to long-term research is one reason that the LTER network has been so successful. However, such commitment also makes change difficult. This balance between continuity and change is a core challenge for the LTER program. The long-term nature of the science means that we need to maintain long-running datasets and not continually reinvent approaches. However, to do current, cutting-edge research, we also need to gradually incorporate new approaches and, in doing so, possibly leave some things behind.

My recommendation to someone wanting to get more involved with LTER research is to capitalize on the site-level data as an asset to start investigations at a given site but to not pass up any opportunity to initiate or be involved with cross-site synthetic work.

One aspect that is missing from the LTER network is a mechanism to conduct standardized experiments across sites. Although many sites happen to do similar experiments (such as nitrogen fertilization), the design and implementation usually takes place without coordination across sites. Issues such as sampling design, plot size, and measurement techniques reduce our ability for synthesis. In addition, having standardized manipulations across sites would elevate our network-level science. This would necessarily mean including sites that may not be focusing on the manipulation in question and installing new experiments that may be repetitive with existing ones. Currently, each site makes its science decision based on site-level needs. Adding some network-level funding for research would allow network-level research as well.

CONCLUSION

I am currently moving to the University of Colorado to take more leadership responsibilities at NWT. I am drawn by the ability to ask really big questions—much bigger than I can do within the confines of just my research group—on a spatial and temporal scale that is essential, given recent changes in our world. I am also drawn by a place where the contributions of so many outstanding scientists present such solid footing to embark on the next steps and to train the next generation of scientists. Amazing things happen when so many remarkable scientists get together to work on one problem: I am honored to be a part of it.

ACKNOWLEDGMENTS

The National Science Foundation (NSF) has supported the NWT LTER program since 1980, most recently with award DEB-1027341. I am grateful for the NSF LTER program, the Andrew Mellon Foundation, and NSF awards DEB-0919569 and DEB-1457827 in supporting my research at NWT. I also thank the many incredible graduate students and postdoctoral fellows who took research in so many interesting directions (I. Ashton, C. Buena de Mesquito, E. Farrer, M. Jabis, A. Miller, J. Smith,

M. Spasojevic, H. Steltzer), field technicians and undergraduate students who worked alongside us, and my colleagues at NWT as well as across the LTER network.

REFERENCES

Cleland, E.E., Collins, S.L., Dickson, T.L., Farrer, E.C., Gross, K.L., Gherardi, L.A., Hallett, L.M., Hobbs, R.J., Hsu, J.S., Turnbull, L., and Suding, K.N. (2013). Sensitivity of grassland plant community composition to spatial vs. temporal variation in precipitation. *Ecology*, **94**, 1687–1696.

Eitzel, M.V., Diver, S., Sardinas, H., Hallett, L.M., Olson, J.J., Romero, A., Oliveira, G., Schuknecht, A.T., Tidmore, R., and Suding, K.N. (2012). Insights from a cross-disciplinary seminar: 10 pivotal papers for ecological restoration. *Restoration Ecology*, **20**, 147–152.

Farrer, E.C., Herman, D.J., Franzova, E., Pham, T., and Suding, K.N. (2013). Nitrogen deposition, plant carbon allocation, and soil microbes: Changing interactions due to enrichment. *American Journal of Botany*, **100**, 1458–1470.

Hallett, L.M., Diver, S., Eitzel, M.V., Olson, J.J. Ramage, B.S., Sardinas, H., Statman-Weil, Z., and Suding, K.N. (2013). Do we practice what we preach? Goal setting for ecological restoration. *Restoration Ecology* **21**, 312–319.

Hallett, L.M., Hsu, J.S., Cleland E.E., Collins, S.L., Dickson, T.L., Farrer, E.C., Gross, K.L., Gherardi L.A., Hobbs, R.J., Turnbull, L., and Suding, K.N. (2014). Biotic mechanisms of community stability shift along a precipitation gradient. *Ecology*, **95**, 1693–1700.

Peters, D.P.C., Laney, C.M., Lugo, A.E., Collins, S.L., Driscoll, C.T., Groffman, P.M., Grove, J.M., Knapp, A.K., Kratz, T.K., Ohman, M.D., Waide, R.B., and Yao, J. (2013). Long-term trends in ecological systems: A basis for understanding responses to global change. USDA Agricultural Research Service Publication No. 1931. Washington, D.C

Spasojevic, M.J., Bowman, W.D., Humphries, H.C., Seastedt, T.R., and Suding, K.N. (2013). Changes in alpine vegetation over 21 years: Are patterns across a heterogeneous landscape consistent with predictions? *Ecosphere*, **4**.

Suding, K.N. (2011). Toward an era of restoration in ecology: Successes, failures, and opportunities ahead. *Annual Review of Ecology, Evolution and Systematics*, **42**, 465–487.

Suding, K.N., I.W. Ashton, I.W., Bechtold, H., Bowman, W.D., Mobley, M.L., and Winkleman, R. (2008). Plant and microbe contribution to community resilience in a directionally changing environment. *Ecological Monographs*, **78**, 313–329.

Suding, K. N., Collins, S.L., Gough, L., Clark, C., Cleland, E.E., Gross, K.L., Milchunas, D.G., and Pennings, S. (2005). Functional- and abundance-based mechanisms explain diversity loss due to N fertilization. *Proceedings of the National Academy of Sciences U.S.A.*, **102**, 4387–4392.

PART FOURTEEN
NORTH TEMPERATE LAKES (NTL) LTER SITE

30 My Evolution as a Long-Term Ecological Research Scientist

John J. Magnuson

IN A NUTSHELL

1. My college education as a fish and fishery ecologist provided a solid base for my evolution to a scientist absorbed by the long-term ecology of lakes in the landscape.
2. Graduate students in the Long-Term Ecological Research (LTER) program and in my course lectures came to represent more disciplines and became more interdisciplinary, often addressing major ecological questions using long-term data.
3. Viewing the dynamics of a time series and spatial maps became strong approaches in the LTER program for communicating with colleagues and the broader community.
4. The LTER program would have failed without the realization and the broad application of collaboration. That is true, of course, for much of what we do.

5. The LTER program is a great way to participate in and learn from a life of science teaching, research, application, and outreach.

PERSONAL OVERVIEW

My association with the LTER program began in the late 1970s when I was a 41-year-old associate professor at the University of Wisconsin–Madison. It continued through the remainder of my professional life to the present; I am now an 80-year-old emeritus professor at the Center for Limnology at the University of Wisconsin–Madison.

I had been the program director for Ecology in the Division of Environmental Biology at the National Science Foundation (NSF) for 1 year (1975–1976) and saw the first movements toward such a program. I participated in all three NSF workshops in the late 1970s to consider and plan an LTER program. At the workshops, I represented the perspectives of limnology and our field site at the Trout Lake Station in northern Wisconsin. Ideas being discussed and planned were of great interest to me. I believed that research opportunities at field stations with this long-term approach were important to the ecological sciences and to biological field stations across the country.

My colleagues and I at the University of Wisconsin–Madison responded to NSF's initial call for proposals; we were one of the first six sites to be funded for a proposal entitled "Long-Term Ecological Research on Lake Ecosystems." I enjoyed leading the North Temperate Lakes (NTL) site in Wisconsin with my colleagues from the start in 1980 until July 2000, when I retired to become an emeritus professor, and Stephen Carpenter became the principal investigator from 2000 to 2009. From the beginning, I participated in network governance as a principal investigator and served as chair of the LTER network in 2006 to 2007. I participated internationally both through the International LTER (ILTER) and through joint research with international colleagues working on lakes and streams. Today, I continue at a much-reduced level to publish and participate in the NTL LTER program, now led by Emily Stanley.

Although I would like to ascribe all the changes that occurred in my evolving life in science to the LTER program, I am struck with the complexities of our lives and the many facets even of our science lives. So, at the onset, I must present the disclaimer that my evolution was amplified by other portions of my life in science and may have developed as I matured and aged as a scientist regardless of the LTER program.

APPROACH TO SCIENCE

Before becoming involved in the LTER program, my research, teaching, and coworkers were in fish ecology and limnology. This work extended from the small lakes of Wisconsin, to the Laurentian Great Lakes, to biological oceanography and fish behavior in Hawaii and elsewhere in the Pacific, and off Cape Hatteras in the Atlantic.

After the LTER program formed, I continued to work in fish and fisheries and in limnology, but the flavor changed, seasoned by the LTER program. My students came from a broader mix of subspecialties within limnology, ecology, and environmental science. The research directions of my colleagues and the foci of scientific meetings that I attended moved away from water and fish a bit. Scientists with whom I interacted were largely those in the LTER program; mostly my colleagues at Wisconsin but also colleagues from other LTER sites located in prairies, forests, and deserts. I became an advocate and practitioner of long-term, landscape-level ecology. I have always believed in and worked in interdisciplinary sciences, but these practices were broadened and realized through the LTER program. The questions I found most interesting became those at landscape, regional, and global scales—and at multiple years to decades or longer when possible.

ATTITUDES TOWARD TIME AND SPACE

Through the LTER program, I became a strong advocate for the importance of time and space in both the philosophy and practice of research in ecology. Our synthesis book (Magnuson et al. 2005) gives some of our thoughts and makes clear that temporal and spatial scales were central to our NTL program. Some surprises became apparent. I will point out three from a paper by Kratz et al. (1995). First, variability in ecological properties and processes across time and space provided a nondimensional way to compare widely divergent parameters and ecological systems, such as between lakes, prairies, deserts, and forests. Second, even though the LTER program seemed to be focused on interyear temporal variability and changes over decades or longer, the variability associated with locations in an LTER site exceeded the variability associated with different years. Almost all the LTER sites existing when we did these analyses chose locations to study within their site based on elevational position from high to low in the landscape. They clearly had a desire to and a sense of how to capture the spatial variability at their sites. Third, often the differences in landscape position were only a meter or so, but locations still exhibited dramatic changes in properties and processes along that gradient; good examples are a catena in the shortgrass steppe or adjacent lakes in northern Wisconsin from high to low in the groundwater flow system. Differences along the flow system would appear to be important in both of these cases.

Our interests in time and space led to our use of new statistical techniques capable of extracting pattern in time and space. These included sophisticated mapping and time series analyses, as well as approaches to visualization.

COLLABORATION

I learned that collaboration came in many forms and for many purposes. On the pragmatic side, we needed to collaborate across disciplines to prepare and succeed with our LTER proposals. We needed to collaborate across LTER sites and with NSF to manage and administer the complex organization we call the LTER program. We also needed to collaborate in the collecting and sharing of data within and among LTER sites and with broader science communities. From the very beginning, our site scientists were preadapted to share data; we collected and managed the data as a group process. Coming from limnology and oceanography, we knew we needed to collaborate in the sharing of resources to accomplish the interdisciplinary questions we wished to answer. We needed to collaborate to broaden our working definition of interdisciplinary to include for example, social sciences, information management, remote sensing, and instrumentation.

The LTER program would have failed without the realization and the broad application of collaboration. That is true, of course, for much of what we do. Networks and partnerships are strong forces that make it clear that much of what we wish to accomplish cannot be achieved as individuals through individual effort.

Some people are preadapted to be collaborative and recognize the benefits from the onset, and others are motivated by narrower self-interests (i.e., need to support their disciplinary research or a student). Some researchers initially attracted by resources, evolve to be strong collaborators and believers in the process. Some never do, but even these researchers can be important to an LTER program for the information and knowledge they provide to solve the overall environmental puzzle. Not everyone needs to be at the table for the same reasons.

APPLIED RESEARCH

The research at our LTER site, including that which involved me directly, had components that were and are relevant to management or policy issues. Examples are climate change and variability, where my colleagues and I used the long-term ice data. This set of continuing lake-ice analyses were used at the global scale in the 1995, 2001, 2007, and 2013 reports of the Intergovernmental Panel on Climate Change—the IPCC reports (I participated in the 1995 and 2001 reports), and at the regional scale through "Confronting Climate Change in the Great Lakes Region" (Kling et al. 2003). In Wisconsin, our data and analyses were used in Wisconsin's Initiative of Climate Change Impacts (Google WICCI). Other examples include an invasive crayfish and acid rain. The rusty crayfish transformed inshore lake communities of several of the primary lakes at the NTL site. Subsequently, we reduced their abundance to determine whether the effects were reversible (Hein, Vander Zanden, and Magnuson 2007). For acid rain, we conducted a major chemical manipulation of Little Rock Lake by dividing it into two parts with a rubber curtain; one side served as a reference, the other we acidified and later neutralized over a series of years (Frost et al. 2005).

Although involvement in the LTER program amplified my beliefs that most management and policy issues required a breadth of disciplines and approaches, most of the strong examples for me personally came from other sources. That was especially true for study committees of the National Research Council such as *Decline of Sea Turtles: Causes and Prevention* (Magnuson et al. 1990) and *Upstream: Salmon and Society in the Pacific Northwest* (Magnuson et al. 1996). Locally, there was the water quality of the lakes at Madison, Wisconsin, and electric power plant impacts and siting. I also relied on major management or policy issues in our teaching of "Limnology, the Conservation of Natural Resources." This course started with the physical, chemical, and biological components of aquatic ecosystems and ended putting them together when we dealt with particular management and policy issues.

COMMUNICATION

Viewing the dynamics of a time series and spatial maps are strong vehicles for communicating with colleagues and the broader community. I began using the approach of displaying a time series at the 10-year review of the LTER program, later published in *BioScience* (Magnuson 1990). I used the analogy of the "Invisible Present" to make the case that long-term research is an important approach to understand as well as inform. Ever since the 10-year review, I have repeatedly used the "Invisible Present" idea and the time series of lake-ice breakup and freeze to communicate the importance of long-term research; each extension of the time series revealed new underlying causes or relationships. The tool has worked well with professional managers, policy-makers, and the public, as well as with students in secondary schools and colleges. In the climate change arena, professionals and the public have told me that my presentation of the "Invisible Present" convinced them that climate change is real and should be addressed.

Another paper in *BioScience* from the 10-year review addressed the challenge of the "Invisible Place" (Swanson and Sparks 1990). Their presentations and discussions made a strong case for why a system cannot be understood or communicated by studying only one meter square plots; study at multiple spatial scales was essential. I have personally found that people generally understand maps as well as time series; carefully chosen maps help them recognize and distinguish larger-scale processes from local phenomena. Audiences have commented that they had not realized that what they thought was a locally generated process actually originated from processes working at much larger spatial scales.

Time and space viewed in simple presentation formats greatly enhance our abilities to communicate by allowing the audience to learn from what they can actually see. The approach provides a view of what one is trying to communicate that many have not seen before. I believe that using various examples of the "Invisible Present" and the "Invisible Place," that is, time and space at multiple scales, is a communication tool that the LTER program has contributed. The approach helps us all understand the logic of Senge's (1990) truism that "a fundamental characteristic of complex human systems: cause and effect are not close in time and space." His statement clearly applies to complex ecological systems or complex human–ecological systems.

MENTORING

Initially, my collaborators in the LTER program were concerned that while long-term ecological research was indeed important, it would not well serve our students in their thesis research. The thought was that thesis research should address a particular process and also use experiments. In contrast, we discovered that our students could address major questions about lake ecological systems using data from the LTER program, supplemented by particular measurements of interest as well as experiments embedded within the LTER framework. Experiments were not always useful, but some were essential and included whole lake manipulations, the use of "limnocorrals," and laboratory research in aquaria or large outdoor tanks (i.e., microcosms and mesocosms).

The long-term changes and dynamics that emerged from our data sets often caused our students, as well as their advisors, to think and postulate about what was going on. Questions were raised that we had not formulated initially. The element of discovery was exciting and motivating. Often, the multiple levels of our long-term data allowed students to unravel puzzles using a broad range of statistical analyses. This was very reminiscent of one of the original rationales we had put forward to justify the LTER program: namely, that complicated system behavior could be understood though time series of multiple components and processes (Franklin, Bledsoe, and Callahan 1990).

Another change was occurring in student publications from thesis research: namely, that multiauthored papers became the rule rather than the exception. The questions they addressed depended on the author's own background and knowledge—but importantly also on the background and knowledge of their student and faculty colleagues. This was an important change that the LTER program helped to bring about, because of its interdisciplinary nature.

One wonders whether new disciplines are emerging through students using interdisciplinary approaches and a more diverse set of mentors in the analyses of time and space data from LTER sites. I think so, but I am not certain. A clearer case of this phenomenon for me comes from my work on the distribution of fishes, where C. Clay and I teamed up to coadvise our students, me from a fish ecology perspective and Clay from an acoustics (underwater sonar) perspective. Neither of us understood the other's science very well, but our students exhibited a practical mix of competencies from our mentoring and understood both disciplines, perhaps as a new discipline of fish acoustics.

SKILL SET

Here I have to be a bit more inclusive. "My" refers to our LTER site, its principal investigators, researchers, and students. We have incorporated so many new tools, approaches, and disciplines to the mix that no one individual can use or manage them all, except through the interdisciplinary mix of colleagues working together. These tools include a rich array of new disciplinary approaches, methods specific to the various subdisciplines, computer and information management tools, remote sensing and sensor technologies on lake buoys, new statistical approaches appropriate to our data and the questions being addressed, new process models and system models, and many more.

For the principal investigator, the challenge is to provide a site management system that allows these new tools and approaches to develop and for individuals to gather around questions that they could not have addressed alone. We learned to be a collective interdisciplinary team and to address new synthesis and science issues by using new tools as they became available or were needed. For me personally, I took a short course sponsored by the LTER program on remote sensing. Later, I used what I learned and coauthored a paper with colleagues (Riera et al. 1998).

PERSONAL CONSEQUENCES

In the early years of the LTER program, I greatly enjoyed visiting the sites during our principal investigator meetings; this exposed me to ecosystems at a level deeper than ecotourism. My life became more interesting, as I learned about new disciplines and accumulated a changing group of friends and colleagues. My personality is such that focusing on the same thing over a career was not an intellectual trait. I also became less of an "I and me" person and more of a "we and us" person.

I met a wonderful group of scientists, researchers, NSF personnel, administrators, students, and staff. Long-term associations with these people blurred the distinction between acquaintance, employee, supervisor, colleague, and friend. A few came and left, others became long-term.

CHALLENGES AND RECOMMENDATIONS

Improvement, of course, is a job that is never complete. Improvement has been continuous in the LTER program, and I believe it will continue. There will always be more we can do and more that we can do better.

What I worry about most is the science. I worry that some LTER sites are losing or never reached a focus on the long-term and broader spatial scales. I always worry that new methods and techniques, especially the really complicated ones, can distract any of us from being scientists and contribute to our losing sight of the major goals of attempting to understand and predict the behavior of the world around us. For example, in the analysis of remote sensing images, the time and energy spent in learning the technique and processing the images can prevent one from asking the ecological questions on spatial heterogeneity or spatial structure that influence function or dynamics. Or, in processing samples of aquatic insects, learning to identify and count dipteran larvae can prevent one from asking questions about their long-term ecology. Some of this can be overcome through interdisciplinary teams with sets of skills working together, but if one is not thoughtful, one can lose sight of the larger picture owing to the detail.

A few things I learned were how to develop a nonthreatening network of partners that is open to contributions from all. In our group meetings, all were encouraged to speak—and we listened. I strove to reduce hierarchy, to maintain creativity, and to empower people to do what they do. Thus, I learned to lead a group process rather than lead a group. Leading a group process moves away from "listen to me and follow me because I know what we should do" to guiding an interactive approach that facilitates the participation of all individuals in the discussion and planning. A common example at an LTER site is in preparing a renewal proposal.

I also learned that at the site and intersite levels that it was critical to have an executive committee of close colleagues to help filter among the many pathways. I learned that as the principal investigator, you are where the buck eventually stops when ideas exceed resources.

CONCLUSION

Participating in the LTER experience changed my approaches to scientific inquiry and my beliefs on how to do science. It also placed me within a mix of scientists and scientific questions at some distance from where my formal education had taken me. I believe it will do the same for others as their "lives in science" evolve. At this time, the lengths of time series on a multitude of variables as well as broad spatial views are available on a diverse array of ecosystems and landscapes. Each year these scientific resources will grow. The mix of individuals who participate in the LTER program will continue to expose and resolve new questions and puzzles and that they will creatively model and test previously hidden system behaviors. I expect these opportunities will continue and greatly exceed where we are today. The information and knowledge coming from LTER scientists at these expanding time and space scales are especially critical to guide humankind as our world spins into an uncertain future.

ACKNOWLEDGMENTS

I thank the NSF for supporting the NTL LTER program for seven grant cycles from 1981 to the present (current grant: DEB-1440297 LTER 2014–2020 from the LTER program, Division of Environmental Biology). I thank the lead principal investigators, S. R. Carpenter and then E. H. Stanley, who creatively led our program after I stepped down in 2000. Over the years, many staff members, students, and postdoctoral and faculty associates energized me and our program, but I especially appreciate B. J. Benson, C. J. Bowser, and T. K. Kratz, with whom I worked closely over my tenure.

REFERENCES

Hein, C.L., Vander Zanden, M.J., and Magnuson, J.J. (2007). Intensive trapping and increased fish predation cause massive population decline of an invasive crayfish. *Freshwater Biology*, **52,** 1134–1146.

Franklin, J.F., Bledsoe, C.S., and Callahan J.T. (1990). Contributions of the Long-Term Ecological Research Program. *BioScience*, **40,** 509–523.

Frost, T.M., Fischer, J.M., Brezonik, P.L., Gonzáles, M.J., Kratz, T.K., Watras, C.J., and Webster, K.E. (2005). The experimental acidification of Little Rock Lake. In J.J. Magnuson, T.K. Kratz, and B.J. Benson, eds. *Long-term Dynamics of Lakes in the Landscape,* pp. 168–186. Oxford University Press.

Kling, G.W., Hayhoe, K., Johnson, L.B., Magnuson, J.J., Polasky, S., Robinson, S.K., Shuter, B.J., Wander, M.M., Wuebbles, D.J., Zak, D.R., Lindroth, R.L., Moser, S.C., and Wilson, M.L. (2003). *Confronting Climate Change in the Great Lakes Region: Impacts on Our Communities and Ecosystems.* Union of Concerned Scientists, Cambridge, Massachusetts, and the Ecological Society of America, Washington, DC.

Kratz, T.K., Magnuson, J.J., Bayley, P., Benson, B.J., Berish, C.W., Bledsoe, C.S., Blood, E.R., Bowser, C.J., Carpenter, S.R., Cunningham, G.L., Dahlgren, R.A., Frost, T.M., Halfpenny, J.C., Hansen, J.D., Heisey, D., Inouye, R.S., Kaufman, D.W., McKee, A., and

Yarie, J. (1995). Temporal and spatial variability as neglected ecosystem properties: Lessons learned from 12 North American ecosystems. In D. Rapport, and P. Calow, eds. *Evaluating and Monitoring the Health of Large-Scale Ecosystems*, pp. 359–383. Springer-Verlag, Berlin.

Magnuson, J.J. (1990). Long-term ecological research and the invisible present. *BioScience*, **40**, 495–501.

Magnuson, J.J., Kratz, T.K., and Benson, B.J. (2005). *Long-term Dynamics of Lakes in the Landscape*. Oxford University Press, New York.

Magnuson, J.J., Allendorf, F.W., Beschta, R.L., Bisson, P.A., Carson, H.L., Chapman, D.W., Hanna, S.S., Kapuscinski, A.R., Lee, K.N., Lettenmaier, D.P., McCay, B.J., MacNabb, G.M., Quinn, T.P., Riddell, B.E., Werner, E.E. (1996). *Upstream: Salmon and Society in the Pacific Northwest*. National Research Council. National Academy Press, Washington, DC.

Magnuson, J.J., Bjorndal, K., DuPaul, B.D., Graham, G.L., Owens, D.W., Peterson, C.H., Pritchard, P.C.H., Richardson, J.I., Saul, G.E., and West, C.W. (1990). *Decline of the Sea Turtles: Causes and Prevention*. National Research Council. National Academy Press, Washington, DC.

Riera, J.L., Magnuson, J.J., Vande Castle, J.R., and MacKenzie, M.D. (1998). Analysis of large scale spatial heterogeneity in vegetation indices among North American landscapes. *Ecosystems*, **1**, 268–282.

Senge, P.M. (1990). *The Fifth Discipline, the Art and Practice of the Learning Organization*. Doubleday/Currency, New York.

Swanson, F.J., and Sparks, R.E. (1990). Long-term ecological research and the invisible place. *BioScience*, **40**, 502–508.

PART FIFTEEN
PALMER ANTARCTICA (PAL) LTER SITE

31 Learning from a Frozen Ocean

The Changing Face of Antarctic Ocean Ecology

Hugh W. Ducklow

IN A NUTSHELL

1. The temporal perspective provided by the Long-Term Ecological Research (LTER) program places individual, limited, short-term observations and experiments in a valuable context. Proper interpretation of short-term results may be incomplete without a longer-term perspective. This often makes me skeptical of individual studies.
2. The remote location and harsh environment of Antarctica place special demands and constraints on research, collaboration, and education. Meeting these challenges is one of the most exhilarating aspects of our LTER project.

3. Keeping the proper balance between maintaining continuity of observations and keeping the research program new and innovative is another key challenge for research in the LTER program. But rather than constraining them, the ongoing nature of the LTER program facilitates and enhances creative observations and innovation.

PERSONAL OVERVIEW

In 2001, I joined the LTER network as lead principal investigator for the Palmer LTER project (PAL), one of two pelagic marine sites in the LTER network. That was my first formal exposure to the LTER program, about midway through my scientific career. After majoring in the history of science in college, I received my PhD in environmental engineering from Harvard in 1977. I was originally trained as an environmental microbiologist and gradually evolved into a biological oceanographer and ocean biogeochemist. Prior to joining PAL, I worked in other large, multidisciplinary, and interdisciplinary ocean science programs. In addition to leading PAL, I study the roles of ocean microbes in the biogeochemical cycling of carbon and other elements in the ice-influenced ocean surrounding Antarctica. As a principal investigator, I participate in planning and guiding the LTER network. Network participation has significantly broadened my perspective on my own personal scientific work. This participation has been one of my more interesting and fulfilling experiences as a scientist. Over the past 20 years, research has shown that the western Antarctic Peninsula is one of the most rapidly warming regions on earth, and we are gradually beginning to understand how the ecosystem is responding to this unprecedented rate of change.

APPROACH TO SCIENCE

Joining PAL changed my life. (Actually, going to Antarctica for the first time changed my life, but the LTER program gave me the opportunity to go there every year.) But what the LTER program has done is to make me appreciate variability, especially at lower frequencies, and the futility of understanding an ecosystem without devoting years of study to it. You can spend a very intensive year investigating a system and write your results, and come back several years later and the system will have changed. I think of the beautiful studies of coral reefs by the Odums (Odum and Odum 1955) and wonder how different their conclusions might have been if they had continued their observations over a few years. Even my own work is vulnerable to this criticism: one of my favorite achievements was constructing a virtual seasonal cycle of bacterial and phytoplankton activity in the Ross Sea, comprising careful, painstaking observations in November 1994, December 1995 to January 1996, and March to April 1997 (Ducklow et al. 2001). For my own PhD dissertation on coral reef microbial ecology most of the experiments were conducted during short field trips to different reef locations and not repeated in different seasons or years. Here I think of John Magnuson's seminal article concerning "the Invisible Present"—without the long-term telescope, it is impossible to place ecosystems observations in their proper context (Magnuson 1990). Much can be learned, of course, through short-term experiments or observations, but as a result of the LTER program, I appreciate the context in which short-term results need to be evaluated.

Our site attracts many potential collaborators, but there are challenges too. Because of the remote location and very constrained space and facilities, we have limited opportunities to bring students and colleagues to our site. The National Science Foundation (NSF) sets an absolute limit on the number of people we can have in our project in any given year, and it even specifies when they can come and go. Access is time-consuming,

FIGURE 31.1 The Palmer Antarctica Long-Term Ecological Research site in the western Antarctic Peninsula is only accessible by ships, such as the Antarctic Research and Supply Vessel *Laurence M. Gould*. (Photo courtesy of Hugh Ducklow.)

and not everyone can fit a field trip to Antarctica into his or her schedule (Figure 31.1). So we have to plan far ahead to set priorities and choose carefully. These extreme factors demand creative thinking to find common denominators with other sites in the LTER network. How is the polar sea ice zone like a forest in Oregon or a coral reef in Tahiti? Or even like the Dry Valleys at McMurdo, our nearest neighbor just across the Antarctic continent? Working to create good science while meeting logistical challenges is, for me, the most exciting and satisfying aspect of research in the LTER program.

ATTITUDES TOWARD TIME AND SPACE

One of the great challenges of research in the LTER program is maintaining the continuity and corporate memory so critical for a long-term project, but at the same time, keeping the program fresh and moving forward. It is difficult to learn that conducting research in the LTER program absolutely does not mean that one just keeps doing what one has been doing, no matter how well one does it. There is a paradox here: it is absolutely necessary (1) to maintain the long-term view and have a backbone of long-term observations to detect variability and change and (2) to test hypotheses about low-frequency processes. But it is also crucial to invigorate the program with new people, tools, and ideas. The long-term infrastructure provides a unique opportunity, not an obstacle, to try out new ideas and develop new ways of looking at decades of prior findings. The long duration of support associated with LTER sites makes innovation possible.

Research in LTER program provides a good example of the interaction between maintaining long-term observations and striking out in new directions. Although I value and appreciate the opportunity to innovate, it has been a challenge to balance continuity and innovation, especially at a remote site in the harsh Antarctic environment, and in a challenging funding environment. Research at PAL began with a central hypothesis stating that sea ice variability was the principal control on ecological interactions in the Antarctic marine ecosystem. By about 1996, investigators recognized that the western Antarctic Peninsula was one of the most rapidly warming places on the planet, and this discovery kicked off a decade-long effort to identify and document ecosystem responses to climate change. This effort has been highly successful (Montes-Hugo et al. 2009; Schofield et al. 2010), but it came with hidden perils: do we keep focusing on long-term change at the expense of other possible research directions? In a fixed funding climate, and in the very finite world of Antarctic logistics, every decision is zero-sum. If we choose to do some particular experiment, or sample a new region, those decisions inescapably limit us from doing something else, or might force us to curtail something we have been doing for many years.

COLLABORATION

Another huge advantage of the LTER program is the way that it helps attract other, non-LTER scientists to work at our site. Of course, Antarctica's general popularity and the LTER infrastructure also help attract researchers. An ongoing sampling infrastructure and rich data resources strengthen outside projects and free them from reinventing the wheel. As our project has matured, we have been increasingly successful at attracting additional scientists to come and work at Palmer Station. This influx of personnel leads to new collaborations and friendships that extend well beyond the lifetime of an individual science project or grant, and it renews one's own perspective on science. Nonetheless, there needs to be a balance between the ongoing needs of our project and the benefits of collaboration.

Another challenge of working in Antarctica is that there is virtually no travel and personnel exchange between stations and sites around the continent, with a few exceptions such as next door neighbor bases like McMurdo Station (United States) and Scott Base (New Zealand). We visit our colleagues at the British Rothera Base for 1 day each year but only because we are on a ship that stops there. Even virtual meetings by Skype are a challenge: the US Antarctic Program does not permit Skype communication as a rule. Most Antarctic stations have limited bandwidth—or may have Internet connectivity only when satellites are directly overhead. A truly international network of scientists collaborating across the continent remains a vision for the future.

APPLIED RESEARCH

I spent about half my career working in and around Chesapeake Bay, so my engagement with applied science is much less now than it was then. But the reason does not have to do with the LTER program itself, as much as with the particular nature of our site. Antarctica has no indigenous population and no society, except for the artificial and barely sustainable scientific presence. In spite of its remote location, Antarctica affects and is affected by societies elsewhere. Think about sea level rise. Antarctica is far from pristine. Of course, the major impact on our site is global climate change. Our site and the regional ecosystem are changing in response to the rapid warming of the climate. This process is the major focus of our research. People always ask, "What good is scientific research that is done so far away?" We believe that because of the large and clear climate change signals we see, Antarctica is an invaluable natural laboratory for studying global change. Even so, our interactions with nongovernmental organizations and other stakeholder groups such as the Antarctic Treaty or Greenpeace are fairly limited. It seems inevitable that sometime in the future, the Antarctic Treaty will break down, at which point we as scientists will have a long list of new challenges to address.

COMMUNICATION

We recognize that every site in the LTER network is different, but we think that a combination of factors makes Palmer Station an especially exhilarating and valuable site for outreach and education. First, our education and outreach program is different from most other sites in that there are no local schools and no citizens—our schoolyard is 16,000 km away! Second, ours is a marine pelagic site: our natural home is the open sea, far from land, and with no fixed reference points for our work. It is nearly impossible to bring K–12 students and members of the public at large to our site. The public can join passenger cruises and visit the Palmer Station at their own (great) expense. We share this challenge with just one other site in the LTER network (McMurdo Dry Valleys). We have a constant challenge to communicate the true nature of our work to students and the public who live and work on land. On the other hand, Antarctica has captured the public's imagination, and we try to exploit everyone's interest to find new teaching opportunities and new ways to educate the public about climate change. We believe that Antarctic examples are particularly engaging tools for teaching the principles of science, technology, engineering, and mathematics (STEM). It is no coincidence that when the national news magazines have articles on climate change, they put penguins or polar bears on the cover. We are now using webcam technology to bring Antarctica to classrooms: see our PenguinCam at https://www.usap.gov/videoclipsandmaps/pal-webcam.cfm.

MENTORING

I came to the LTER program a bit late in my career, so my attitudes and practices about mentoring students were pretty well established. Therefore, I do not believe that I mentor students differently as a result of my participation in the LTER program, but I make sure that they know about it and have every chance to participate in it. I really appreciate the LTER network as a great venue for introducing undergraduate and graduate students, postdoctoral fellows, and new young faculty to a large and diverse scientific community. For example, I have had undergraduate interns who have already worked at two or more other LTER sites.

On the other hand, my experience in the LTER program has made me a little bit intolerant. When I review proposals or papers on microbial processes in oceans or lakes that report a series of observations or experiments made within a year or even over several years, I cannot help but think (1) what was it like last year? and (2) what will it be like 5 years from now? I try to excuse PhD research from this criterion because graduate students have limited time. Most NSF projects are designed to fall short in this regard; a 3-year award has one or maybe two field seasons, after which you have to start all over again. In most of oceanography, we are lucky to get two cruises in a single award. One of the great enduring beauties of the LTER program is the opportunity to go back again and again and get the research right. We owe a huge debt of gratitude to the visionary program managers who created the LTER program and to those who have maintained it over the past 35 years.

SKILL SET

I did have to develop some new analytical skills to fulfill some of the core data requirements of the LTER program. But with a few special exceptions, such as learning how to measure nutrients and carbon dioxide, which I inherited when I joined PAL, I think

I probably would have developed most of the same new skills anyway, even if I had never joined the LTER program. Science progresses and you have to keep up, or you become irrelevant. Similarly, I have been involved in network-based science for a long time, including information management. I was already familiar with a federated information management structure, but the size and diversity of data holdings in the LTER program were unique in my experience.

PERSONAL CONSEQUENCES

How has participation in the LTER program affected me as a person? That is a difficult question. Being a principal investigator in the LTER program has certainly had a major effect on me professionally—I think about LTER matters every day. Of course, my participation has affected where and when I travel. I would not have been to Antarctica 20 times in the past 12 years otherwise—and possibly still would never have been to South America. I cannot say that research in the LTER program has affected my approach to other cultures or ethnic groups, because there is no indigenous population in Antarctica—only scientists. I definitely have higher visibility as an Antarctic scientist than I would otherwise. I had about a dozen requests for interviews during the recent government shutdown, about the possible effects of a hiatus in sampling on the long-term integrity of our project.

CHALLENGES AND RECOMMENDATIONS

Do not let research, education, and outreach activities at your site get old! When I joined the LTER program, my understanding was that not just the research, but also the groups were long-term. It was certainly true of PAL. In 2001, the majority of the researchers had been involved with the program since it started in 1990. To be sure, there had been some turnover (and not without turbulence, just like the ocean), but several of the founding scientists were still active leaders in the project. Gradually, I came to understand that turnover—sometimes orderly, sometimes not so much—was the rule rather than the exception in the LTER network. In our program today, only two researchers remain from 2001 when I succeeded Ray Smith as principal investigator. One of the achievements in my career that I am most proud of is the creation of our current team, a group of excellent scientists who are generous, open-minded, and really fun to be with in the field. If I were to recommend an extreme measure to ensure the freshness and vitality of the LTER program, I would set a limit of two proposal cycles (12 years) on the amount of time someone can lead a given LTER site. I say that as I am working on my third 6-year LTER proposal. Researchers should be able to stay longer—but not too long.

CONCLUSION

I think I was preadapted to work in the LTER program. As an oceanographer, I have always been fairly collaborative and interdisciplinary. By interdisciplinary, I mean a microbiologist working with physicists, geochemists, and modelers, not just other biologists (but maybe working with penguin and whale scientists counts, too). Oceanography has always seemed inescapably collaborative and interdisciplinary to me. Organisms and processes occur in a moving fluid medium; space and time are intertwined in a complex web extending across a wide range of scales; and everything we see now is somewhere else a few minutes, hours, or days later. We cannot use the meter-quadrat approach of the "real" ecologists (Pound and Clements 1898). Physical–biological interactions dominate the simplest observations. Besides the nature of the ocean itself, working on ships also brings us closer together. We are in a very intimate, close-knit environment and society aboard a research vessel. Think about standing at the rail of a small ship in a heaving sea, throwing up, with your advisor, student, technician, or colleague standing at your side, sick as a dog. One of the things I always loved about oceanography is the way it forces your life and work together, literally in the same boat. That experience fosters collaboration like nothing else.

I had already been working in Antarctica for about 8 years when I joined PAL. I had been involved for over a decade in another large, interdisciplinary, centrally planned, NSF-sponsored, big science project, the Joint Global Ocean Flux Study (JGOFS). JGOFS made me even more interdisciplinary and collaborative than I already was. Furthermore, as a decade-long program with two important time series stations of its own (Karl and Michaels 1996), it made us all appreciate longer time scales, and the blessing and curse of doing time series research. Those two stations are still in operation, one of the crown jewels of US oceanography. Joining the LTER program after my long experience in oceanography was familiar in some ways—still doing oceanography, still being collaborative and interdisciplinary; but quite different in others: the scientific and social contexts in which we work were greatly expanded. Oceanography has always been regarded as an exotic offshoot of ecology, but the LTER program enables us to enrich our research through contacts and collaborations within and outside the network. So the LTER program has allowed me to come full circle, back to the central core areas of ecology, where I began as an undergraduate in the 1970s.

ACKNOWLEDGMENTS

I gratefully thank all the many Palmer LTER colleagues, students, technicians, interns, and volunteers who have contributed to our project through 25 years of working together. I also acknowledge support from polar contractors Raytheon Polar Services and Antarctic Support Consortium for help with logistics and field and shipboard operations, often under terrible conditions. I appreciate support from NSF award ANT-0823101 from the Antarctic Organisms and Ecosystems Program, and all the other support, tangible and otherwise, provided by the LTER network and sister sites. Finally, Palmer LTER is supported by NSF grant PLR-1440435 in the Antarctic Integrated and Systems Science Program of Antarctic Sciences.

REFERENCES

Ducklow, H.W., Carlson, C., Church, M., Kirchman, D., Smith, D., and Steward, G. (2001). The seasonal development of the bacterioplankton bloom in the Ross Sea, Antarctica, 1994-1997. *Deep-Sea Research II* **48**, 4199–4221.

Karl, D.M., and Michaels, A. F. (1996). Preface: The Hawaii Ocean Time Series (HOT) and the Bermuda Atlantic Time Series (BATS). *Deep-Sea Research II* **43**, 127–129.

Magnuson, J.J. (1990). Long-term ecological research and the invisible present. *BioScience* **40**, 495–501.

Montes-Hugo, M., Doney, S.C., Ducklow, H.W., Fraser, W.R., Martinson, D.G., Stammerjohn, S.E., and Schofield, O. (2009). Recent changes in phytoplankton communities associated with rapid regional climate change along the Western Antarctic Peninsula. *Science* **323**, 1470–1473.

Odum, H.T., and Odum, E.P. (1955). Trophic structure and productivity of a windward coral reef community on Eniwetok Atoll. *Ecological Monographs* **25**, 291–320.

Pound, R., and Clements, F.E. (1898). A method of determining the abundance of secondary species. *Botanical Gazette* **25**, 382–383.

Schofield, O., Ducklow, H.W., Martinson, D.G., Meredith, M.P., Moline, M.A., and Fraser, W.R. (2010). How do polar marine ecosystems respond to rapid climate change? *Science* **328**, 1520–1523.

PART SIXTEEN
PLUM ISLAND ECOSYSTEM (PIE) LTER SITE

32 Mysteries in the Marsh

Anne Giblin

IN A NUTSHELL

1. I feel as though my graduate student experiences "preadapted" me to become involved in long-term ecological research. I already enjoyed collaborative research and instantly felt comfortable in the Long-Term Ecological Research (LTER) program when I first had the opportunity to work in it.
2. Working on large, collaborative projects offers a great number of opportunities for students and postdoctoral fellows, but their mentors need to ensure that they develop intellectually independent ideas. Giving students and postdoctoral fellows the long-term collaborative view of science while having them develop as fully independent scientists is a balancing act that I try to always keep in mind.
3. The LTER program has led me into an increased level of communication with the public, students, and local and regional level managers and policy-makers. I have found that at every level people are hungry for scientific information, and my interactions with all of them have been extremely rewarding—although challenging. It

has forced me to expand my communication skills and work with others who have the gift of science translation.
4. There are costs and benefits to scientific collaboration that change with the size of the project and with one's level of involvement in the project. Entraining young scientists is a challenge for large-sized projects, such as those in the LTER program.

PERSONAL OVERVIEW

It was 1975 when I and several other beginning graduate students first walked down a short path through the woods to the Great Sippewissett Marsh in Falmouth, Massachusetts. Ahead of me marched my major professor, Ivan Valiela. As we explored the marsh, Ivan pointed out numerous circular plots staked in the grass. These, it turned out, defined the bounds of his fertilization experiments. The grass within some of the plots was distinctly greener and taller compared to others. Ivan began explaining the marsh fertilization experiment that he had begun 5 years earlier with John Teal. He described how the responses of the marsh seemed to differ with the amount of added fertilizer. The community composition of the vegetation had been changing over time. Shifts in species composition were occurring within the plots that received the highest doses of fertilizer, and Ivan wondered whether all the fertilized sites would eventually converge. Or would the community shift in some entirely different way? To test this, he explained that they had started new plots with an even higher dose of fertilizer. "Of course," he said, "they may continue to change again and it may be many years, even decades, before we know the answer." "Cool," I thought, and then with a bit of panic, I wondered, "just how long does he expect me to be here?"

Like many of the "older" members of the LTER network, I had my first experiences in collaborative long-term ecological research prior to the founding of the LTER program. At Great Sippewissett Marsh, I worked side by side with students examining everything from bacteria to fish. Later, I sampled Mirror Lake before it was part of the Hubbard Brook LTER site. I made my first trip to work at Toolik Lake in 1985 before the founding of the Arctic LTER site. At the time, I did not fully appreciate how rare such sites were. In every case, I was struck by the power of long-term observations and experiments. I came to the conclusion that long-term data were needed to address many critical ecological questions, but I did not fully appreciate what it took to collect those data. But looking back, I can say that in all cases that these research efforts were led by one or two scientists who had an encyclopedic knowledge of all of the work done at the site—and who somehow managed to maintain enough funding to keep the research active for many years. The leaders assembled a team of collaborators with complementary skills and expertise. Data management consisted of trying to make sure that walls of notebooks, papers, and PhD dissertations were not lost through hurricanes, fire, or flood. A secure backup meant that there were copies of critical data somewhere in a dry basement. Outreach was usually limited to occasional press releases or talks in the local library.

APPROACH TO SCIENCE

It is impossible to separate out my experiences in the LTER program from my experiences on other large projects. I came to The Ecosystems Center at the Marine Biological Laboratory in Woods Hole, Massachusetts, because my graduate work convinced me that I liked working on collaborative projects that were examining an ecosystem from an interdisciplinary perspective. The LTER program formed just as I was starting my career and I grew into it, rather than making a conscious decision to seek out research opportunities at an LTER site. Within a few years of starting work at Toolik Lake with John Hobbie, Gus Shaver, and others, the site had become the Arctic LTER site (ARC). A decade later in 1997, a research project on a salt marsh north of Boston where I was working became the Plum Island Ecosystems LTER site (PIE). My associations with LTER sites has only strengthened my opinion that to understand how an ecosystem works, and how it will respond to perturbations, we need information from many disciplines collected at the same place for long periods. I also saw first-hand the power of long-term experiments and cross-system comparison with modeling. At ARC, I saw fertilization experiments still yielding surprising results 20 years after they were started, just as they had at Great Sippewissett Marsh. It took more than a decade to see the impact of sea level variation on salt marsh primary production and two decades for Jim Morris to see an 18.2-year cycle in sea level variation. I also learned how strong collaborations could foster better science and expand my horizons. It was a real privilege, although somewhat intimidating, as a young scientist to spend hours in a room with experienced scientists hashing out ideas and experiments. Of course, this does not mean that there is no place for smaller focused studies or individual research in environmental research. Well-framed questions at all scales are needed to make progress. But personally I have found collaborative, interdisciplinary research more exciting.

Large projects force one to do science outside of one's immediate discipline and learn more about what your collaborators are doing. I certainly believe that the LTER program has broadened my understanding of the role of social science in understanding ecological change. At PIE, humans are major players in the hydrologic cycle in the watershed and major predators on a number of key species in both the watershed and the estuary. Social scientists have brought important insights on water use and species management that inform our predictions of how the system will respond in the future. The LTER program is debating whether or not social science should be integrated with the ecological studies. In some cases, predicting how the system will respond to perturbations requires that we understand human responses as well as natural response to changes.

ATTITUDES TOWARD TIME AND SPACE

The LTER program has not changed my attitude about the importance of understanding how ecological processes vary over time and space. Yet with the reality of shrinking science budgets comes the question as to how to best manage an LTER program. The lure of collecting that next data point is not easy to resist. Yet when is long-term data long enough? What criteria do we use to decide to cut off an experiment at 6 years as opposed to continuing it for 24 years? When do we decide that we have answered a question and move on to a new one? I have not found it easy to answer these questions. I do worry that by asking LTER sites to identify new themes every 6 years LTER sites will move away from asking the more difficult long-term questions. We do not want LTER sites to become scientific supertankers—unable to change course on short notice when necessary—but we do want them to address questions that cannot be examined within the time frame of standard grants.

COLLABORATION

Although collaboration clearly benefits science, does it always benefit the scientist? How big is too big and when do the costs outweigh the benefits? I have considered this question often and frequently look back on an article my colleague, Bruce Peterson wrote (Peterson 1993). Bruce framed his argument from the point of view of project size but also pointed out that within large projects smaller networks exist. Because of this, the experience one has with a large project depends on one's role with the project. Working on a large, interdisciplinary project can be tremendously valuable for students. Students working together learn from one another and often develop friendships and professional relationships that can last throughout their careers (not to mention the occasional marriage). Cramped field houses and challenging field experiences build camaraderie, and students learn to appreciate how their work fits into a bigger question. Postdoctoral fellows also benefit because broad projects such as those in the LTER program are large enough that they can begin to develop the cooperative and leadership skills that are needed for successful collaborations later. However, once a scientist becomes an assistant professor, involvement in a project at an LTER site becomes more complex. At this stage in their careers, scientists are intellectually ready to begin really launching collaborative projects of their own and participating in collaborative synthesis efforts. But many academic institutions believe that a scientist getting tenure needs to show independence. This often means papers published with only the scientist and his or her students as authors, decreasing enthusiasm for collaboration. The expectation that a scientist will publish several papers a year to get tenure does not favor research on long-term studies or experiments. Heavy teaching loads coupled with high research expectations make it far less likely that a young scientist will take on service roles within the project. As a result, it can be difficult to engage and keep young scientists involved in core activities at an LTER site. In contrast, mid-career and later career scientists are more inclined to feel comfortable in a project where credit is shared and projects take longer; then collaborations can be very intellectually rewarding. Peterson (1993) suggested that when compared to a single-researcher grant, group creativity in large projects is still rising, even though the role for individual creativity has declined significantly. I agree with this analysis. Thus, the challenge becomes managing the internal networks of an LTER program in such a way that the needs of scientists at all career stages can be met.

APPLIED RESEARCH

During the early part of my career, there were endless debates about the value of "pure" versus applied science, and young scientists were advised to stick to basic questions. Ironically, good long-term data were being collected within management agencies, but this was often referred to as "mindless monitoring." The data were seldom published, and their value was not recognized. Applied questions were viewed as specific problems that could be addressed by local consulting firms with engineering solutions or by regulations. Scientist who did become involved in applied issues and who spoke out about them often saw their careers threatened. As a graduate student, I was inspired by *Silent Spring* but also saw Rachel Carson's credentials and interpretations questioned. Claire Patterson's pioneering work on lead pollution was still being attacked well into the late 1970s. There were exceptions of course, but students were generally advised to choose their questions carefully, and young scientists were warned against becoming too involved with applied issues. The distinction between basic and applied science is seldom discussed now. The LTER program has been very much involved in helping achieve this change. Strong partnerships between federal agencies and academic scientists did not begin with the LTER program, but the colocation of five LTER sites in US Forest Service Experimental Forests helped cement these relationships. Increasingly, agency and academic scientists began to plan projects together. Although the groundbreaking work in acid rain by Gene Likens began at Hubbard Brook more than a decade before the LTER program was founded, it later became closely associated with the Hubbard Brook LTER site. The Andrews LTER site was the location of major studies on the spotted owl, an iconic symbol of the Endangered Species Act. Through the 1970s and into the 1990s, it became clear that "applied" questions were large in scope and complicated and that solving them required fundamental knowledge. Now many LTER sites are concerned with climate change and invasive species.

COMMUNICATION

Sometimes the general public is skeptical that climate change and sea level rise are really happening, but the managers and regulators we work with through PIE are not. The agencies and nongovernmental organizations in our region are hungry for information that they can use to manage watersheds and marshes in our region in the face of these changes, and the general public is increasingly interested.

I find working with all of these groups to be very rewarding, but it is challenging as well. I credit the Leopold Program with teaching me some of the skills that are needed to more clearly communicate science to diverse audiences, but I still find that without outside input, I slip into scientific jargon and obscure acronyms. For me, as for many others, the most difficult issue is to put the proper emphasis on uncertainty, and convey what is known and what is not known with the proper balance. The K–12 coordinator at PIE, Liz Duff, regularly has scientists from the LTER program present to students in the Schoolyard Program. This activity is a good reality check, and Liz and the students do not hesitate to tell us when we are not making sense. My experiences have brought home to me the importance of science communication and the need to provide both training and opportunities for young scientists to be involved in outreach. Fortunately, I am seeing increased opportunities for scientists to become trained to be better communicators both within and outside the LTER program.

MENTORING

The LTER program, and my association with other large projects, has had an impact on how I mentor students. I believe that when working on large projects, it is important to structure things so that the student's core thesis project is distinct and is a major piece of work where they can take intellectual ownership. In the best possible cases, the student also collaborates on other projects or incorporates site data into their work to gain a valuable long-term perspective. However, the thesis needs to be largely generated by the individual student. I give postdoctoral fellows a bit more flexibility, although for their own success they still need to show that they have made a distinctive contribution, so their network within a large collaboration is often still somewhat limited.

SKILL SET

Through the LTER program, I become much more aware of the challenges of good data management. It would be a stretch to say that this is now a skill in my tool kit. To be perfectly honest, I have sat through entire meetings where I did not know the meaning of most of the acronyms being discussed, and could probably still not get some of them correct even if there was a gun to my head. But because of the LTER program, I have thought much more about the importance of creating and sharing long-term data sets that are useful to the entire community, not just to the researcher who collected them. I am concerned about the increasing burden that is being put on the scientific community to not only make the data available but also to create easy systems for manipulating and compiling it for use by others.

PERSONAL CONSEQUENCES

I have certainly enjoyed my experiences in the LTER program. Seeing many of the LTER sites and learning how they are approaching their questions have been tremendously informative to my own work. When I took over as principal investigator of PIE, I talked to many other principal investigators about their experiences and management styles and received much helpful advice. Other members of the LTER program have always been generous with their time and many have become good friends.

CHALLENGES AND RECOMMENDATIONS

It can be difficult for young scientists to participate in large projects because of the academic expectations that they will show independence. Fortunately, many young scientists find a way to get through this period, but a greater institutional recognition of the benefits of collaboration is needed to ensure that the best and brightest young scientists want to become involved in large projects in general and the LTER program in particular.

Research at LTER sites is a little bit different from research in most academic environments in that they often have a number of highly trained, "long-term" staff. Providing professional opportunities for the staff is equally important to the success of projects at an LTER site, but somewhat more difficult to do. With the exception of the information management community, workshops, meetings, and training programs are often restricted to students and faculty. I believe that this is a challenge in professional development that agencies seem to handle a little bit better than does the LTER program, which tends to focus on more classical academic advancement.

Finally, recalling Peterson's (1993) essay, as projects become larger, both fiscal and intellectual maintenance costs increase. This trend creates a challenge to the management of an individual LTER site and to the ability of sites to participate in LTER network activities. The LTER program has taken on a diverse array of activities in education, data management, and outreach that have considerably increased the scope of the LTER sites. All of these activities are worthwhile, but the sheer volume and scope of them may be spreading the human capital within the LTER program too thinly. I think that the LTER program needs to carefully consider what should be core activities and what should be pursued with partners.

CONCLUSION

I occasionally still go back to the Great Sippewissett Marsh to lead a class field trip. Last September, I ran into Ivan Valiela as he was putting on his boots to once again map the vegetation in the fertilized plots. He enthusiastically described the recent changes that he had observed in the plots and how they might be related to changes we were seeing in the marshes at PIE. I share his enthusiasm and I no longer worry that I will still be working at the same sites for decades on end; in fact I look forward to it. Even after more than 30 years, the marsh is only slowly giving up its secrets.

ACKNOWLEDGMENTS

I am grateful for support from the National Science Foundation (OEC-1238212).

REFERENCE

Peterson, B.J. (1993). The costs and benefits of collaborative research. *Estuaries,* **16**, 913–918.

33 Perspectives on a 30-Year Career of Salt Marsh Research

James T. Morris

IN A NUTSHELL

1. A hallmark of my career has been the development of a model of the responses of salt marsh vascular plants to changes in sea level. This discovery would not have been possible without long-term support from the National Science Foundation (NSF) Long-Term Ecological Research (LTER) and Long-Term Research in Environmental Biology (LTREB) programs.
2. The LTER and LTREB programs have provided platforms for student research that would have been difficult or impossible to duplicate. Most of my students have benefited from the background of data, which stimulate a never-ending source of thesis topics and from the logistical support.
3. My communication skills have been improved by LTER-sponsored workshops with journalists. I also have had an opportunity to share my enthusiasm for fieldwork with primary school students and teachers.
4. Many of my numerous collaborations are consequences of novel, long-term data that emerged from research supported by the LTER and LTREB programs.
5. There are important environmental trends that develop slowly in response to climate or that reveal themselves infrequently, such as disturbance responses, thresholds, and

tipping points. These require long-term, place-based observation of the kind that the LTER and LTREB programs are designed to facilitate.

PERSONAL OVERVIEW

My history with the LTER program began in the late 1970s. As a Yale graduate student working at The Ecosystems Center, Marine Biological Laboratory (MBL) at Woods Hole, I participated in a workshop organized by Dan Botkin to develop a rationale for a long-term ecological monitoring program (Botkin 1978). After a 2-year postdoctoral fellowship, I moved in 1981 to the University of South Carolina (USC), which had sponsored one of the first LTER sites, North Inlet (NIN). North Inlet was the perfect place for starting a research program in salt marsh ecology, and my research there eventually was supported by the NSF LTREB program. I owe a great deal to NSF for that.

My early career benefited enormously from infrastructure at USC's field laboratory and support by the NIN LTER program, which I did not fully appreciate at the time. The NIN LTER would fade into history, to be replaced in part by the National Oceanic and Atmospheric Administration's National Estuarine Research Reserve Program, and my career continued with NSF core program and LTREB grants. I would not become a scientist in the LTER program until mid-career, when former colleagues at The Ecosystems Center invited me in 1998 to join the Plum Island Ecosystem (PIE) project, which is administered through the MBL. My role has been to investigate the environmental factors controlling primary production in the intertidal marshes (Figure 33.1). Primary production is one of the "five Callahans" and for those who are unfamiliar with the term, the Callahans are the core questions that every LTER site is required to study.

APPROACH TO SCIENCE

I was a disciple of long-term interdisciplinary research before my affiliation with the LTER program, and I have been a strong proponent of place-based research since graduate school,

FIGURE 33.1 Technician Karen Sundberg (right) and postdoctoral associate Katherine Renken in a Plum Island salt marsh attempting to right a "marsh organ" after it was toppled by ice during a harsh winter. The marsh organ is a bioassay that measures the responses of plants to relative elevation.

when Hubbard Brook was becoming an icon of intensive place-based research and when Bormann and others (Bormann et al. 1968; Bormann, Likens, and Melillo 1977) were unraveling the details of the nitrogen cycle and role of disturbance. At about this same time, people such as Peter Vitousek were demonstrating the value of an extensive network of research sites (Vitousek 1977). Both approaches are needed if we are to unravel the generalities of ecosystem organization and function. The LTER program has reinforced my views about this.

One of the hallmarks of my career is the discovery of a response by salt marsh primary producers to anomalies in mean sea level (Morris and Haskin 1990). Primary production at NIN was greater during years of higher than average sea level. The next question was: would we see the same response in other estuaries such as Plum Island Estuary? The answer is, conditionally, yes; primary production responds positively to a rise in sea level in marshes perched at an elevation greater than that which is optimal for the vegetation (Morris, Sundberg, and Hopkinson 2013). Ultimately, this line of research resulted in a model of how marshes respond to sea level rise (Morris et al. 2002), which I refer to as marsh equilibrium theory, and this has certainly had an impact on site-specific and cross-site understanding in estuarine ecosystems. These discoveries would not have been possible without a long series of place-based observations.

ATTITUDES TOWARD TIME AND SPACE

Ecologists want to understand the dynamics of environmental change and make predictions. There are many environmental trends that develop slowly or that reveal themselves infrequently, such as disturbance responses, thresholds, and tipping points. These require long-term observation, a perspective not shared by all. I am reminded of a comment by a former dean and scientist that the LTER program consisted of mindless data gathering. I think not, but explanations of the changes we observe require experimentation, hypothesis testing, and modeling.

New discoveries lead to more questions. The questions will dictate the appropriate scales of time and space. Fresh ideas will never cease, and I do not believe that the core monitoring requirements of LTER sites, in the right hands, will inhibit discovery. On the other hand, I would argue that flexibility and adaptation are essential. There should be enough freedom (1) to adapt to the new discoveries and methodologies and (2) to address novel questions.

COLLABORATION

The LTER program attracts scientists who are collaborative by nature. I have many collaborators inside and outside of the LTER community. All of these collaborations have been facilitated by work I have done as a LTER and a LTREB scientist, and they have been shaped by the workshops and meetings I have attended. Although collaboration is a prerequisite for success in the LTER program, it is not a prerequisite for success as a scientist. Collaborators fall along a continuum. At one end, there are "service providers" who may not contribute intellectually. At the other end, there are individuals whose leadership and intellectual contributions are substantial and who push the frontiers of science. I suspect that collaborators representing both ends of the spectrum are necessary for the success of a large, multidisciplinary program.

APPLIED RESEARCH

My experiences in the LTER and LTREB programs have led to applications of practical importance. For example, I am knowledgeable about the causes of wetland loss and of the solutions to this problem. I do not practice marsh restoration, but the basic knowledge I have gained from field and modeling studies is informative to those who do and to those who want to forecast different scenarios. Consequently, I have significant interactions with governmental and nongovernmental organizations. However, I would like to think that my research has value because it has pushed the frontiers of theory in my discipline. General principles that have predictive power emerge from theory. Applied science tends to be case-specific.

I have learned to appreciate the importance of social science and the role that people play and have played (legacy effects) in shaping the environment. For example, analyses of sediment cores extracted from Plum Island salt marshes and modeling demonstrate that land clearing and decimation of beaver populations following the European colonization have had a dramatic effect on marshes and their responses to rising seas. With the abandonment of widespread agriculture in New England, reforestation of the landscape has reduced sediment loadings into the estuary and now limits the response of marshes. Coastal development poses another challenge for marshes. Rising seas force the migration of marshes inland, but the extent of the migration is governed by local topography and barriers imposed by people. Constructing a barrier to protect property is costly in terms of the actual capital expense and future ecosystem services lost, but there are benefits as well as costs. The decision-making process involves politics, economics, public policy, psychology, social justice, and more. My exposure to the LTER program stimulated my interest in this very complex subject. Social science has an important role in an LTER program charged with observing and explaining environmental change, and it should be adequately funded.

COMMUNICATION

The LTER program has created opportunities for me to improve my communication skills and to reach members of the community both directly and through the media. There have been workshops with science writers designed for networking and communication. I participated in one that improved by ability to respond to the media and share my ideas about science. I learned the value of the metaphor. The outreach done by LTER programs through the Schoolyard LTER Program is also highly effective. It provides a mechanism for scientists to interact directly with the public. I have been able to kayak Plum Island Estuary with students and teachers under the auspices of the PIE Schoolyard LTER Program—what a pleasure and a privilege. This is a great way of sharing the joy of science and beauty of nature, and I am convinced of the benefits. I need to make more time for this.

MENTORING

The LTER and LTREB programs have provided platforms for student research that would have been difficult or impossible to duplicate. Most of my students have benefited from the background of data, which has provided a rich source of thesis topics and logistical support. The rewards associated with mentoring students include intellectual stimulation, camaraderie, and often increased research productivity. The students also benefit from exposure to an extended network of students. Also, there is the personal satisfaction of training the next generation of scientists. But the cost-to-benefit ratio has not always been favorable, and I think that most academics recognize that committing to a new student is a risk that occasionally fails. There is an emotional toll from those that do fail, and the research program suffers from the waste of resources and productivity.

Supporting a graduate student is a never-ending quest for funding, which now must cover tuition and fees as well as overhead, and it is a financial shell game. I sometimes question why the scientists must bear this burden.

SKILL SET

My scientific tool kit has grown as a consequence of collaborations within the LTER network. At meetings, I learn what others are doing and discover common interests that sometimes blossom into new research, approaches, or collaborations. For example, Bob Christian gave an excellent short course on the topic of network analysis at a workshop I attended at an LTER All Scientists Meeting. This piqued my interest in the subject and led to the publication of a theoretical paper on the information content of computer-generated food webs varying in size and complexity (Morris, Christian, and Ulanowicz 2005).

Generally speaking, I have put new skills to good use, even in ways that transcend ecology. For example, what I learned about networks also changed my views about social organizations. Like food webs, social networks have attributes or metrics that can be quantified (e.g., connectivity, flow efficiency), although the currency of flow is information, social contact, communication, an so on. I cannot support this with data, but I believe that the success or failure of a social network (e.g., an administrative unit) can be correlated with these metrics. I look at social networks now in terms of who is highly connected, who is not, and who should be. Maybe this has made me a better administrator.

PERSONAL CONSEQUENCES

My profession has affected my personal life in several ways, but the greatest effect has been the time required for travel. This certainly is not a problem unique to scientists, but time spent traveling for meetings and research at remote field sites can take a toll on family life. Every excursion to the PIE site and meeting takes the better part of a week. On the positive side, there are LTER workshops in interesting places (airport hotels excluded!) and meetings at which I present the results of my research. As an environmental scientist, I have been fortunate to witness parts of both the constructed and the natural environments that most people cannot access or even imagine. In each case, whether field research, workshops, or scientific meetings, there is the cultural dimension, and a sharing of ideas and knowledge to which I have been privileged to both receive and contribute. Balancing the pros and the cons, I have to say that my travels have been richly rewarding.

CHALLENGES AND RECOMMENDATIONS

The LTER program was inspired by a need for long-term, place-based observation, and it serves that function very well. One of its challenges is how to utilize this extensive network of sites to test hypotheses that transcend the boundaries of individual ecosystems. The extensive dimension of the LTER program is the domain of intersite research and is encouraged by NSF. Some of these efforts have been successful, such as the Long-Term Intersite Decomposition Experiment (LIDET) (Harmon et al. 2009). I asked Mark Harmon why he thought this particular project was successful. He speculated that its simple design and minimal commitment of time required of participating sites were important. Also, LIDET fit the mission of the LTER program, that is, it was a long-term study of decomposition and one of the core elements, a "Callahan." Mark added that LIDET was not without problems, including a lack of support for data management.

Perhaps there needs to be a physical or virtual center along the lines of an observatory network or national center dedicated to interecosystem research and theory that uses the extant network of LTER sites and other sites of intensive fieldwork.

Another challenge of the LTER program is how to broaden participation, and I mean this in the widest sense. The LTER program is perceived by some as a "closed community," but the reality is that core monitoring and data archival activities, designed to continue in perpetuity, severely limit how the funding can be allocated or reprogrammed. I think these limitations force the LTER community to act defensively and, consequently, like a closed club. But, however club-like the program may seem, it maintains a large membership. A visit to an All Scientist Meeting will confirm this.

Participation is open to any scientist who submits a grant to one of the NSF core programs for research at LTER sites. Leveraging the funding at LTER sites should be an advantage to grant applicants, but I do not know this to be true. What is true is that support from core programs at NSF is so low that it discourages young people from pursuing careers in science. It may be time to rethink how funding is awarded in core programs. A hybrid lottery system is one possible alternative. Peer review panels, still the gold standard, could separate proposals into three pools: those rare gems that are truly extraordinary and must be funded, those that are worthy of funding, and those that are not. The gems would be funded and the worthy proposals subjected to a lottery, all subject to the availability of funds. A lottery would level the playing field and mitigate the stochastic effects of panel composition and expertise.

The LTER program strongly favors interdisciplinary collaboration, and I cannot imagine how it could succeed otherwise. However, collaborations can be challenging for junior faculty members, when promotion criteria emphasize first authorship of grants and publications. Consequently, I would advise junior faculty members to establish themselves as independent and productive scientists before committing significant time to what might be construed as a support role. But weigh this advice against the numerous benefits of forming collaborations that can last a lifetime. The best advice is to seek a balance.

Outreach is a desirable goal, with important societal benefits, but it should not be forced. I fear that the ever increasing escalation of broader impacts now required of grantees will crowd out the socially awkward, mad geniuses who wish to be sequestered in a laboratory to work on their science. It can be a very personal experience. The long-term impact on the nation's competiveness in science and engineering fields of broader impacts has not been established. Scientists are not uniformly the most effective communicators, and most do not have access to the outreach opportunities available to LTER scientists.

CONCLUSION

My exposure to place-based research began prior to the formation of the LTER network and prior to locating to the USC in 1981, home to one of the first LTER sites, NIN. Unfortunately NIN was short lived, but NSF would reward the work I was doing on primary production in salt marshes with an LTREB grant. I am an advocate of place-based research, and the discoveries coming out of long-term observations of primary production at NIN helped when colleagues in Woods Hole recruited me to the PIE site in 1998. Both programs, LTREB and LTER, have been instrumental in shaping my career. The LTER program has not been my major source of support, but I think the program was never intended to be a dominant source of support for individual scientists. It was intended primarily to support long-term measurements of key ecosystem processes. Now that an extensive network of sites has been established, one of the challenges is how to utilize them to test broad hypotheses that cross ecosystem boundaries. This would be an appropriate role for a national center or observatory. For faculty beginning a career and interested in joining the LTER community, I would recommend that they take advantage of LTER infrastructure and collaborate, but establish a record as a successful, independent scientist. Establish a reputation as a science leader, not as a science service provider. Independence is more difficult today because funding from NSF's core programs is especially tight. Given the extreme pressure on resources, I suggest that NSF explore alternative funding mechanisms, including a hybrid lottery system that uses a peer review system to identify proposals that are worthy of funding. Finally, there has been an escalation of expectations for broader impacts, one of the two criteria used by NSF in the merit review of proposals. This is dilutive of the science, and I would argue that the first priority should be the intellectual merit.

ACKNOWLEDGMENTS

I am grateful for support from the NSF (OCE-1238212, DEB-1052636).

REFERENCES

Bormann, F.H., Likens, G.E., Fisher, D.W., and Pierce, R.S. (1968). Nutrient loss accelerated by clear-cutting of a forest ecosystem. *Science* **159**, 882–884.

Bormann, F.H., Likens, G.E., and Melillo, J.M. (1977). Nitrogen budget for an aggrading northern hardwood forest ecosystem. *Science* **196**, 981–983.

Botkin, D.B., ed. (1978). Long-term Ecological Measurements: Report of a Conference, Woods Hole, Ma, March 11-18, 1977, NSF National Technical Information Services, Springfield, Va., 26 pp.

Harmon, M.E., Silver, W.L., Fasth, B., Chen, H., Burke, I.C., Parton, W.J., Hart, S.C., Currie, W.S., Long-Term Intersite Decomposition Experiment (2009). Long-term patterns of mass loss during the decomposition of leaf and fine root litter: An intersite comparison. *Global Change Biology* **15**, 1320–1338.

Morris, J.T., Christian, R.R., and Ulanowicz, R.E. (2005). Analysis of size and complexity of randomly constructed food webs by information theoretic metrics. In A. Belgrano, U.M. Scharler, J. Dunne, and R.E. Ulanowicz, eds. *Aquatic Food Webs: An Ecosystem Approach*; pp. 73–85, Oxford University Press, Oxford.

Morris, J.T. and Haskin, B. (1990). A 5-yr record of aerial primary production and stand characteristics of *Spartina alterniflora*. *Ecology* **71**, 2209–2217.

Morris, J.T., Sundareshwar, P.V., Nietch, C.T., Kjerfve, B., and Cahoon, D.R. (2002). Responses of coastal wetlands to rising sea level. *Ecology* **83**, 2869–2877.

Morris, J.T., Sundberg, K., and Hopkinson, C.S. (2013). Salt marsh primary production and its responses to relative sea level and nutrients in estuaries at Plum Island, Massachusetts, and North Inlet, South Carolina, USA. *Oceanography* **26**, 78–84.

Vitousek, P.M. (1977). The regulation of element concentrations in mountain streams in the Northeastern United States. *Ecological Monographs* **47**, 65–87.

PART SEVENTEEN
SANTA BARBARA COASTAL (SBC) LTER SITE

34 Evolution of an Information Manager

Margaret O'Brien

IN A NUTSHELL

1. I work in the realm between environmental research and information technology. My academic training provided background in environmental and ecological research. The Long-Term Ecological Research (LTER) program's culture of data sharing necessitates understanding formats, structures, and vocabularies required to effectively manage research products.
2. I do not "educate" per se, but I advance the field of data management by mentoring and by training student assistants. Because of the LTER program's early adoption of data management plans, I am able to review and help write these for researchers.
3. The diverse data of the LTER program are widely regarded, with standardized, complete metadata records, so they are sought after by programmers creating sophisticated software tools. I provide an example of real-world data management

for those in informatics research and translate concepts between programmers and environmental researchers.
4. The broad array of tasks means that collaboration among information managers is essential. Collaboration works best when groups are small, problems well defined, and when we already have a similar approach and technology.
5. When I started in the LTER program, cataloging and publishing data represented a novel concept. Now, data management is no longer an afterthought. The LTER program has fostered that.

PERSONAL OVERVIEW

I am not a research scientist; I am a data manager. The role is a bridge between research and information technology, and my role in the LTER program is to increase the visibility and awareness of data science for ecology. For most of the last decade, I have been the information manager for the Santa Barbara Coastal site (SBC) at the University of California (UC) Santa Barbara (Figure 34.1). My primary responsibilities are to facilitate

FIGURE 34.1 Margaret O'Brien and colleagues on the beach at Santa Barbara, California. (Photo courtesy of John Porter.)

research at SBC and to assemble data packages for cataloging. This means that I work with system administrators on hardware needs and user accounts and with scientific staff to keep time series data up to date. I design data products and metadata systems for sharing within and outside the group, design and maintain dynamic websites, and devise the software to hold the "information management system" together. SBC shares scientists and technical personnel with the Moorea Coral Reef site (MCR). For practical reasons, the two sites' data management practices are closely aligned. This arrangement provides an effective environment for collaboration, allowing each to leverage the other's work. Within the LTER network, I recently served as the cochair of the Information Management Committee (IMC), and because of considerable recent attention paid to all aspects of data availability, that position has provided additional experience and exposure.

I came to the LTER program as a scientific programmer with an academic background in biological oceanography. In my previous job, I generated data products for a national repository designed for satellite calibration and validation. Generally, I worked alone after consulting on the requirements and specifications, and I knew the data intimately. Those skills were transferable to the LTER program, but that is where the similarities ended. Within the LTER program, I found a diverse corpus of data, and a community of site information managers with a wide variety of backgrounds and points of view. Previously, I had worked to provide data products for a specific use, and the pathway was clear. For the LTER program, the eventual use of the data that I post is not always known, and its diversity makes data management strategies equally complex.

Several times at social gatherings, I have been asked to describe what an information manager does, and the answer is something like: "It is analogous to being a librarian, but we work with data instead of books, and we have to write our own software. Usually we also edit 'the books.' Sometimes we even write 'the books' and build the 'bookcases.'" My role does not fit into any of the traditional definitions for ecologist, researcher, librarian, editor, programmer, system administrator, or into any other job category that currently exists. It requires some knowledge of all of these fields and sometimes ad hoc, in-depth knowledge. Researchers value my ability to cross between the scientific and software realms.

APPROACH TO SCIENCE

The goal of SBC's information management system is to create well-described, high-quality data, and my day-to-day tasks usually involve writing the code to accomplish that. Data management software is generally rare, and if it does exist, is often not easily adapted to local needs, which are driven by the policies of both SBC and the LTER network in addition to available hardware support. The issues that I encounter now with ecological data from the LTER program are not the same as I used to encounter. The three facets of "big data" are "volume, velocity, and variety." Of these, volume and velocity get the most attention because of their sheer visibility. But the third axis—variety—is more insidious and is what characterizes most ecological data.

Data diversity, plus the need to be integrated with several researchers' laboratories, has required me to develop flexible pathways for data handling and more complex vocabularies for categorizing it. I have had several experiences working with software developers, both as contractors and while engaged in informatics research. These are successful when there is a high degree of coordination and communication, that is, with "joint development" where the programmers work very closely with us, the clients. For example, SBC and MCR recently outsourced a coding project to export metadata from a relational database for multiple uses. I estimate that while the programmer worked full-time on the actual code, nearly half my time was invested in discussion and coordinated activities. Clearly, this was not a timesaver for me during development, but the product is now tuned, tested, and viable.

ATTITUDES TOWARD TIME AND SPACE

I came to the LTER program from a culture of preserving data for the long-term, so my attitude about this basic need has not changed. In my pre-LTER position, data were collected to advance scientific knowledge but also with predefined plans for their reuse. In the LTER program, however, two things are different. First, data collections are designed primarily to address specific location-based scientific questions, and reuse is of secondary importance. Although archiving is still important, reuse is not necessarily planned for. Secondly, data are extremely diverse, and encompass many forms: experimental, observational, instrument, laboratory, and satellite. Long-term time series make up about half of SBC's data holdings, which implies that their observations are consistent despite changes in methods or instrumentation. The added complexity combined with potential future uses means that SBC data require extensive curation, much the same way that museum specimens require not just basic descriptions but also records of their context. In all, it has meant I need to have several different management strategies. SBC's data have four to six pathways for curation, with some overlap.

The long-term nature of investigations in the LTER program means that software requires maintenance, just as data do, especially in an environment where technology is always changing. So every piece of code that I have will be either maintained and improved or abandoned and replaced. Our small, local code projects include plans for upkeep. A research group the size of SBC cannot afford to replace customized code continually, and we try to use inexpensive labor when we can (e.g., students). However over time, new languages become popular (e.g., Python, R), and it becomes more difficult to find young, inexpensive labor fluent in older or proprietary languages (e.g., Perl, MATLAB, SAS).

COLLABORATION

The broad array of tasks that represent an information manager's activities means that collaborating to solve common software problems is essential. One person rarely has both the time and breadth of knowledge to get everything done. Several years ago, I introduced MCR's recently hired information manager to the LTER network. Since then, she has become my own closest collaborator, and this has been a huge advantage. Each site has a dedicated, primary information manager, but our complimentary skill sets and deliberate use of similar computing architecture allow us to efficiently divide work. For example, she is the database administrator for both SBC and MCR, and I take care of the transformations to website content.

As in any field, software collaboration involves trial and error, but for me, it has been one of the most rewarding aspects. Software collaboration tends to work more smoothly if everyone already has a similar strategy and approach to data management. For the most part, some early choices by the LTER program have made this simpler. For example, the choice of a granular XML specification to exchange metadata has meant that we have a common vocabulary. We also can design tools individually that can be shared broadly. In addition, we can translate all metadata to other specifications that might be appropriate.

Quite often, software collaboration simply is not possible, even when two (or more) sites have the same data management needs. First, the diversity of research goals and technology among LTER sites can mean that skills and priorities do not always match. Secondly, each person joining a project changes the scope, so what could have been a short-term project becomes a large investment. Participants have to choose between meeting their sites' immediate needs and the potential long-term benefits of the larger project. Most often, the immediate needs prevail.

LTER network also stresses the importance of integrating data management with science, which means that I have opportunities to work with scientific programmers as well. I have worked with several laboratory groups on code for quality control and data cataloging. As with collaborations among LTER sites, this collaboration is always easier if we already have a similar approach and goals as well as a shared technology.

APPLIED RESEARCH

I participate in applied research mainly by providing examples of real-world data management (called "use cases") for software developers exploring the practical use of new informatics concepts. Interaction with the target community is essential for any software development project. My time is occasionally funded by an outside software development project, but requests for volunteered help are much more common. Some of these new technologies are likely to eventually benefit both my site and the LTER network. I would like to have time to explore prototypes, but usually these are not necessary for day-to-day data management and are sometimes hard to justify.

Innovative scientific data management tools are usually funded through the grant process, and to prove useful in the long term, they also require adoption by the community and eventually, maintenance. Practically speaking, however, neither adoption nor maintenance occurs within the 2- to 3-year time horizon of a typical grant. Consequently, even if there is adequate interaction with the target community during the grant period and a successful product emerges, adoption and identification of potential maintenance requirements cannot happen until much later, after funding is terminated. Like scientific endeavors, sometimes software projects do not yield the expected results. But if a negative result is seen as failure, rather than increased knowledge (as is the case for scientific experiments), it could be that the scientific grant process is not the right mechanism for software production. Software developers are not supposed to "experiment," they are supposed to "produce."

COMMUNICATION

Sadly, my field of expertise does not have a component that is particularly interesting to the public or to K–12 communities. Eventually it will—because data literacy is likely to be an essential skill of the twenty-first century (e.g., National Science Foundation, 2007). Until then, there are ample opportunities for service in other sectors. Because it is a relatively new field, scientific data management has an ill-defined scope and set of practices. When I joined the LTER program, I was surprised to learn that many outside research groups consulted us for advice. LTER's IMC has responded to this need by hosting interagency workshops on specific topics; accommodating a data management session at the large triennial All Scientists Meeting; reorganizing our own committee's annual meeting to hold regular large, open, conference-style meetings (Gries and Jones 2008; Jones and Gries 2011); or joining the Earth Science Information Partners (ESIP 2014). Attending these meetings, and occasionally organizing sessions for them, has exposed me to a much larger community of practitioners and potential collaborators. Locally, our university library is considering establishing a data curation center and has started an advisory group, of which I am a member.

MENTORING

My own position does not include teaching, although occasionally I am able to hire undergraduate student assistants. Generally, they are technologically unskilled, but they have broad experience as computer and Internet users, and they already expect data and other resources to be easily available online. Almost inherently, they appreciate the need for completeness and structure in a way that scientists of my generation do not. Student assistants welcome learning about the day-to-day tasks that make data sharing a reality. Mentoring and teaching them is a pleasure, even when they arrive in the office with very small skill sets. Ideally, some of them might take up information management as a career.

Earlier, I mentioned introducing a second information manager at UC Santa Barbara to the network. I do not consider this mentoring per se. Although that person was a new hire, she came to the job with the necessary and (fortunately for me) complementary skills. My role was more of guide or consultant: the one to explain acronyms, elucidate arcane practices, and warn about the pitfalls.

SKILL SET

When I started with SBC, my background was scientific, and even though I considered myself fluent in some programming languages, I had essentially no formal software training. I think my situation is typical of most people trained in the sciences: scientific software is usually closely tailored to the immediate problem, and self-taught skills are adequate. I have since expanded my knowledge of the same languages I used before to keep up with the growing volume and diversity of SBC's data. I became fluent in our metadata exchange specification and related standards and have had to modernize the code I use for website construction. But still, none of my training is formal; it is all "on the job." But, by supporting the community of information managers with missions similar to mine, the LTER program has fostered the environment that helps all of us find the fastest route to new solutions. For example, all LTER sites create metadata according to Ecological Metadata Language, so the IMC is the first place I go to discuss its usage.

Diverse data require fairly sophisticated and flexible software. Data management is still a fairly young field, so most tools are simply not fully mature. Fortunately, the National Science Foundation has been generous with supplementary funds for information management, and we use these to contract out some software development. To be ensured of software that was robust and to take advantage of the skills of trained developers, it became imperative that I understand the software development process and use it to our advantage. The process is more complicated than it looks. Large companies have departments dedicated to acting as the interface between the code and the user. LTER information managers such as myself have had to learn to be that interface ourselves.

PERSONAL CONSEQUENCES

I travel two or three times a year for work related to the LTER program, usually to workshops or meetings. Many of these trips are deliberately held at LTER sites, an experience that has immensely broadened my understanding of the LTER network's research diversity. Being associated with the LTER program, and especially in my leadership role as cochair of a standing committee, has also allowed me to reconnect with friends I knew before my move to UC Santa Barbara. The IMC is a close-knit group, perhaps in part because many people are not fully aware of what our jobs involve, so we value each other's understanding. Now I count other site information managers among my friends.

CHALLENGES AND RECOMMENDATIONS

My position at UC Santa Barbara is in an academic track called "specialist," and of the career tracks available it is probably the best fit. A specialist is reviewed with the same framework as tenure-track professors and professional researchers. However, my main responsibilities are to facilitate ecological research, which does not foster growth as a researcher in the field of ecological informatics, because that field is peripheral to the focus of data needs of the LTER program. There are new technologies available that are likely to eventually benefit both my site and the LTER network. I would like to have time to learn from these, but usually these prototypes are not necessary for day-to-day data management and seldom fall within my scope of work.

The role of scientific information manager must become better defined. The field is still new to research programs, and there is little agreement on how the data management roles should be integrated or how the performance of its practitioners should be measured. Clarifying these parameters will help the LTER program's information management community evolve from a group of people with collections of ad hoc solutions to an actual profession. For data managers themselves, the same components that are missing from research scientists' training are missing from ours: mainly training in management techniques and good software practices.

CONCLUSION

It is not enough to simply say that my involvement in the LTER program has affected my professional identity over time; I would have to say that it has enabled my development. The skill sets and competencies that information and data scientists use vary widely, and most practitioners are hybrids. For decades, however, the LTER program has had a culture of curating and sharing data and has created the environment that fosters the development of careers such as mine.

ACKNOWLEDGMENTS

First, I wish to thank the scientists, students and technical staff of the SBC LTER site. Under D. Reed's leadership, they continually have recognized the importance of sustainable data practices, even though the dividends may not be immediate. Second, many of the ideas presented here came from discussions with members of the diverse and creative LTER IMC, whom, after years of interaction, I happily count among my friends as well as colleagues.

REFERENCES

ESIP. (2014). Earth Science Information Partners, http://commons.esipfed.org. Last accessed on April 15, 2014.

Gries, C., and Jones, M.B., eds. (2008). Proceedings of the Environmental Information Management Conference (EIM 2008). Available at https://eim.ecoinformatics.org. Last accessed on April 15, 2014.

Jones, M.B., and Gries, C. eds. (2011). Proceedings of the Environmental Information Management Conference (EIM 2011). Available at https://eim.ecoinformatics.org. Last accessed on April 15, 2014.

National Science Foundation. 2007. Cyberinfrastructure Vision for 21st Century Discovery. http://www.nsf.gov/pubs/2007/nsf0728/nsf0728.pdf. Last accessed on April 15, 2014.

PART EIGHTEEN
SEVILLETA (SEV) LTER SITE

35 From the Long-Term Ecological Research Program to the National Science Foundation and Back

A Personal History of Long-Term Ecological Research Science and Management

Scott L. Collins

IN A NUTSHELL

1. I have been formally associated with the Long-Term Ecological Research (LTER) program since 1988, spanning most of my research and teaching career.
2. The LTER program provides the infrastructure to offer a variety of research experiences for undergraduate students. Also, through LTER programs, I have worked with many graduate students in collaborative research projects and helped train my own students about the value of studying long-term changes in ecological systems.
3. Because LTER sites are highly visible regionally and nationally, I have had many opportunities to communicate the value of long-term research to the public through a variety of mechanisms, such as field trips, visitor center interactions, and public lectures.
4. Because it is a functioning network, the LTER program fosters collaboration. Some of my most satisfying research experiences have been collaborating with friends and colleagues on cross-site synthesis research.

PERSONAL OVERVIEW

I have been involved in the LTER program for more than 25 years. I started in 1988 as one of the senior scientists at the Konza Prairie (KNZ) site, collecting core vegetation data. In 1992, I moved to the National Science Foundation (NSF), first as a program director in Ecology and then as the program director for the LTER program for 6 years. In 2003, I moved to the University of New Mexico (UNM) to become principal investigator at the Sevilleta (SEV) site. By the end of 2014, I will have served as principal investigator at the SEV site for 12 years. I have also served on the LTER Executive Council, and after a change in governance structure, I served as the chair of the LTER Science Council and Executive Board from 2011 to 2014.

Currently, I am a professor of biology at UNM. My formal training was in plant community ecology, primarily focusing on grassland ecosystems, but through diffusion, trickery, and persistence, my interests have also drifted toward interactions between plant community structure and ecosystem processes. By drifting in this direction, I have become increasingly interested in how technology can provide high-resolution environmental data sets via sensor networks and wireless technology (Collins et al. 2006). This mix of interests has allowed me (1) to participate in a variety of cross-site syntheses using both low-resolution (Suding et al. 2005) and high-resolution (Vargas et al. 2010) LTER and non-LTER data and (2) to promote the value of long-term research and collaboration as a scientific process.

APPROACH TO SCIENCE

I think my favorite approach to ecological research is through long-term field experiments, although I have certainly benefited enormously from short- and long-term observational and experimental studies as well. Several experiences early in my career led me to blend these approaches. First, I did my PhD research on the plant ecology (habitat structure) of the wood warblers (Family Parulidae) in central Maine and Minnesota. I used an observational approach that combined the concept of gradient analysis and niche differentiation in co-occurring species. Toward the end of my PhD program, I met with a distinguished ecologist who was on campus to give a seminar in our department. When our laboratory group met with him, he asked me to tell him about my PhD research. Following a brief summary, he said to me something to the effect that working with bird communities was not a good idea because "you cannot do experiments."

Evidently, he believed that manipulative experiments are the pinnacle of science. After 35 years, I finally have my retort, "Oh yeah? So are you saying that climate scientists are not scientists?" I managed to publish my "nonscientific" observational dissertation studies (Collins, James, and Risser 1982; Collins 1983) despite my purported inferior approach to research. But the statement about the value of experiments stuck in my mind, and I wanted to work in systems where I could use experimental approaches when appropriate. Working in grasslands has been a tremendous advantage for me because they are so amenable to experiments, and I would argue grasslands have become model ecosystems for experimental hypothesis testing and generating ecological theory.

The nice thing about the LTER program is that you can just remain a core scientist and do your research and training "thing"—or you can assume various leadership roles, ranging from workshop organizer to chair of the LTER network. It is a fact that large research networks require a certain level of travel, meetings, bureaucracy, and overhead to function effectively and to maintain some degree of coherence, coordination, and general level of communication. Site- and network-level committees, synthesis working groups, and ad hoc groups require volunteers to participate and someone to function in a leadership role. Since escaping NSF, I have been elected chair of the LTER Executive Committee and Science Council. I have also been elected president of the Ecological Society of America. Evidently, people think my time is better spent in conference rooms rather than out among the research infrastructure at the SEV site that I do not know how to operate.

I have long believed in sharing data. When I completed my PhD research, my dissertation was written as manuscript chapters, like most are today. My dissertation seemed so thin that I was somewhat embarrassed by it. So, I typed up all my raw dissertation data and included it as an appendix to fatten up my dissertation. By doing so, anyone could have access to my dissertation data via interlibrary loan. I remain committed to data sharing and open data access, preferably with no restrictions except in rare but justifiable cases (e.g., PhD research). Open access to data is an expectation for LTER programs and is encouraged by the data access policy of the LTER network, which I strongly endorse. Access is provided via LTER site websites, and wider access will eventually occur through the Provenance Aware Synthesis Tracking Architecture and the LTER Network Information System data portal. Providing the metadata necessary for others to use data can be tedious and time consuming, but it is worth the effort. Like most others, I struggle with this responsibility; however, the sophistication of LTER data management systems and especially LTER information managers is making the job easier and easier over time. Share your data! You will expand your collaborations and might even get involved in working groups pursuing new and exciting synthesis research.

My connections with the KNZ site and time at NSF exposed me to the need for experimental tests of the likely ecological impacts of global environmental change. Stepping in to lead an LTER program in New Mexico opened up many new avenues for research and presented multiple opportunities to expand my research interests and approaches. Based on my time at NSF, I think it is much easier to take over an existing LTER program than it is to start a new one. When I arrived, the SEV LTER site was struggling a bit, but it had all the elements of a successful LTER program, including an excellent group of congenial scientists and very compelling ecosystems. In particular, it had a lot of gorgeous desert grassland suitable for new climate change experiments! New Mexico also afforded an opportunity to develop some of my own ideas about the impact of climate variability on arid ecosystems in collaboration with several of my new colleagues at UNM.

I believed that moving from NSF to New Mexico to be the principal investigator of the SEV site put me under a rather intense spotlight regarding expectations from research staff, departmental colleagues, and my former colleagues at NSF. Perhaps some of this pressure was self-imposed, but I was determined to help the SEV program grow and gain exposure through outstanding long-term research. Quite frankly, I think my colleagues and I have been pretty successful. Several new long-term experiments are ongoing or planned, and SEV scientists continue to successfully compete for research funds and publish articles in excellent journals. Despite all of those accomplishments, the LTER program director at NSF has recommended that funding for our highly successful research program be terminated.

ATTITUDES TOWARD TIME AND SPACE

This is a rather odd category to address in this context. One of the challenges with a life in the LTER program is that you can become somewhat callous about "short-term" research. The vast majority of ecological research is conducted on 3-year time intervals associated with typical grant cycles. Because we so often see different trends in our data long after 3 years, it is difficult not to dismiss such research as "short-term" and therefore less informative or reliable. I think this has certainly colored my evaluations of journal articles more so than research proposals. We all understand the time and creativity constraints imposed on everyone by limited resources and the 3-year funding cycle. On the positive side, an emphasis on collaboration has greatly changed how I evaluate junior colleagues. I consider it to be an asset for young scientists to have several multiauthored papers that demonstrate a willingness to collaborate and share data. Collaboration is not for everyone, or for every research context. But it is great when it works.

Basically, as a scientist in the LTER program I am committed to conducting long-term research. Long-term does not imply anything spatial. However, I think we all wonder if any of our research is scalable. Indeed, I was fortunate to be involved in what turned into a relatively long-term project (circa 8 years) comparing grassland response to fire, grazing, and climate variability at the KNZ site and at two relatively comparable sites in South Africa. Our goal was to determine if our fundamental knowledge gained from years of experience and research at the KNZ site was applicable to similar systems in South Africa. The project forced us (1) to list both the similarities and differences (contingencies) between the two study areas and (2) to use those factors to help explain why responses were similar or different. Surprisingly, many aspects of community structure and ecosystem processes were, in fact, trending in similar directions (Buis et al. 2009; Koerner et al. 2014). That finding was rather satisfying because it suggested that knowledge gained from one system might be useful to understand, and perhaps even manage, other systems with different evolutionary histories that were subjected to similar environmental drivers.

COLLABORATION

My first exposure to the LTER program was in 1986 when I visited my former graduate student officemate at the University of Oklahoma, David Gibson, who had recently accepted a postdoctoral fellowship at the KNZ site. I did not quite understand the concept of LTER program at the time, but I was impressed by the research opportunities at the KNZ site as we drove around the research area before the bison fence was built. I knew that my department at the University of Oklahoma was not going to become a powerhouse in ecology, certainly not based on my research record, so by working at the KNZ site, I could get my "big-time" ecology fix. About 1988, I had the chance to formally join the KNZ program by collecting their long-term data on plant species composition each summer after David Gibson took a faculty job at the University of West Florida.

This greatly expanded my opportunities to think about the value of long-term research and to work with other colleagues interested in different aspects of grassland ecosystems.

At first I was not particularly collaborative. Instead, my students and I worked within the context of the KNZ site, and we used some of the long-term data sets to test hypotheses on factors driving the spatial and temporal distribution and abundance of grassland plants (Collins and Glenn 1991). Because of my interest in plant communities, I was invited by other KNZ scientists to participate in experiments where species composition and community structure were both going to be key response variables. This greatly helped to expand my own research program and interests. However, I think the first really collaborative work I did with KNZ colleagues was research resulting in a paper published in 1998 analyzing the interaction of fire and grazing on plant species diversity (Collins et al. 1998). It was about that time that I began to appreciate how collaboration improved research, in general, and certainly improved the research I was doing. That collaboration led to multiple opportunities to share ideas and work on new projects with colleagues at the KNZ site.

Surprisingly, my integration into the KNZ program really flourished after I moved from the University of Oklahoma to the NSF in 1992. The plan was to work at NSF for 1 year and then resume my faculty position at the University of Oklahoma. That 1 year turned into a very slow spin of 10.5 years instead. Fortunately, NSF allows scientific staff to use 20% of their time for "independent research and development (IRD)." I used my IRD time to stay engaged in research, continue my long-term vegetation surveys at the KNZ site, and to go to the National Center for Ecological Analysis and Synthesis (NCEAS) once a year to work on projects that generally developed around KNZ data sets (e.g., Collins 2000). I also served as the NSF program officer for NCEAS for several years, and that experience emphasized the value of data sharing, collaboration, and synthesis. Because program directors are not allowed to compete for NSF research funds, I lived a vicarious research career at that time through the KNZ program, adjunct appointments at Kansas State University and the University of Maryland, and trips to NCEAS, where it became abundantly clear to me that collaboration facilitated creativity, integration, and generality.

It is quite likely that my time at NSF allowed me to embrace collaboration more than I might have through a traditional academic pathway. First, as an NSF employee I needed existing research infrastructure to maintain a research program. But I think the fact that I did not have to publish and compete for funds made it easier to be integrated into research projects. I did not have to create "the Collins laboratory." I wanted to continue to do research and publish primarily because I enjoyed it, and I knew it would be my only ticket back into academia. But I also knew that if I maintained some scientific credibility while at NSF, it could potentially have a positive impact on my interactions with principal investigators applying for, and being denied, NSF grants.

APPLIED RESEARCH

Given the pervasive nature of global environmental change, I am not convinced that a distinction exists between "basic" and "applied" research. Certainly the boundary between these somewhat artificial research categories is rapidly fading. NSF emphasizes fundamental science but requires broader impacts. I think LTER programs are well suited to integrate knowledge and application. After all, LTER sites are located in US Forest Service Experimental Forests, cities, agricultural systems, and coastal zones. We all hope our research is relevant to conservation and environmental management. The core components of the SEV program are located in a National Wildlife Refuge. We occasionally choose some research projects to explicitly address a local to regional management needs (e.g., Ladwig et al. 2014). I believe that we owe that to those who host our research and to the public in general.

COMMUNICATION AND MENTORING

I think outreach is a fundamentally important obligation of any LTER program. We have an amazing Schoolyard LTER Program at the SEV site, called the Bosque Environmental Monitoring Program (BEMP), led by Dan Shaw of the Bosque School in Albuquerque. This program brings hundreds of precollege students into the field each year to collect long-term data on riparian ecosystems up and down the middle Rio Grande. BEMP also trains undergraduate interns and provides a lot of outstanding in-class activities. In addition, I have worked with several colleagues to create a variety of training and outreach opportunities tied to the SEV site, such as K–12 and Research Experiences for Undergraduates (REU) sites programs. We work especially hard to attract minority students to our summer REU program. An LTER program has the advantage of offering a broad suite of research activities that can attract a diverse set of students, and we have access to a large pool of mentors to work with students each summer, including faculty members, staff members, postdoctoral fellows, graduate students, and affiliated scientists. In addition, many sites have staff with backgrounds in outreach, communication, and education to lead these kinds of activities.

SKILL SET

Moving to a new system and interacting with a new group of collaborators was quite invigorating. My work in Oklahoma and at the KNZ site mostly involved quadrats and stake wire flags. I was a core scientist in the long-term Rainfall Manipulation Plots at the KNZ site, but excellent technicians and postdoctoral fellows managed the technology and I just dealt with my quadrat data. Since moving to the SEV program, I am now the proud owner of several data loggers, multiplexors, irrigation systems, rainout shelters, passive warming infrastructure, and wireless radios, along with dozens and dozens of expensive soil moisture probes, temperature probes, and CO_2 probes. Bright graduate students, postdoctoral fellows, and SEV staff introduced most of this technology to me. I still do not know how any of these things work, but because of the SEV program I am fortunate to be surrounded by excellent technicians, postdoctoral fellows, and colleagues who do. On the downside, I now spend much less time using quadrats to conduct vegetation surveys.

PERSONAL CONSEQUENCES

Being part of the LTER program for more than 25 years, including positions at several sites (KNZ, SEV, and to some extent Central Arizona–Phoenix) has had a huge impact on the research that I do and how I do it. When I was a beginning faculty member at the University of Oklahoma, I struggled to compete for research funds, and I was working under the traditional premise that I needed to establish a personal research identity. Therefore, I needed to focus on a particular theme that would distinguish my research from that of others. My dissertation research had nothing to do with grassland ecology, but being in Oklahoma, I was surrounded by nice examples of both tallgrass and mixed-grass prairie. I was also noticing in the literature that people were doing more "field" experiments and publishing them regularly in important journals such as *Ecology*.

I wanted to publish in *Ecology* before I came up for tenure. Interest in the role of natural disturbances and disturbance regimes was a growing area of inquiry at this time, as well. It did not take long for me to figure out that I could do replicated field experiments in grasslands at relevant spatial and temporal scales using technological tools such as shovels and stake wire flags. These systems responded quickly to experimental and natural disturbances, and no one seemed to mind if I dug holes and created fake gopher mounds in burned or unburned, and grazed or ungrazed grasslands.

My primary assignment at NSF was as a program director in the Ecology Program. However, 3 years after I arrived at NSF, Tom Callahan and Mike Allen tricked me into volunteering to be the program director for the LTER program. Like the Ecology Program, LTER is a core program in the Division of Environmental Biology and a flagship program in the Directorate for Biological Sciences. Working at NSF changed my life because it exposed me to the breadth of ecology as well as to many other related disciplines. But, being the program director for LTER really changed my life because LTER principal investigators and site scientists are not shy about letting you know what they think or what they want. There is an LTER culture! Also, thinking broadly about LTER research programs helped to feed my growing interest in blending community and ecosystem perspectives in my own research. During my time as LTER program director, I was fortunate to be involved in several competitions for new LTER sites, eventually helping to expand the network from 18 to 24 sites. Overseeing the LTER program did have its challenges. Site visits could be a bit contentious and trend toward confrontational at times. Managing renewal panels was a far more active and involved process than the more passive role of program directors in other core programs. And, putting sites on probation at renewal time, sites run by members of the National Academy of Sciences, for example, was somewhat sobering. I was also involved in a somewhat contentious renewal competition that resulted in moving the LTER network office from the University of Washington to the University of New Mexico in 1996 to 1997. All in all, I believe the LTER program benefits greatly from the process of oversight imposed by NSF during site visits and during the evaluation of renewal proposals.

CHALLENGES AND RECOMMENDATIONS

Despite the many positive aspects about the LTER program, there are many negative aspects, too. In no particular order, these include the need for new site and network leadership, maintaining a creative and dynamic research program, the need to bring new people and perspectives into an LTER research program, the need for consistent and appropriate management by NSF, and the National Ecological Observatory Network (NEON).

Indeed, one current challenge for the viability of the program is that NSF program managers need to develop a vision for LTER that is not narrow, prescriptive, and constraining, but instead is responsive to the goals of the LTER community. To me, a good program manager at NSF is far more reactive to the needs of the scientists and not overly proactive in a way that limits scientific creativity. At the site level, changes in leadership can have both positive and negative impacts on an LTER program. Someone needs to have a vision for the project as a whole and the energy to corral independent-thinking scientists into doing collaborative science. We need to do a better job of preparing others during their careers to step into leadership roles at the site and network levels. Yet, who wants to impose bureaucratic time-sinks on our productive young colleagues? How do we engage new people into an LTER program when resources are already committed? I recommend using a more flexible centralized funding mechanism at the site level. In this way, resources can be directed toward core activities and some funds can be reserved for new projects and participants. NEON is the new "big dog" on the scene. Initially, NSF made a huge mistake, in my opinion, by driving a wedge between NEON and the LTER program during the planning phases of NEON. I think I can say that because I was the first program director for NEON, and in that role it was my job to make sure the two programs remained distinct. During the planning stages of NEON, one high-level administrator in the Biological Science Directorate at NSF told me that, "If more than half of the NEON observatories are at LTER sites, then NEON will be a failure." I found that perspective rather appalling. Why not build on our strengths? Of course, scientists in the LTER network, along with many others, were actively involved in the planning of NEON from the get-go, and as it turns out several NEON core or "relocatable" installations are partnered with LTER sites. The two programs are clearly complementary: NEON is by necessity a top-down, highly coordinated infrastructure program and LTER is a bottom-up research program built around common core areas of investigation. We all have a stake in the success of NEON and want it to be wildly successful. Yet, because of this start, we now must actively work to bridge the two programs.

CONCLUSION

Given my background and experiences, I would urge young scientists to get involved in the LTER program as soon as possible. Working within the context of a research platform, such as that in the LTER program, has tremendous benefits, including abundant background data; research infrastructure; and a ready supply of collaborators, technicians, and students. I struggled to get funding on my own when I was an assistant professor, but by engaging with other scientists and working at KNZ I found a mechanism for success and a constant supply of new questions to be answered. I think with the right mindset, anyone can be successful within the context of the LTER program.

A second piece of advice is: do not be afraid of change. My moves from academia to NSF and back to academia were challenging. Giving up the security of tenure was extremely risky, especially to move to a bureaucracy such as NSF where research opportunities might be limited. LTER program allowed that transition to work for me. Leaving behind a very secure, highly paid, science-oriented federal job to resume an academic position and to once again compete for funding was also quite worrisome. But each time I did a major "reboot" of my life, I found the new systems, colleagues, and challenges to be quite invigorating. So, big change can be a strong source of positive motivation, excitement, and especially creativity.

The LTER program has had an enormous impact on my career. It has allowed me to meet and even work with some of the most accomplished ecologists in the United States. It has provided me with opportunities to think completely differently about how to do ecological research. Indeed, we operate under the luxury of doing long-term studies without having to justify long-term funding. Working in the LTER program also poses considerable advantages for training undergraduate and graduate students, and for building important understanding that is relevant to management and the general public. For a long time, the LTER program was the only game in town. Now we have the good fortune to witness the development of technologically advanced observational networks that will eventually provide national-scale information on ecosystem status and trends. Such networks are the perfect complement to the LTER network, which focuses on understanding process and mechanisms driving status and trends. Together, as these networks gather complementary and spatially extensive long-term data, they will eventually form a powerful and essential tool for understanding and managing the consequences of global environmental change. We must work together now to integrate these observational and experimental networks to address a number of the large-scale environmental challenges facing humanity today.

REFERENCES

Buis, G.M., Blair, J.M., Burkepile, D.E., Burns, C.E., Chamberlain, A.J., Chapman, P., Collins, S.L., Fynn, R.W.S., Govender, M., Kirkman, K., Smith, M.D., and Knapp, A.K. (2009). Controls of aboveground net primary production in mesic grasslands and savannas: An inter-hemispheric comparison. *Ecosystems*, **12**, 982–995.

Collins, S.L. (2000). Disturbance frequency and community stability in native tallgrass prairie. *American Naturalist*, **155**, 311–325.

Collins, S.L. (1983). Geographic variation of habitat structure for the wood warblers in Maine and Minnesota. *Oecologia*, **59**, 246–252.

Collins, S.L., Bettencourt, L.M.A., Hagberg, A., Brown, R.F., Moore, D.I., Bonito, G., Delin, K.A., Jackson, S.P., Johnson, D.W., Burleigh, S.C., Woodrow, R.R., and McAuley, J.M. (2006). New opportunities in ecological sensing using wireless sensor networks. *Frontiers in Ecology and the Environment*, **4**, 402–407.

Collins, S.L., and Glenn, S.M. (1991). Spatial and temporal dynamics in species regional abundance and distribution. *Ecology*, **72**, 654–664.

Collins, S.L., James, F.C., and Risser, P.G. (1982). Habitat relationships of wood warblers (Parulidae) in northern central Minnesota. *Oikos*, **39**, 50–58.

Collins, S.L., Knapp, A.K., Briggs, J.M., Blair, J.M., and Steinauer, E.L. (1998). Modulation of diversity by grazing and mowing in native tallgrass prairie. *Science*, **280**, 745–747.

Koerner, S.E., Burkepile, D.E., Burns, C.E., Eby, S., Fynn, R.W.S., Hagenah, N., Matchett, K.J., Thompson, D.I., Wilcox, K.R., Govender, N., Collins, S.L., Kirkman, K.P., Knapp, A.K., and Smith, M.D. (2014). Plant community response to removal of large herbivores differs between North American and South African savanna grasslands. *Ecology*, **95**, 808–816.

Ladwig, L.M., Collins, S.L., Ford, P., and White, L.B. (2014). Chihuahuan Desert grassland responds similarly to fall, spring, and summer fires during prolonged drought. *Rangeland Ecology & Management*, **67**, 621–628.

Suding, K.N., Collins, S.L., Gough, L., Clark, C.M., Cleland, E.E., Gross, K.L., Milchunas, D.G., and Pennings, S.C. (2005). Functional and abundance based mechanisms explain diversity loss due to nitrogen fertilization. *Proceedings of the National Academy of Sciences U.S.A.*, **102**, 4387–4392.

Vargas, R., Baldocchi, D.D., Allen, M.F., Bahn, M., Black, A., Collins, S.L., Yuste, J.C., Hirano, T., Jassal, R.S., Pumpanen, J., and Tang, J. (2010). Looking deeper into the soil: Environmental controls and seasonal lags of soil CO_2 production and efflux across four vegetation types. *Ecological Applications*, **20**, 1569–1582.

36 The Long-Term Ecological Research Stimulus

Research, Education, and Leadership Development at Individual and Community Levels

James R. Gosz

IN A NUTSHELL

1. Through the Long-Term Ecological Research (LTER) program, I have learned to appreciate the complexity of environmental dynamics when they are analyzed at multiple time and space scales.

2. My experience as a postdoctoral fellow and in the LTER program facilitated much of my understanding of interdisciplinary research because of access to multiple disciplinary approaches and accumulation of long-term and multiple-scale information.
3. My teaching of science benefited through recognition of the need for a combination of a deep understanding of each discipline's role in an issue (reductionist approach) and the collaborative need for integrating disciplines to fully understand complexity. No single discipline can answer the complexity in an environmental question.
4. I have improved my communication with the public through the combination of teaching and research reporting. The challenge is to develop the information in ways that can be communicated: free of scientific jargon, containing only essential data, and developed in scenarios that are recognized as real-life situations. The public has many forms and levels of understanding—there are K to gray and ordinary citizens and policy-makers; consequently, communication needs to be targeted appropriately.
5. I value the role of collaboration; there is tremendous satisfaction and reward from working in teams that can accomplish so much more than can an individual. This collaboration requires compromise, interaction, and time, but those that strive for this approach to science are well recognized.
6. I am fortunate in being in positions that have created opportunities for sustaining a long career in stimulating interdisciplinary and collaborative science.

PERSONAL OVERVIEW

I had a traditional forest management and soil science education (Michigan Technological University and the University of Idaho). However, my entrée into ecosystem science was set up by my very valuable postdoctoral fellowship at the Hubbard Brook Experimental Forest under the guidance of Gene Likens from 1969 to 1970, before the formation of the LTER program. The Hubbard Brook experience, quite literally, educated me about systems thinking, with the watershed approach to understanding integrated responses from complex, multifactor interactions and influences of forest management as disturbances. I was fortunate in having a dissertation topic at the University of Idaho dealing with nitrogen inputs to forest systems in an agricultural landscape in Idaho, which is likely why I was offered the postdoctoral fellowship. Although there were a number of publications from my Hubbard Brook work, the attempt to develop a total system energy analysis was my first experience with systems ecology (Gosz et al. 1978). Following this "new education" in ecology, my employment at the University of New Mexico in 1970 allowed the development of watershed approaches to quantify hydrologic and nutrient dynamics of southwestern conifer forest ecosystems. That was augmented by a sabbatical in New Zealand studying forest dynamics and nutrient cycling associated with exotic forests and nitrogen-fixing plant communities.

My interactions with the LTER program have been extensive, starting with my service on the advisory committees for the Konza and H. J. Andrews programs in the early 1980s. This followed some interactions with programs during the US International Biological Programs (US IBP) that contributed to large scale, integrated, and team-based approaches to scientific questions. When the US IBP and subsequent LTER programs were initiated, these were additional interactions that tremendously impressed and further "conditioned" me for (at that time) thinking about "big science" questions and how both place-based and interdisciplinary team-based approaches were critical to advance the information and understanding of complex systems. Those early years did not focus on comparisons among sites because it was before the pressure to network among sites; however, they did help demonstrate how collaborative and at least multidisciplinary, if not interdisciplinary science, was critical in properly understanding ecosystems.

FIGURE 36.1 International Long-Term Ecological Research (LTER) Chairman James Gosz presenting workshop goals for international long-term ecological science, developed at the LTER All Scientists Meeting, September 23, 2003, in Seattle, Washington. (Photo courtesy of LTER Network Office.)

In 1988, we were fortunate to get funding that started the Sevilleta (SEV) LTER program. I was the principal investigator for the first two 6-year awards but was aided by a great team that developed the science for this biome transition zone. That also set the stage for being elected to chair the US LTER network and the later International LTER (ILTER) network—all experiences that furthered my appreciation for interdisciplinary, place-based research efforts (Figure 36.1). I am now a professor and special assistant to the dean in a College of Natural Resources and special assistant to the vice president for research at a land grant institution (University of Idaho), where the applications of science to resource and environmental management are important issues. My experiences with the LTER program have clearly been instrumental in carrying out my current position because of the reputation and respect given to LTER science.

APPROACH TO SCIENCE

The LTER program influenced my thinking on environmental research but, more than that, it also helped clarify what I perceive as the distinction between reductionist approaches and team-based collaborative and ecosystem approaches. Both are essential, and my understanding of each evolved as I became more comfortable with collaborative efforts and avoided criticizing each of their characteristics. Arguments by proponents of each approach reinforced my awareness of the myopic nature of a single approach to science and now influence how I review programs, literature, and proposals.

A more recent evolution of system science has added socioecological factors and the coupled human–natural ecosystem approach. We addressed this in a number of ways during the 1990s, from a minimalistic approach (inviting a social scientist to meetings) to gradually including socioeconomic scientific involvements in developing scientific questions. No doubt, the awarding of two urban LTER programs greatly stimulated this direction and the current development of approaches in synthesis themes involving social and ecological dynamics.

ATTITUDES TOWARD TIME AND SPACE

The development of my thinking over 40 years has gone from addressing a particular question or hypothesis in a short time (dissertation period or 1- to 3-year award) to understanding that the time period for the research determines the results. To truly answer many questions, one needs multidecadal data. Only when these longer term studies demonstrated such timescale dependency did this lesson sink in. For example, studies at the Central Plains site in Colorado showed that the relationship between net primary production and precipitation changed after drought; previous models did not work after this change to the system and multiple decades were required to understand the dynamics of the relationship. Also, the ability to compare across space opened questions of spatial scale dependency. Finally, theory that linked scale dependencies of time and space were tremendous changes to how we approach research questions. Not all such insights came from research in the LTER program, but it certainly was a contributor. I believe that the LTER program is unique in the academic research community in balancing the need for a core set of measurements to demonstrate time dependency while adding new questions that spin off of core areas of research. The successful sites do both very well, but it continues to be a significant challenge. A site cannot be perceived as a place just for doing monitoring. However, when monitoring is fundamental to understanding new questions or the basis for asking new questions, it is clearly an important requirement.

COLLABORATION

I now focus entirely on how I can stimulate collaboration through team-based approaches to large projects and funding opportunities. Although I appreciate the value of investigator-initiated research, I know my faculty colleagues are able to do that and that I can add a dimension to the research portfolio of the institution by creating opportunities for larger programs and awards. Experiences in the LTER program have led to this and made me comfortable in focusing my efforts along these lines. An example is work done between the Jornada Basin and SEV programs, where intersite collaboration with multiple investigators has allowed the development of robust conclusions based on both large-scale and long-term studies (Peters et al. 2006). My institution allows and encourages my synthetic activities. I normally help create the appropriate team for a program and develop the leadership teams.

The domestic experiences in the LTER program have served as a valuable template for the development of the International LTER network. The lessons learned in the US LTER program have been successfully applied to the development of similar programs in many countries. Frankly, it was an easy sell, once we had the communication techniques, language barriers, and cultural differences understood. The timing also was appropriate as long-term research was a relatively new approach compared to the normal focus on conducting science in reductionist ways over short time spans and by single investigators.

APPLIED RESEARCH

My background and training in the LTER program of using an ecosystem approach has been greatly enhanced by developing communication and interactions with stakeholders and resource managers. I learned some of this during the early days at the H. J. Andrews site, when local US Forest Service managers were involved in the science discussions about research needs. We now know stakeholders have to be at the table when "starting" a project or proposal to help define the appropriate "use-based research" that benefits management and societal interests. Fortunately, funding agencies also appreciate this relatively new approach that includes human factors.

More recently, I have been involved with developing new directions in the Idaho Experimental Program to Stimulate Competitive Research (EPSCoR) that changed the culture associated with previous EPSCoR awards and have stimulated the combination of a collaborative ("OneIdaho") and Social-Ecological-Systems (SES) science effort. This SES thinking also involves bringing many stakeholders as well as resource and policy decision-makers into the development of the needed science. That approach was also critical to our award for a Research Coordination Network (RCN) with the theme of Complex Mountain Landscapes and the development of a Mountain Social-Ecological Observing Network. Meetings and projects are developed with a balance of researchers, stakeholders, resource managers, and policy-makers.

COMMUNICATION

My teaching is essentially communicating science and research results from the LTER program, not only from SEV but from other place-based programs. I use these results to illustrate the complexity of human–environmental systems. My efforts to communicate to others about science also involve stakeholders, community leaders, outreach activities, and university administrators in discussions. Starting with a good "elevator speech" and building the story involved with understanding the complexity of a system is critical. In the past decade, with the emphasis on dealing with the science of coupled natural–human systems, tailoring presentations appropriate for the stakeholder audience has proved challenging. However, the decades of teaching ecosystem science from the perspectives of the LTER program have been very instructive. I have not been involved directly in the K–12 and Schoolyard work at SEV but was very supportive of developing that aspect—or working with those who were involved directly. The National Science Foundation (NSF) was very keen on developing this capacity and although at times it seemed to be an unfunded mandate, it evolved into a great program and is viewed as very successful. Sites often had to complement the modest funding that came from NSF with other sources to achieve the goals of making local schools appreciate ecosystem science and long-term trends in environmental measurements.

MENTORING

We need to stress the importance of understanding multiple approaches to science, when each is appropriate, and the consequences of doing each. This lesson is most difficult when mentoring graduate students and new faculty members. There is no reason why new faculty members, for example, cannot focus on particular questions, develop investigator-initiated proposals, have sole-authored publications, but also participate in team-based, collaborative efforts where their expertise is critical to the success of the work. The key is a balance among the products from both approaches (if the institution involved can appreciate multiauthored publications as a strength). Some of the best examples of this dual capability come from researchers with clear strengths in a particular discipline but interest in putting that disciplinary strength into a team-based approach involving many disciplines. Evidence in the LTER network that shows the clear advantage of a multipronged approach, is the increasing numbers of publications, both sole and multiauthored, that take advantage of data developed from long-term and large-scale studies and from multiple disciplines. The challenge in mentoring new faculty members (and graduate students) to achieve this result lies in the recognition that one must obtain the right balance between the added time commitments of collaborations and the net benefits of such interactions to career-long productivity. For me, an excellent example of mentoring has been associated with NSF's Integrative Graduate Education and Research Traineeship, or IGERT, program. At least at institutions with which I have been associated, the students see the challenges and rewards of teamwork, interdisciplinary approaches, place-based studies, and information management. At the University of Idaho, it also is possible to have a multiple-authored chapter in a student's dissertation. Additional efforts are needed for mentoring teachers, especially in K–12 systems. The Schoolyard LTER Program has been a key asset to educational programs at LTER sites, and a great deal of credit needs to go to NSF for continued support and stimulation of those activities. Other efforts at training teachers exist, but it is likely that more support would allow much more to be accomplished.

SKILL SET

Early in my career, my tool kit probably did expand as I learned of techniques that could be applied in my research at SEV. More importantly, my association with others and seeing applications of their techniques greatly educated me on how these techniques could be applied to questions about which I had an interest. I do not claim to be an expert on these techniques but I do know the importance of doing them and when they are appropriate. At SEV, we aggressively worked to acquire new technologies such as geographic information system technology, remote sensing, and wireless communication and networking of instrumentation that would help develop innovative approaches to traditional field ecology. Perhaps the most lasting or best appreciated skill set is the development of appropriate information management. The LTER program was recognized early in its development as a leader in demonstrating the value of managing long-term data. Many discussions were held early in the development of the LTER program on the proprietary nature of data and the best protocols for making data widely available, but in the past decade, a great deal of support has developed for data sharing. This support is now widely shared by NSF and other federal agencies that fund research, but it took time for the research community to accept this philosophy and execute it in an efficient manner. The LTER program still plays a leadership role in information management, both at site and network office levels. Although not a sophisticated information manager, I do build this requirement into the various programs with which I am involved.

PERSONAL CONSEQUENCES

My experiences have greatly altered my approach to science and my appreciation of other disciplines, languages, cultures, jargon, perceptions, and interpersonal skills! Team building, both locally and internationally, requires at least recognition of the variabilities in these attributes among collaborators, but more importantly, the value they bring to the team and projects. My networking skills have developed through my association with communities in the LTER and ILTER programs. In fact, it was clear that I needed to develop networking skills to be successful. I have many friends and colleagues in the larger community in the LTER network who are extremely valuable to me in my sunset years in academia. In addition to my scientific efforts, I believe that my involvement with the LTER program was largely the springboard for my invitation to become a program officer at the NSF in the mid-1980s. That invitation was a tremendous benefit to my professional career and allowed me an opportunity to understand science from a perspective not acquired in any other way. It also led to the development of SEV, a first in terms of a large-scale collaborative research effort in the biology department at the University of New Mexico. That success helped me with my role in the Sustainable Biosphere Initiative (led by Paul Risser and Jane Lubchenco), followed immediately by an invitation for me to become division director for Environmental Biology at NSF. This position led, in turn, to many more lessons and connections with the scientific community. The NSF experiences led to a third NSF position in the EPSCoR Office. I have no doubt that my training, experience, and connections in the scientific community were initiated in large part through my involvement with the LTER program.

CHALLENGES AND RECOMMENDATIONS

The key is continued promotion and funding of LTER-like science, whether at LTER sites or in other programs (e.g., Critical Zone Observatories, agency research stations). I firmly believe that the values demonstrated by long-term research, interdisciplinary team approaches to complex system dynamics, system comparisons to understand principles behind those dynamics, scale-related issues, and coupled human–natural approaches will be the keys to improved science in the future. The LTER sites are like the libraries we have in the United States. There cannot be one on every corner but there must be some that develop the place-based knowledge that is required to understand the totality of issues involved in complex system dynamics.

My experiences over many decades allowed this "gradual" change in how I approached mentoring, education, and communication of science. The LTER program needs to continue to challenge itself to develop these attributes to the fullest extent possible among all age levels from kindergarteners to graying scientists, and from new scientists to policymakers. Long-term research has always been a challenge to sell to funding agencies, as science changes and societal issues create new demands. In the future, it will be essential to maintain the productivity of the LTER program, hopefully expand the network, create innovative ways of doing science through interdisciplinary approaches, and continue to demonstrate the great value of long-term data.

CONCLUSION

The place-based approach taught me the value of accumulating knowledge at a place and how it stimulated synthesis, interactions among disparate studies, new insights, collaborations, and new publications that would not likely have occurred among independent studies. However, it also was sobering in the sense that this collaborative behavior to do new science was a cultural change from single investigator-initiated, and isolated research efforts. This approach required communication, compromise, tolerance to different ideas and jargon, and learning the true meaning of what is required to do team research. Although difficult at times, I believe those early years at Hubbard Brook as well as my experiences in the LTER program from SEV and from a national and international network of sites changed my impressions of what science is, or should be for these complex issues. That has been my motivation for much of what I have done in the subsequent decades and continues in my present efforts. Aspiring ecologists and students should not take these challenges lightly because time and interpersonal skills are important requirements as well as disciplinary strength. The LTER program is an excellent opportunity to meet these challenges and create a successful career.

ACKNOWLEDGMENTS

My scientific career was greatly influenced by many colleagues, most notably by the mentoring of G. Likens and the late P. Risser. A combination of federal funding from NSF, US Department of Agriculture, US Fish & Wildlife Service, US Forest Service, and the US Department of Energy National Laboratories provided sustained support since the 1970s, but the major influence was the NSF funding that started the SEV LTER program in 1988. Renewals of the SEV LTER in 2000 and 2002 following my return to New Mexico from NSF and continued the research goals of the program along with the numerous supplements for equipment and Research Experiences for Undergraduates support throughout the funding period for LTER. The International Program at NSF provided numerous awards for international travel for LTER and international researchers and students.

REFERENCES

Gosz, J.R., Likens, G.E., Holmes, R., and Bormann, F.H. (1978). Energy flow in a forest ecosystem. *Scientific American*, **238**, 92–102.

Peters, D., Gosz, J., Pockman, W., Small, E., Parmenter, R., Collins, S., and Muldavin, E. (2006). Integrating patch and boundary dynamics to understand and predict biotic transitions at multiple scales. *Landscape Ecology*, **21**, 19–33.

PART NINETEEN
SHORTGRASS STEPPE (SGS) LTER SITE

37 Long-Term Ecological Research and Lessons from Networked Lives

John C. Moore

IN A NUTSHELL

1. The Long-Term Ecological Research (LTER) program has affected how I conduct and evaluate ecological research. Working with the LTER program has given me a greater appreciation for the complexity of the natural world and has provided a framework to study it.
2. The LTER program has provided the best possible venue to connect ecological research with classroom instruction, mentoring, and professional development.
3. Translating our science to the public is a challenge. My experiences in the LTER program have provided multiple opportunities to work with the public, K–12 and

college or university students, and professionals in different fields. This process has honed my communication skills.
4. The ideas that emerge from true collaborative science cannot be understated. The work at an LTER site and within the LTER network works best when we collaborate.

PERSONAL OVERVIEW

I received my undergraduate training in ecology at the University of California (UC) Santa Barbara. At UC Santa Barbara in the 1970s, the ecology program focused largely on populations and communities. Field observations, laboratory studies, manipulative field studies, and equation-based modeling were the norm. I recall the first set of litter and soil samples of arthropods that I sorted were extracted using Tullgren funnels and thought at the time that a person would have to be insane to pursue this type of work as a career. Two years later, I was in the graduate program at Michigan State University working with Dr. Richard Snider where I studied the impacts of herbicides on soil arthropods in no-till corn. At Michigan State, I learned the importance of species life histories, behaviors, and tolerances to environmental variation.

My first exposure with the LTER program started in 1979, during my first year of graduate school at Michigan State University. A National Science Foundation (NSF) program officer was visiting the university to promote the concept of the LTER program and the first round of competition. Being 22 years old at the time, it was difficult for me to appreciate discussions about a program that would potentially operate over several decades. As a graduate student, it was a lesson in the planning, extended time frame, and other programmatic logistics of collaborative science.

I started working at the Shortgrass Steppe (SGS) LTER site in 1982 when I was a doctoral student working under Dr. David Coleman. My studies focused on soil invertebrates and microbes in natural and managed ecosystems and evolved into the study of the structural and dynamic properties of food webs and ecological networks. This was about 2 years before the SGS was designated a site within the LTER network.

During the 30-year tenure of the SGS as an LTER site from 1984 until it was decommissioned in 2014, I participated in SGS research, educational, and administrative activities as a graduate student researcher (1984–1986), research scientist (1986–2002), coprincipal investigator (2002–2010), and principal investigator (2010–2014). I am also a senior scientist with the Arctic LTER site (ARC). I was invited to join ARC in 1998, as part of an effort to expand studies to include soils and soil ecology. I am now investigating the impacts of nutrient additions, warming, and herbivory on soil food webs within and across several biomes.

APPROACH TO SCIENCE

My experiences in the LTER program have clearly influenced my ideas about how environmental research should be conducted. I approach my research and education activities from a systems perspective that requires a longitudinal view of processes and outcomes. Part of this approach comes with experience and age, learning to be patient and to take the long view, but the impact of the LTER program is clear. Through my engagement within the LTER community I have been exposed to different approaches, techniques, and ways of thinking about topics. I have gained a greater appreciation for the qualitative research approaches used in education and social science research.

Although the experiences that I have had with the LTER program have influenced my approach to science and education, they need to be viewed in the context of other experiences. What follows provides this context. The title I chose for this chapter is a variation of the title of a keynote talk I gave mid-career at a banquet to faculty and administrators at the University of Northern Colorado (UNC) in recognition of that year's college and university scholars. In what may seem to have been a literary "shaggy dog" approach, I used the opportunity to summarize my research and education activities and presented the idea of forming a center that focused on precollegiate activities. The basis of my argument was to connect research activities to ongoing K–12 teacher professional development and K–12 student enrichment programs. For context, this was the mid-1990s, before Science, Technology, Engineering, and Mathematics (STEM; and its unfortunate precursor acronym SMET) was part of our lexicon, before the early years of the NSF push for "broader impacts," and before the establishment of the Schoolyard LTER Program. The idea of linking activities that were largely the purview of the academic affairs side of campus with those of the student affairs side of campus was rare.

I started the talk by summarizing my research program in soil ecology and food web theory. Much of that work included studies that I was involved in since graduate school at the Central Plains Experimental Range (CPER, or SGS) and agricultural sites in eastern Colorado and elsewhere. I began by noting that *soil* and *dirt* were not synonymous, and that *dirt* was a four-letter word. I described soils as a living system and provided examples of biota that resided in soils, discussed their natural histories, the interactions in which they engaged, and their importance to soil formation, nutrient dynamics, and plant growth. Using our description of the soil food web for the CPER (Hunt et al. 1987), I shifted the talk to food webs and food web theory, emphasizing themes of the LTER program that included the need to study networks using information from multiple scales of organization—melding principles from autoecology, synecology, and ecosystem ecology.

The second part of the talk focused on K–12 education and engagement activities. I focused on my experiences leading to our newly US Department of Education–funded mathematics and science Upward Bound and K–12 teacher professional development activities. The program included regular advising during the academic year and a 6-week summer resident program for between 40 and 60 high school students that included courses, advising, and cultural activities. Our activities to promote professional development of teachers included research internships, a teacher-in-residence program, and the establishment of demonstration plots on our campus that replicated a nutrient addition study at SGS. In both cases, trips to SGS, interactions with scientists in the LTER program, and access to resources and funding opportunities associated with the LTER network were to become a regular component of our education and engagement activities.

Reflecting back on the keynote talk at mid-career from the current perspective of beginning the latter part of my career helped me answer how my LTER experience affected me as a scientist and educator. Trying to separate the impact that my LTER experiences had from the different groups of people and institutions that I have worked with was another matter. My experience at the Natural Resource Ecology Laboratory (NREL) at Colorado State University (CSU) as a graduate student, research associate, and more recently, as a senior scientist and director have influenced me as a scientist and educator. Like the LTER program, the NREL has a rich history in ecosystem science. Many of the scientists at the NREL were proponents of and participants in the LTER program and the International Biological Program (IBP) before it (see Coleman 2010). The same could be argued for my work in education. My tenure as a faculty member in the Department of Biological Sciences at UNC (1991–2006), with its roots as a normal school and strength in teacher preparation, and my experiences with Upward Bound at CSU and UNC shaped my appreciation for K–12 engagement. In each case, my professional identity evolved over time as a consequence of the interactions and experiences I gained from each institution and my involvement in the LTER program. In the process, these experiences also helped to shape the LTER program.

ATTITUDES TOWARD TIME AND SPACE

The LTER community has been a leader in promoting ecosystem science; the notions of complex adaptive systems, hierarchical organization, and scale (spatial and temporal); and in the evolution of the discipline of ecology, to include the social sciences through the lens of the socioecological perspective (Collins et al. 2011). In time, as data from the long-term studies at different sites accumulated, patterns began to emerge that would not be apparent from short-term studies, affecting my attitudes toward time and space, and their role in ecological studies. Transient versus asymptotic dynamic behavior were no longer abstract concepts. The spatial heterogeneity of landscapes and the impacts it has on populations, communities, and ecosystem-level processes could not be ignored. The legacy effects of past events or states, and variation in current states, are recognized as integral components of current states and their future trajectories.

Coming to understand each of the aforementioned concepts has changed how I conduct research for the better, but also has created inherent conflicts in terms of what constitutes credible research. Manipulative field experiments were long the mainstay of a standard 3-year grant, yet now I question the utility of the approach or relevancy of the results, even when the approach is appropriate. Observation-based science and basic natural history studies have been deemed passé, unless new technologies (read molecular techniques) are used. Models are viewed with suspicion. Statistical techniques and approaches are conflated with rigor and elegance. I do not mean to sound cynical about the state of ecology, but rather my sense is that ecology is in a transition period where these new understandings and the appropriate approaches have yet to be sorted out. How ecology evolves in an era of large network-based monitoring and "big data" will be interesting. LTER science and scientists will no doubt have an important role in working through this transition.

COLLABORATION

My experiences in the LTER program have affected the national and international focus of my research and education collaborations by providing circles of colleagues and friends in grassland ecology, polar biology, soil ecology, food web theory, and education. I had phenomenal mentors, and, with their guidance as my career progressed, I began to appreciate their contributions on several levels in ways that affected how I observe and formulate questions, conduct studies, and communicate results. The LTER program has provided me with the opportunity to experience a liberal arts career of sorts. I have worked with scientists from different disciplines, mathematicians and statisticians, science and mathematics educators, social scientists from a variety of fields, educational psychologists, journalists, and artists. These interactions have been both rewarding and, as discussed below, challenging.

The LTER program has influenced my assessment of the nature of collaboration and the way in which I collaborate with others. As ecologists, we are inculcated to embrace collaboration. By its nature, ecology and in particular ecosystem science, draw principles from the humanities and mathematics that are vital for understanding complex environmental systems. Students of the philosophy of science immediately recognize that different disciplines vary in what motivates them, their methodological approaches and practices, and the process of confirmation that defines the nature of meaningful evidence. Working with individuals with different backgrounds on a common set of questions within the LTER program has been stimulating, efficient, and productive—the operative phrase being "a common set of questions."

The collaborations that I have been involved in that work are the ones in which the group has interest in a topic or challenge, defines a common set of questions to address, and collectively develops a framework from which to operate. Too often, individuals from different disciplines are invited to participate on a project rather than participate in the development of the project. I experienced elements of this practice when the LTER program first promoted education and outreach activities in the late 1990s and again with the early collaborations with the social sciences from 2004 to 2006. In each case, the collaborations struggled until they started to focus on the topics and the challenges that were of mutual interest.

APPLIED RESEARCH

The LTER program has afforded me the opportunity to work with a diverse group of individuals, institutions, and agencies from the private and public sectors. Many of these efforts involve basic research, but most are often classified as applied research, extension, or more recently as engagement activities. These types of activities are central to what we do as scientists in the LTER program. Our project had several scientists from the US Agricultural Research Service (ARS) as co–co-principal investigators and senior personnel, was colocated on land administered by the ARS, and regularly interacted with local trade associations. Should I consider colleagues who engage in applied research or extension activities as valued colleagues? I never gave this a second thought. What was clear was that an appreciation for different stakeholders was not universally shared; not in ways that were hostile or negative, but rather, in ways that viewed such engagement as not being at the center of what we do as academicians. Asking us to comment on this in some ways hints at the persistence of this dichotomous approach to research—either it is applied or it is not.

A recent 2-day road trip organized by the upper administration for department heads and administrators at CSU emphasized the aforementioned dichotomy for me. The theme of the trip was "engagement and outreach" with diverse stakeholders. The administration was interested in highlighting different activities. However, it also wished to showcase how this type of work aligned with the mission of the university and how this work constituted scholarship that should be valued and evaluated for promotion and tenure considerations. During the trip, we were asked to fill out surveys that questioned our understanding of engagement activities, whether we valued these types of activities, and how we would evaluate faculty and students that were involved in them. My initial reaction to the questionnaires was to challenge the premise that these activities were not valued, until I realized that had it not been for my experiences in the LTER program and the appreciation for engagement and outreach gained through them, I might have reacted in a very different way.

COMMUNICATION

For teaching and outreach, I initially used the data and publications from the LTER network as sources of examples for lectures and professional development workshops. Later, I worked with local school districts and conducted surveys of teachers on the types of professional development activities that would serve them best. The feedback I received revealed the need for authentic research experiences that connect to local and regional scientific issues. The Schoolyard LTER Program has broadened the impact of our K–12 outreach by providing a local research venue and a network from which to develop and manage these activities. My recent learning progression-based research has guided the development and revision of courses and degree programs in this area.

MENTORING

The LTER program has affected my teaching and involvement in outreach activities, as well as in my current administrative roles as a department head and laboratory director, how I value and prioritize teaching and outreach when allocating resources, conducting annual evaluations, and making promotion and tenure considerations. My association with the LTER program has affected the way in which I mentor students and junior faculty with respect to the costs and benefits of collaboration in multidisciplinary research and with respect to education and engagement activities. The LTER program offers abundant research opportunities, often with several investigators collaborating. As stimulating and rewarding as this environment is, it does present challenges for students and junior faculty in balancing their needs to collaborate with needs to develop an independent research agenda or line of scholarship. Contributing to group efforts is important but not when it occurs at the expense of taking the lead and publishing as senior author.

Involvement in education and engagement activities presents its own set of issues. We sometimes forget that not all academic cultures share the same appreciation for research, education, and engagement activities, and prioritize them differently (Hermanowicz 2005). I was invited to a leading research institution to give a seminar and meet with faculty, postdoctoral fellows, and students about my work on soil food webs. My last office visit of the trip was with a postdoctoral fellow. After being dropped off at the laboratory and introduced, we exchanged the usual niceties. Then his visage became very serious, he looked over my shoulder at the office in the back of the laboratory and around the laboratory, closed the door, sat down, and asked about the education and engagement activities that I had been involved in. How was I able to balance these activities with my discipline-based research? How did my colleagues view these activities? Were they valued? These were activities that this person valued and wanted to become involved in, but felt that they would somehow negatively affect his career advancement. It seems that the fellow's mentor and the faculty of the department as a whole viewed education and engagement activities as important but discouraged them when balanced against primary research, views that I shared at one time and had to reconcile.

Through my work with Upward Bound I developed an appreciation for education and engagement activities, but initially viewed them as something separate from my faculty duties, in part because they did not fit neatly into the traditional scholarship, teaching, or service categories of my job description. Ultimately, I was able to align these activities with my appointment and legitimize them with my colleagues through my association with the LTER program that focused on balancing basic research with education and engagement. My advice to junior faculty is to pursue the scholarship and engagement activities about which they are passionate and to strike the balance between research, teaching, and service that is appropriate for the setting that they are in.

SKILL SET

Have my experiences in the LTER program expanded my professional tool kit or afforded new skills? As with many of the questions we were asked to address, this one was difficult to answer directly. The LTER program provided training on how to conduct research and offer education programs within a site and across a network of sites. Separating the impact of the LTER program from other such arrangements is difficult. During my career, I have worked at two national parks—Rocky Mountain National Park, Colorado, and Wind Cave National Park, South Dakota. Both of these provided infrastructure, colleagues to work with, and logistical and grant support. In the education realm, I worked in and directed Upward Bound Programs, which were part of a broader association of hundreds of Upward Bound programs.

PERSONAL CONSEQUENCES

My involvement with the LTER program has provided me with a network of colleagues that share common interests in research and education. Visiting the different LTER sites has highlighted for me how common challenges require nuanced approaches that adapt to the idiosyncrasies of place. Some of my most valued friendships and collaborations were initiated by a site visit or by discussions of data generated from a particular site. I found the cross-site initiatives to be particularly rewarding, many of which have developed into long term collaborations.

CHALLENGES AND RECOMMENDATIONS

Are there impediments in the LTER program that need to be overcome to catalyze the development of participants (e.g., senior personnel, postdoctoral fellows, graduate students, undergraduate students) as research scientists, educators, or communicators to the public? I will address three observations and concerns—and provide recommendations where I can. First, there is a perceived exclusivity of the LTER program from colleagues outside the community, who at times see the network as being impenetrable and difficult to navigate. Whether this is real or not, or whether we agree or disagree with this observation, I have heard this complaint too many times for it to be ignored. Second, the inclusion of social science research, and education and engagement activities as formal components of the LTER program were important steps in the evolution of the LTER program. Having said this, I think that the social science and education components, while appreciated have yet to be fully embraced by the community as a whole as part of the LTER mission. Finally, much of the discussion about data has centered on data access and meeting the requirements set forth by the NSF that data be made available in a timely manner. With the amount of data that is being generated, it seems that the network is missing an opportunity. The cross-site efforts that have worked with common data sets have yielded some important results and insights. Rather than being reactive to data policy, the network could be proactive in this area. For example, data management is an important service, yet it also can serve as a venue for scholarship.

CONCLUSION

I end my essay as I started it, by restating that my experiences in the LTER program have shaped the way that I conduct research, the way that I approach education, and the way that I engage in outreach to the public. My experiences with the LTER program have greatly enriched my career. My research, education, and outreach activities do not exist in different career silos but rather are inextricably interrelated. After more than 30 years of involvement in the LTER network, trying to separate its impact from other aspects of career that have shaped my professional identity would be impossible.

ACKNOWLEDGMENTS

The author wishes to thank the scientists, students, and staff of the SGS LTER and ARC LTER programs, and the staff, students, and parents of the UNC COSMOS program for their collaborations, support, and fellowship over the years. This work was supported in part by grants from the US Department of Education (P047M 990093-00), NSF (DGE-00086443, ESI-0119786, DEB-0423385, DEB-0832173, DEB-0823405, and DEB-1027319), CSU, and UNC.

REFERENCES

Coleman, D.C. (2010). *Big Ecology: The Emergence of Ecosystem Science*. University of California Press, Berkeley.

Collins, S.L., Carpenter, S.R., Swinton, S.M., Orenstein, D.E., Childers, D.L., Gragson, T.L., Grimm, N.B., Grove, M.J., Harlan, S.L., Kaye, J.P., Knapp, A.K., Kofinas, G.P., Magnuson, J.J., and William, H.M. (2011). An integrated conceptual framework for long-term social–ecological research. *Frontiers in Ecology and the Environment*, **9**, 351–357.

Hermanowicz, J.C. (2005). Classifying universities and their departments: A social world perspective. *The Journal of Higher Education*, **76**, 26–55.

Hunt, H.W., Coleman, D.C., Ingham, E.R., Ingham, R.E., Elliott, E.T., Moore, J.C., Rose, S.L., Reid, C.P.P., and Morley, C.R. (1987). The detrital food web in a shortgrass prairie. *Biology and Fertility of Soil* **3**, 57–68.

PART TWENTY
VIRGINIA COASTAL RESERVE (VCR) LTER SITE

38 Networking

From the Long-Term Ecological Research Program to the National Ecological Observatory Network

Bruce P. Hayden

IN A NUTSHELL

1. As a scientist, the Long-Term Ecological Research (LTER) program has been on my mind for more than three decades.
2. As an educator, I have served in the classroom for 41 years. The merger of the physical and the ecological sciences was at the core of my teaching philosophy.
3. As a science communicator, I informed the general public on issues of climate and climate change.
4. As a collaborator, I found that understanding strengths and weaknesses in collaborative partnerships best ensures success.

5. As a science leader, I served at the National Science Foundation (NSF) as the Director of the Division of Environmental Biology (DEB), established the Schoolyard LTER Program, and launched the National Ecological Observatory Network (NEON).

PERSONAL OVERVIEW

My disciplinary background includes formal graduate education at the University of Wisconsin in meteorology, climatology, and paleoclimatology, as well as in oceanography and biology (mycology, botany, zoology, and genecology). As a postdoctoral fellow, my scientific identity was on track to culminate as a paleoclimatologist. As an assistant and associate professor, my identity morphed to include coastal geomorphology (Hayden et al. 1995). Finally, my experiences in the LTER program have vectored my career toward the interactions of climate and vegetation (Hayden 1998).

My affiliation is with the Virginia Coast Reserve (VCR) site in the LTER program (1986–2014). As one of the founding principal investigators of the VCR site, I have served in subsequent renewals as its principal or co-principal investigator. Our site-based research plan focused on the Virginia Coast Reserve on Virginia's eastern shore with a focus on the dynamics of the chain of 14 barrier islands, bounded by the entrance to the Chesapeake Bay to the south and Assateague Barrier Island to the north. This peninsula is 100 km in length by 20 km in width. Only the islands fronting the Mississippi delta are more dynamic in both the temporal and spatial domains.

Prior to joining the LTER program, my research was hemispheric to regional in scope, and it focused on the environmental dynamics of the Atlantic Coast from Florida to Cape Cod at 50-m intervals (Fenster and Hayden 2007). The time domain that I explored ranged from decades to centuries. Currently, I am professor emeritus in environmental sciences at the University of Virginia. I continue to work on storminess over North America using records for all of the LTER sites in the contiguous states (Hayden and Hayden 2003).

APPROACH TO SCIENCE

My approach to science is basically empirical, and this science includes the use of existing data banks and the development of new data. My perspectives have always been multidisciplinary, as a result of my graduate studies and interests in zoology, botany, biometeorology, and atmospheric sciences. This breadth of academic training and experiences was well suited and preadaptive to my subsequent leadership of and research contributions to the emerging VCR LTER program.

When I first joined the LTER program, of necessity, my research became much more place-based and model-based (Hayden 1994); it was consequently less global and more local, with a strong focus at the VCR site. Nonetheless, I have come to believe that my research has become more insightful, more synthetic, and more comparative in nature as a consequence of my involvement in the LTER program. In contrast, I do not think that I have become more theoretical as a consequence of my experience at the VCR site. My academic and research careers have been highly multidisciplinary for a very long time.

ATTITUDE TOWARD TIME AND SPACE

Most of my early research was at regional to hemispheric spatial scales and at decadal to millennial time scales. Topically, my research interests lie at the junction between climate and vegetation at varying time and spatial scales. My approach is to treat the vegetation as thermodynamic wicks. In my view, temperature and vapor pressure are the key state variables. Precipitation has proved wanting in explaining plant biogeography. These proclivities I trace back to my postdoctoral studies. At regional and decadal scales, research questions involve storminess and coastal geology (Fenster and Hayden 2007) and coastal climatology (Hayden and Hayden 2003).

COLLABORATION

Collaboration requires a tolerance of the epistemological norms of the participating collaborators. These norms address the basic question: "How do I come to believe what I believe?" Intolerance of competing beliefs systems renders collaboration unlikely. For example, scientists using the case study approach may have a difficult time collaborating with someone using a survey methodology or a first principle, theoretic modeling approach. Among the impediments to tolerance are language, mathematics, cultural norms, national affiliations, ethical traditions, age, institutional standards, and academic disciplines. It is not required that collaborators adopt a common epistemological norm. However, respect of the belief system or norms of others is critical, as is a welcoming environment. Collaboration within disciplines is usually somewhat free of epistemological conflicts, as colleagues broadly know approaches to the problem at hand.

Collaboration between scholars from different disciplines is more likely to suffer epistemological impediments. Of the many possible impediments, language, cultural norms, and nationality are difficult for me. When epistemological compatibility is the case, the tool set is such that each collaborator benefits. A kind of "ecological" mutualism is a desired outgrowth. All members of the collaboration should benefit. In one sense, I am a poor collaborator because I am impatient and eager to address the task at hand early and finish on time. Collaborations usually require that time be set aside by each of the participants. In a sense, time management may become an impediment to collaboration. An important downside of collaboration is its impact on professional advancement in the junior ranks. Many institutions, mine included, place high value on single-authored papers, as well as on synthetic and comparative publications. Junior faculty face impediments associated with the publish-or-perish imperative. With academic advancement, scientists in the LTER program can more freely take advantage of the extensive opportunities in network science. Prior to my experiences in the LTER program, I had a long-time (circa 16-year) collaboration with a two-investigator model of operation. We researchers knew each other's strengths and weaknesses. There were no conflicts of vision.

APPLIED RESEARCH

The VCR program, in partnership the Virginia Institute of Marine Sciences and The Nature Conservancy, is engaged in a major restoration of the bays and lagoons of the Delmarva Peninsula. In the 1930s, sea grasses became extinct on the peninsula. A small sea grass colony was found in the past decade and is being successfully encouraged by a seeding program. We expect widespread colonization and a conversion from a heterotrophic to autotrophic ecosystem. With good fortune, we hope that the basic science of the current program of restoration will be viewed as applied science. However, applied science is not a requirement of the LTER program at the National Science Foundation (NSF).

COMMUNICATION

On June 1, 1992, I began an LTER-electronic newsletter on the Web. The Web language for the newsletter was Gopher. It was followed by Mosaic, and both text and images could "ride" the Web. Next came Netscape, and communications carried their forms with them. My LTER newsletter is now communicating via blogging software called Drupal; it carries whole databases over the Web. To date, more than 150 short stories focused on science have been posted on these venues. The last edition of the newsletter was published on August 2, 1996.

The next 2 years I spent as director of the Division of Environmental Biology (DEB) at NSF. While there, I initiated the LTER Schoolyard Program and identified the funding stream for it that has endured to this day. I also proposed the NEON program in 1997. Bill Michener and I, along with 12 collaborators and with help from more than 700 other scientists, completed the planning and design process. Nearly a third of the NEON planners were LTER scientists and educators. Today, NEON is in the construction phase. NEON was a major collaboration.

I have taught in the classroom for 41 years. In those years, I have taught climatology and meteorology to most of the University of Virginia graduate students supported by the VCR program. For me, and I hope for the students, this has been an enriching activity. Teaching young ecologists the physics of the atmosphere at the graduate level is a challenge. Part of that challenge is teaching across discipline boundaries. The downside is the collision between travel associated with the LTER program and the inflexible scheduling of classroom teaching. Without colleagues to occasionally stand in for me in the classroom, it would be impossible to meet all the demands of participation in the research enterprise embodied by the LTER network.

MENTORING

The most important mentoring in which I have participated is the preparation of a colleague to take over the leadership of an LTER site. Although there is no hard and fast rule for how long a principal investigator should serve, two 6-year terms is common. Among the management and administrative skill sets that a new principal investigator needs to master are budgeting, people management, proposal crafting, and long-term planning. With proper lead-time, hands-on experience and focused mentoring will help the transfer of leadership to a new principal investigator.

SKILL SET

As with most people in my age cohort, I have witnessed and participated in the transition from mainframe computation to the desktop, characterized by increasing portability and miniaturization. Keeping up with technocrats, including the continual upgrade of hardware and software, is taxing and sometimes frustrating. The cyber network itself morphed several times: Gopher, Mosaic, and Netscape search engines, to say nothing of the Geographic Information System and Global Positioning System. Statistical and modeling software as well as social media software demand our time and attention.

PERSONAL CONSEQUENCES

I have always considered myself to be on the shy side. I realize that there is little evidence in support of this self-appraisal, but nonetheless it stands. Friendships and interpersonal relationships are time-dependent and mature or decay with events of the day. Remembering names and faces is not one of my strong suits. In contrast, structured, professional interactions have always been easy for me. For example, structured activities in the LTER program related to travel are a pleasure and account for 90% of my total travel. My extensive trips on LTER business are only possible due to the support and understanding of my family. Friendship circles within the LTER community might be characterized as a giant, multidimensional Venn diagram. It is a science hyperspace, if you will. These many intersecting circles of science friends, and science colleagues play a role in advancing ecological sciences in the broadest sense. A subset of these circles might be classified as science leadership circles. Each LTER site has at least one leadership circle. My site, VCR, has several. Given the 6-year funding cycle of LTER sites, it is not surprising that science leadership cycles (1) play a critical role in moving science questions and hypotheses forward and (2) define or redefine the path of discovery for the site. The power of intermeshing cycles is witnessed at the All Scientists Meeting held every 3 years.

Diverse cultures and ethnic groups are underrepresented at most LTER sites and in the LTER network in general. To a significant degree, this is a pipeline problem. The Research Experience for Undergraduates and the LTER Schoolyard Programs are a modest effort to improve the situation. It is also true that students often consider nonecological professions as pathways to fulfilling careers. At the University of Virginia, nearly half of the students envision a medical career. Next in line are careers as lawyers. Ecological and physical sciences are not a major part of student planning. We need to earn our majors.

CHALLENGES AND RECOMMENDATIONS

LTER sites cannot be all things to all people. Mission creep, or the growth of expectations, is and has been an impediment to the research planning process. A prominent leader of one of the sites stated, "I am sick and tired of all these opportunities being foisted on us." In the 1980s, John Brooks warned the leaders of the LTER sites that if LTER did not work and behave as a network, it would not long survive. This imperative by Brooks set in place a natural tension between site-based elements of the LTER network and expansion of scientific scope and promise of new discovery. Monetary resources were earned by proposal writing, and we institutionalized the LTER network and the network office. It was mission creep of the first order. Cross-site research, information management, social science dimensions, and additions of coastal and urban sites into the LTER network enriched the opportunities available to participants. In this sense, mission creep is both a blessing and a curse.

In my view, mission creep has enriched the network and the sites. The mission of the LTER network office (LNO) is, on the one hand, to realize Brooks' vision for the LTER program. On the other hand, the LNO must garner support and motivate the leadership of all sites to be successful. Unlike the LTER program, NEON is, by design, the product of an incorporated Board of Directors and its Chief Executive Officer; research sites are all alike as platforms for scientific investigation, with a common contribution to the whole. Unlike NEON, LTER is a confederation of semi-independent sites, each with its own research mission. Each LTER site is funded by proposals to NSF. The network office, currently in Albuquerque, New Mexico, is likewise funded by a proposal submitted to the NSF. The continual application to NSF creates a potential conflict between funding long-term observations and addressing new research questions. This conflict is addressed by each site as it writes its own proposal and by NSF when it decides which proposals to fund. It is therefore a conflict of values, not a conflict that affects the overall vision of the LTER program. As a final thought—perhaps someday a new John Brooks will come along, and motivate and sell the merger of LTER and NEON!

CONCLUSION

I have been a scientist in the LTER program for more than a third of my 73 years. Now, I look forward to the coming decade so that I can continue to be a part of the LTER community, to complete my work on the history of storms, and to remain as an LTER-science outreach blogger on climate ecosystem dynamics. Given my standing in the LTER community, I was encouraged to apply for the post of director of the DEB at the NSF. There, I was able to propose and help develop the concept of NEON. With Dr. William Michener, we crafted the successful proposal for the development of NEON. More than 700 scientists contributed to the plans for NEON. This is but one example of LTER scientists taking time out of their careers to serve NSF and the ecological community. As to launching an LTER career, I suggest maintaining geographic mobility, building a strong academic record, enriching your tool kit, and developing a can-do attitude.

ACKNOWLEDGMENTS

I would like to acknowledge the organizers and participants of this book. Their support is greatly appreciated. I am currently supported by NSF grant DEB-1237733.

REFERENCES

Fenster, M.S., and Hayden, B.P. (2007). Ecotone displacement trends on a highly dynamic barrier island: Hog Island, Virginia. *Estuaries and Coasts*, **30**, 978–989.

Hayden, B.P. (1994). An overview of biological models: A physical scientist' perspective. In P.M. Groffman and G.E. Likens, eds. *Integrated Regional Models*, pp. 13–35. Chapman & Hall, New York.

Hayden, B.P. (1998). Ecosystem feedbacks on climate at the landscape scale. *Philosophical Transactions of the Royal Society of London B*, **353**, 5–18.

Hayden, B.P., and Hayden, N.R. (2003). Decadal and century-long changes in storminess at Long-Term Ecological Research sites. In D. Greenland, D. Goodin, and R. C. Smith, eds. *Climate Variability and Ecosystem Response at Long-Term Ecological Research Sites*, pp. 262–285. Oxford Press, New York.

Hayden, B.P., Santos, M.C.F.V., Shao, G., and Kochel, R.C. (1995). Geomorphological controls on coastal vegetation at the Virginia Coast Reserve. *Geomorphology*, **13**, 283–300.

39 Sharing Information
Many Hands Make Light Work

John H. Porter

IN A NUTSHELL

1. The Long-Term Ecological Research (LTER) program has shaped almost every aspect of my scientific career. It has enabled me to pursue ecoinformatics, a new and growing field, while allowing me to build on my training as an environmental scientist within the context of an intelligent, vibrant, and dedicated team of researchers and collaborators.
2. Skills that I learned initially as part of workshops sponsored by the LTER program— on Geographical Information Systems (GIS), ecological information management, and wireless sensor networks—are now the skills I teach to others in a variety of formal and informal educational settings, including graduate and undergraduate classes.
3. As a leader in sharing scientific data, the LTER program provides a strong positive and dynamic example of how data can be shared to enable new scientific syntheses.

I communicate widely within the LTER network and with the larger community regarding the ethics, techniques, and values of data sharing.
4. Collaboration, with researchers and other information managers, is a critical aspect of successfully promoting the sharing of ecological data and the important new discoveries that arise from such sharing.

PERSONAL OVERVIEW

I started my work with the Virginia Coast Reserve (VCR) project in the LTER program at its inception in 1987. I started work at VCR site immediately after completing graduate school, as a postdoctoral fellow (1988–1991), then subsequently as a co–principal investigator and eventually as principal investigator. Although my primary contribution to the project has been as an information manager, I also engage in a variety of landscape, environmental sensing, and population-related research. Also, I briefly served as principal investigator (1997–1998), when the former and subsequent principal investigator (Bruce Hayden) did a rotation at the National Science Foundation (NSF). Within the LTER network, I have been very active in the Information Management (IM) committee and served on the LTER Executive Committee (1997–2002) and as a cochair of the Network Information System Advisory Committee. In addition, I served as a part-time program director in Biological Databases at NSF (1993–1994). Academically, I am a research associate professor at the University of Virginia, where in addition to my research, I teach courses on GIS.

My graduate training was in environmental sciences, focusing on population and community ecology, with a heavy emphasis on statistics. This background, coupled with interests in computers and technology, has allowed me to participate in, and help lead, a revolution in the way that researchers in the LTER program, as well as others, now archive and share data.

APPROACH TO SCIENCE

Unsurprisingly, given my role as an information manager, working in the LTER program has led me to appreciate how the depth and diversity of data available at an LTER site increases the range of ecological questions that can be answered rigorously. John Magnuson's conceptualization of the "invisible present" (i.e., how time series of data allow us to interpret present observations) and Fred Swanson's and Richard Sparks' "invisible place" (i.e., how data from a variety of locations are needed for general understanding) have a strong resonance with me (Magnuson 1990, Swanson and Sparks 1990). However, I would extend those concepts to include the "invisible moment," arguing that understanding what is happening at a specific moment, at a specific site, may depend on an array of factors, many of which a particular researcher is ill-equipped to measure. Thus, the wealth of data available to the researcher from at a LTER site can play an important role in understanding patterns and processes. For example, at one point we were mystified by a groundwater level sampler on a barrier island dune; it kept reporting increases in the height of the water table *before* rainfall events, a seeming violation of causality. However, inspection of tidal records indicated that higher-than-normal tides had preceded the storms, and it was actually the groundwater pressure generated by those tides that was raising the groundwater level. This provided new insights into the hydrologic environment experienced by barrier island plants, where soil water availability is tempered by both rainfall and tidal cycles.

At the cross-site level, I continue to be amazed by the progress in information management in the LTER program. Despite modest support (funds for an annual meeting and targeted workshops) the IM committee has generated a series of joint projects, ranging from codevelopment of Ecological Metadata Language, to climate and hydrologic databases, to LTER program-wide bibliographies and data catalogs. The LTER network office provides resources that create a "home" for products derived from the IM group. I believe that the diversity of approaches in the IM group has made it possible to create more robust and general solutions than would otherwise be possible.

ATTITUDES TOWARD SPACE AND TIME

My PhD research focused on the spatial and temporal dynamics of small mammal populations. The LTER program has allowed me to extend my research into other domains through training on advanced geographical analysis tools and wireless sensor networks and by providing the computational and equipment resources needed to implement new systems.

Early on, I bought into the LTER premise that understanding ecological systems cannot be based only on short-term data; however, deciding exactly which measurements at a site need to be long-term was a challenge. Interestingly, the ultimate resolution of the challenge was more bottom up than top down. Some measurement programs designed at the start to be core and long-term were abandoned when they proved to be logistically unwieldy and scientifically uninformative. At the same time, other monitoring programs, initially planned to be of limited duration, grew to be long-term core data sets as they proved to be increasingly valuable, especially in light of new questions and conceptual themes. This sort of cost–benefit analysis with respect to both new and old data sets is an essential part of the LTER program.

COLLABORATION

The LTER program created a new career path for information managers that is intensely collaborative. Indeed, it cannot exist without collaboration. During my graduate training, there was a strong emphasis on individual achievement and single authorship. Since that time, I have observed a strong movement toward collaborative research, wherein a team of researchers interact to produce a series of papers that incorporate the unique contributions of each of the participants. As a result, when I look at junior, or potential colleagues, I consider their capacity to constructively engage in collaborative research, as well as their individual intellectual achievements.

My collaborations are not only within the US LTER network. At the international level, I have engaged in several International LTER and Organization of Biological Field Stations training activities in the United States, Eastern Europe, Latin America, and Africa, but mostly with Taiwan and in the East-Asia Pacific region (Figure 39.1). Some of the activities are best described as "planting seeds." Typically the participants in those activities were investigators or graduate students, who might later be responsible for overseeing or participating in information management, but because they had "other jobs" were unlikely to fully exploit the tools and skills with which they were exposed. Other activities, such as work with the Taiwan Ecological Research Network, had a strong commitment in place, with resources to support people whose primary job had a large IM component. This allowed us to jointly produce tangible products, such as analysis and mapping tools.

APPLIED RESEARCH

My experiences in the LTER program have had an impact on my perceptions of the role of applied science. Although ecoinformatics theory plays an important role in information

FIGURE 39.1 Discussion of a biodiversity database in Taiwan during 2005. Facing the camera from left to right are J. Porter, S.-S. Lu, and C.-C. Lin. (Photo courtesy of Meei-ru Jeng.)

management, the ultimate measure of success is the archiving and reuse of ecological data, which is an inherently applied process. In addition, our LTER site collaborates on a variety of projects that have management implications with The Nature Conservancy.

COMMUNICATION

It is difficult to remember that at the inception of VCR there was no e-mail, no Web, and certainly no videoconferencing. Today, we could not operate without all those technologies. Nonetheless, face-to-face meetings still play an important role. The VCR website provides a primary outlet for sharing data and other information about the site. In addition, outreach is a key element of successful information management in the LTER program. Even within the LTER network, investigators need to be convinced to "buy in" to the model of data sharing dictated by the LTER data policy (Porter 2010). This is even truer for non-LTER scientists, who need to be involved if we want to build a reward system for data contributors based on coauthorship, citation, and acknowledgment (Porter and Callahan 1994; Duke and Porter 2013). I have been involved in a large number of workshops and publications aimed at providing that outreach to the larger community.

MENTORING

Apart from the role of data sharing in enabling new science, some level of mentoring is inherent in passing on best practices for managing information. I encounter too many students who believe that complex spreadsheet structures or the form of tables in journals constitute best practices for archiving data. Much of my mentoring of students and investigators is informal, with discussions over a table with a pad of paper. In addition,

I provide materials on the Web, such as PowerPoint presentations (with audio) that lay out best practices. I have been involved in a large number of IM workshops at both the national and international scale, with participants from within and outside the LTER network. These workshops varied from the general (e.g., general informatics training in Eastern Europe) to the very specific (e.g., scientific workflows for processing data from wireless sensor networks). Helping to formulate and teach the best practices of ecoinformatics has helped clarify my own approaches to managing and analyzing data.

My work with the LTER program has also led to the teaching of graduate-level courses at the University of Virginia and the University of New Mexico on managing environmental data. Such courses are challenging to teach because of the wide diversity of tools that I introduce to students and the skills that are required to master them. Ideally, a student would learn how to use database management systems, a statistical language (e.g., R, SAS), and a programming language (e.g., Python, JAVA). Fully mastering any one of those can require a semester in itself, so these courses remain a work in progress as I and my coinstructors winnow the essentials from the plethora of options.

I have benefited greatly from mentoring by principal investigators and senior personnel at VCR. They have helped develop my scientific perspective; shown me better ways to work with groups of researchers; and shown me how to balance research, teaching, and administrative demands. Outside VCR, interaction with the LTER IM committee has affected me profoundly. Most of what I learned about how to organize groups and meetings came from observing information management colleagues who headed the LTER IM committee during my early tenure. I count many of my fellow information managers among my close friends, and I value my long association with them. They are also a fount of information on a wide variety of technical challenges that I encounter on a daily basis.

SKILL SET

Opportunities provided by the LTER network have largely changed the type of science in which I engage. In 1988, I received my initial training in GIS from a short course sponsored by the LTER network. Subsequently, this training allowed me to adopt a landscape perspective that is particularly valuable at a site that changes as fast as the VCR (where 90% of the land of a major study site was deposited subsequent to 1900). In turn, my knowledge of GIS has provided the opportunity to collaborate with other researchers who wish to incorporate spatial elements into their work. I teach university-level courses on GIS that are attended by many of the graduate students in the LTER program, which allows me to pass on that legacy to others.

Similarly, in 1989, an Internet workshop at the National Center for Supercomputer Applications provided grounding in network-based tools and concepts that helped make the LTER program a leader in bringing ecologists on to the Internet. More recently, I have participated, sometimes as a student, sometimes as an instructor, in large number of ecoinformatics workshops and short courses that have helped me keep abreast of technological developments.

My exposure to wireless sensor network technologies came as result of outreach by David Hughes's "Biological Science by Wireless" project for LTER sites. Interactions with David and his associate, Tom Williams, provided equipment and knowledge that have allowed us to connect our remote barrier island via wireless networks. This technology provided me with a basis that has allowed VCR to transform many aspects of data collection at the site and to write numerous papers on sensors and sensor networks (Porter et al. 2005, 2009; Porter, Hanson, and Lin 2012).

PERSONAL CONSEQUENCES

What would my life be like if the LTER network had not existed? It would certainly be different! My graduate program prepared me for a career studying local populations of small mammals in the context of a biology or environmental sciences department. However, thanks to my participation in the LTER program, I am engaged in developing new, data-rich approaches to ecological research across a wide variety of domains that takes me all over the world, both virtually and physically. Instead of collaborating with a few colleagues on projects over small space and time scales, I number my research collaborators in the dozens and operate at a much wider array of spatial and temporal scales.

As an LTER information manager, I bring together a wide variety of technical and scientific approaches, melding them to produce systems that are capable of supporting new types of scientific enquiry. The LTER program's early requirement for data management provided the space and motivation for experts in the discipline of ecology to acquire the "tools" needed to effectively manage data (Michener et al. 2011). Notably, these tools are not necessarily those used by businesses to manage data. My graduate training as a population and community ecologist, with a background in statistics and computational biology, provided an invaluable perspective to be applied in developing ecological information management systems. LTER information systems need to be usable by ecologists, not computer scientists.

My association with the LTER program has built my profile within the network and in other research communities. In addition to LTER-centric activities, I frequently serve on advisory groups for various projects outside the LTER program. Over time, this has included the Organization of Tropical Field Stations Science Committee, User Working Groups for the National Aeronautics and Space Administration Global Change Master Directory, and the Oak Ridge National Laboratory Distributed Active Archive Center. I have also participated in reviews of the US Geological Survey Biological Resources Division, the Department of Energy Atmospheric Radiation Measurement program, and the NSF Division of Biological Infrastructure. In many cases, I am viewed as a liaison or at least an informal spokesperson for the LTER program as a whole or for the LTER IM group.

Another significant interaction fostered by the LTER program has been with Chau-Chin Lin and his ecoinformatics team at the Taiwan Forestry Research Institute. In 2003, that team approached five different LTER sites regarding informatics techniques and ultimately ended up sending four team members for 90-day sequential working visits to VCR. In turn, I have been involved in a wide variety of informatics activities in the East Asia Pacific region of International LTER that have given me an opportunity to field-test approaches to data integration and analysis in ways that were not feasible in the United States (Lin et al. 2008a, 2008b; Porter et al. 2010).

CHALLENGES AND RECOMMENDATIONS

One underrecognized feature contributing to the success of LTER sites is that through long-term association with one another, researchers become more efficient at working together because they are more knowledgeable about each other's interests and capabilities. In contrast, LTER cross-site efforts tend to be short-term, compounding the difficulties associated with working over distance, with the challenges of organizing new groups. As mentioned previously, I have seen the effectiveness of the IM committee operating over the long-term. I would love to see a long-term program aimed at hypothesis-based, multisite research that could provide independent support for standardized experiments across sites. Data from the National Ecological Observatory Network has some potential application in this area, but as an infrastructure program it lacks the hypothesis and theory-driven element so critical to the success of the LTER program.

Effectiveness of cross-site interactions would also be strengthened by a program of personnel exchanges. One of the keys to the successful interactions I have had with the Taiwan IM group was that visits were extended (e.g., 90 days), so that we had a chance to collaborate in-depth on projects and develop strong personal ties that have persisted to this day (Lin et al. 2011; Porter and Lin 2013). Support for researchers taking a sabbatical at different LTER sites or with external groups would help to foster similar projects and ties across the LTER network and beyond.

CONCLUSION

The LTER program has shaped almost every aspect of my scientific career. It has enabled me to pursue ecoinformatics, a new and growing field, while allowing me to build on my training as an environmental scientist within the context of an intelligent, vibrant, and dedicated team of researchers. The LTER program has given me access to technical resources such as data systems and sensor networks that would never be available to me as an individual researcher. Other sites and the LTER network office have provided partners for my efforts in ecoinformatics that are unparalleled in ecological science. Truly, many hands make light work!

ACKNOWLEDGMENTS

This material is based on work supported by the NSF under BSR-8702333-06, DEB-9211772, DEB-9411974, DEB-0080381, DEB-0621014, and DEB-1237733.

REFERENCES

Duke, C.S., and Porter, J.H. (2013). The ethics of data sharing and re-use in biology. *BioScience*, 63, 483–489.

Lin, C.C., Kassim, A.R., Vanderbilt, K., Henshaw, D., Melendez-Colom, E.C., Porter, J.H., Niiyama, K., Yagihashi, T., Tan, S.A., Lu, S.S., Hsiao, C.W., Chang, L.W., and Jeng, M.R. (2011). An ecoinformatics application for forest dynamics plot data management and sharing. *Taiwan Journal of Forest Science*, 26, 357–369.

Lin, C.-C., Porter, J.H., Hsiao, C.-W., Lu, S.-S., and Jeng, M.-R. (2008a). Establishing an EML-based data management system for automating analysis of field sensor data. *Taiwan Journal of Forest Science*, 23, 279–285.

Lin, C.-C., Porter, J.H, Lu, S.-S., Jeng, M.-R., and Hsiao, C.-W. (2008b). Using structured metadata to manage forest research information: A new approach. *Taiwan Journal of Forest Science*, 23, 133–143.

Magnuson, J.J. (1990). Long-term ecological research and the invisible present. *BioScience*, 40, 495–501.

Michener, W.K., Porter, J.H., Servilla, M., and Vanderbilt, K. (2011). Long term ecological research and information management. *Ecological Informatics*, 6, 13–24.

Porter, J.H. (2010). A brief history of data sharing in the U.S. Long Term Ecological Research Network. *Bulletin of the Ecological Society of America*, 91, 14–20.

Porter, J.H., Arzberger, P., Braun, H.-W., Bryant, P., Gage, S., Hansen, T., Hanson, P., Lin, F.P., Lin, C.-C., Kratz, T., Michener, W., Shapiro, S., and Williams, T. (2005). Wireless sensor networks for ecology. *Bioscience*, 55, 561–572.

Porter, J.H., and Callahan J.T. (1994). Circumventing a dilemma: Historical approaches to data sharing in ecological research. In W.K. Michener, S. Stafford, and J.W. Brunt, eds. *Environmental Information Mangagement*, pp. 193–203. Taylor and Francis, Bristol, Pennsylvania.

Porter, J.H., Hanson, P.C., and Lin, C.-C. (2012). Staying afloat in the sensor data deluge. *Trends in Ecology and Evolution*, **27**, 122–129.

Porter, J.H., and Lin, C.-C. (2013). Hybrid networks and ecological sensing. In S.C. Mukhopadhyay and J.A. Jiang, eds. *Wireless Sensor Networks and Ecological Monitoring*, pp. 99–124. Springer-Verlag, Berlin.

Porter, J.H., Lin, C.-C., Smith, D.E., and Lu, S.-S. (2010). Ecological image databases: From the webcam to the researcher. *Ecological Informatics*, **5**, 51–58.

Porter, J.H., Nagy, E., Kratz, T.K., Hanson, P., Collins, S.L., and Arzberger, P. (2009). New eyes on the world: Advanced sensors for ecology. *BioScience*, **59**, 385–397.

Swanson, F.J., and Sparks, R.E. (1990). Long-term ecological research and the invisible place. *BioScience*, **40**, 502–508.

PART TWENTY ONE
ANALYSIS AND SYNTHESIS

40 Coda

Some Reflections on the Long-Term Ecological Research Program

William H. Schlesinger

IN A NUTSHELL

1. Ecology has a history of long-term studies that offer great insight to ecosystem processes.
2. The advent of the Long-Term Ecological Research (LTER) program institutionalized long-term studies with some core measurements at a selection of sites across North America.
3. The most successful LTER sites are those that have an energetic leader with a clear vision, who has guided the work over many years.
4. Several LTER sites have established successful education programs for K–12 and college-age students, as well as for science policy-makers.
5. Implementation of more and better cross-site work would be welcome.

The various essays in this volume reflect a broad range of experiences among participants in the LTER program. Nearly all are positive: only mad dogs bite the hand that feeds them. All authors appreciate the advantages of long-term funding for their research and lament that funding of the LTER program by the National Science Foundation (NSF) is so limited. There are numerous testimonials for how the LTER program has changed and broadened participation in collaborative science. The real question is whether the LTER program has allowed science to proceed faster, deeper, broader, and with more critical insight than if the program had not been created. To answer that question, I offer a few personal reflections on the LTER program.

First, we must note that long-term research existed well before the LTER program. Edmondson began his long-term measurements of exogenous phosphorus in Lake Washington in the early 1950s (Edmondson 1991). Across the country, Herb Bormann and Gene Likens began long-term studies, now in their 50th year, of forest biogeochemistry at Hubbard Brook in 1963 (Likens 2013). Each of these long-term studies enjoys ample coverage in every text of introductory ecology. The advantages of long-term research are undisputed among those who are funded for it. Indeed, NSF embraces a wide variety of decade-long studies with its Long-Term Research in Environmental Biology (LTREB) program.

The authors of several chapters recall how Howard Odum's early work focused their attention on the connections between large units of the landscape. For many, the writings of Howard Odum border on religion, but his influence on our science remains strong. From its beginning, the LTER program embraced Odum's sense of scale: large scale. Many sites placed special emphasis on hydrologic transport that linked large units of the landscape.

Long-term studies were also fostered by the International Biological Program (IBP), a well-funded, biome-based study of ecosystems from 1964 to 1974. I think it is fair to say that much of the LTER program grew out of the IBP, or was at least strongly influenced by major IBP players. The meld of long-term and large-scale science was a compelling basis to propose an ambitious new program at NSF. Even today, the LTER sites devote three of their five focal areas to the stocks and flows of energy and materials, carrying a distinct flavor of the earlier IBP efforts (Callahan 1984). Even as they embrace socioecology, the new urban ecosystem studies have produced nitrogen and phosphorus budgets for urban areas (Baker et al. 2001; Metson et al. 2012).

The LTER program also tried to be inclusive. The "E" in LTER was purposely chosen to be ecological and not ecosystem. At the onset of the LTER program, some of the best long-term research derived from population studies of desert shrubs at Tumanoc Hill, Arizona (Goldberg and Turner 1986); bird populations at Hubbard Brook (Holmes 2007); moose and wolves at Isle Royale (Nelson et al. 2011); and desert rodents at Portal, Arizona (Brown and Zeng 1989). In the formation of the LTER program, NSF desperately wanted to include such work—and perhaps even have room for a few closet taxonomists as well. It was hoped that population and community ecologists would feel welcome, although many subsequently chose not to be included. Among the chapters here, there is scant mention of long-term population studies benefiting from the context of the larger ecosystem processes also monitored at LTER sites. Nevertheless, the focal area in population dynamics shows the legacy of the desire to include the broad ecological community at LTER sites. Perhaps the trends in the penguin populations at the Palmer Station LTER site best show how population studies can be linked to larger scale, long-term changes in ecosystem function—in this case, in the coastal waters off the Antarctic Peninsula (McClintock, Ducklow, and Fraser 2008).

Large-scale experimentation in ecology began earlier than the LTER program, with the deforestation experiments at Hubbard Brook (Likens et al. 1978), the experimental exposure of the oak–pine forest at Brookhaven National Laboratory to gamma radiation (Woodwell 1967), and manipulation of nitrogen and phosphorus availability in whole-lake experiments in Ontario (Schindler 1974). With these studies, the behavior of ecologists changed from simply describing landscapes to organizing realistic experiments in them.

From the beginning, a few LTER sites also focused on experimentation, as exemplified by David Tilman's work on the diversity, productivity, and stability in grassland ecosystems at the Cedar Creek LTER site in Minnesota (Tilman et al. 1997). His field experiments are among the most widely cited in the history of ecology. Many chapters here comment on the importance of experimental work at other LTER sites. Given that disturbances are expected in nature, studies of the response of forests to fire, drought, and ice storms were an essential part of any holistic ecosystem study. Studies of disturbances are conducive to experimentation in the field, and in a fifth focal area, LTER sites embraced studies of disturbance. Human disturbances due to clear-cutting, agriculture, and lake management have been central at a few LTER sites (Franklin, Bledsoe, and Callahan 1990).

In hindsight, the LTER program would have been more powerful if it had embraced the biome coverage of the earlier IBP and the now-evolving National Ecological

Observatory Network (NEON) efforts. Instead, proposals were accepted from anywhere, with the best 11 selected as the first cohorts for the program. In more recent years, there has been some backfilling of the network by requests for proposals that target particular ecosystems in Antarctic, coastal, or urban settings. All sites are place-based. Sometimes the place is large, but it is seldom regional. Several sites (e.g., Coweeta and North Temperate Lakes) extended the purview of their activities to include both natural and human-impacted systems.

The place-based nature of the LTER sites made it difficult for the LTER program to respond to the challenges of global change research that emerged soon after the LTER program was established (National Research Council 1990). The LTER effort responded with a call for cross-site comparisons, but at the time there had been little standardization of methods or protocols. In the early years of the LTER program, most sites had hunkered down to build long-term observational data sets on key ecosystem metrics, especially net primary production (NPP). For example, Huenneke et al. (2002) show the long-term variations in NPP at the Jornada Basin LTER site in southern New Mexico, as related to El Niño conditions. The need to standardize methods among sites for cross-site comparisons led to several handbooks for soils (Robertson et al. 1999) and NPP (Fahey and Knapp 2005). Even so, few of the chapters in this volume mention any efforts or successes from cross-site synthesis, and the 30-year review of the LTER program stresses the need to develop cross-site experiments more broadly.

One of the most successful cross-site syntheses came from the work of Knapp and Smith (2001), who sought to ascertain the response of NPP to variations in rainfall across LTER sites. Huxman et al. (2004) carried this analysis further, showing a convergence in rain-use efficiency across biomes. Other cross-site comparisons stemmed from funding independent of the core LTER efforts, notably the cross-site litterbag (Long-Term Intersite Decomposition Experiment, or LIDET) experiments organized by Mark Harmon from Oregon State (Harmon et al. 2009). Harmon provided an important component of experimentation in the work that included transplants of litter between sites.

Even today, most studies of the response of ecosystems to the impending global changes in climate and biogeochemistry are largely found outside the LTER network, although they may occasionally involve LTER sites (e.g., the Free-Air CO_2 Enrichment, or FACE; Lotic Intersite Nitrogen Experiments, or LINX; and International Tundra Experiments, or ITEX). For example, with independent funding, Jerry Melillo established the landmark soil-warming experiment at the Harvard Forest LTER site in Massachusetts, which shows the likely response of soil carbon pools to the changes in temperature that will accompany global warming (Melillo et al. 2011).

In some areas, better continental data sets for global change studies are available outside of the LTER network. For instance, the National Atmospheric Deposition Program (NADP) has a 35-year record of rainfall chemistry from 261 sites across the United States. Some sites are also LTER sites, but the larger NADP compilation would be the first port of call for anyone looking for data on regional patterns or long-term changes in deposition. Similarly, nearly all LTER sites have some record of bird populations, but the Breeding Bird Surveys maintained by the Cornell Laboratory of Ornithology are widely used as the definitive record of change in bird populations in North America. In addition, in the face of National Oceanic and Atmospheric Administration's (NOAA) nationwide network of weather stations, I can think of no serious student of climate change who would begin by using the climate data collected at LTER sites.

During the years that I led the effort at the Jornada Basin LTER site, we spent a fair amount of time debating whether LTER was a club or a network, expecting the latter to show positive synergistic results that were more than the sum of the parts. I am pretty sure we never achieved closure on the argument. Many agreed that the All Scientists Meetings are important and influential, but one thing that struck me is how little was mentioned in the chapters of this book about the role or impact of the network office on the function or productivity of the LTER program in general.

Friendships among scientists are important, and all of us who have spent time in field stations during the summer fondly remember our camaraderie in nasty field conditions and in after-dinner conversations on all matter of topics. Now, when I see these workers, we always find time to reminisce about good times and memories of our fieldwork. Nearly every author of chapters in this volume comments on how important collaborations have been to their intellectual development and career. As the LTER program has embraced the inclusion of a wider range of disciplines in its current teams, ranging now into art and ethics, its researchers have a chance to talk to those whose training, aspirations, and career metrics are far different from their own. Who might have envisioned just a few years ago that social scientists would be assessing the value that urban residents place on the biodiversity of vacant lots!

Mentoring and educational programs have flourished as well, including the Schoolyard LTER Program for youth and the Science Policy Exchange, which carries the work from several northeastern LTER sites to policy-makers and the media. Several times, agroecosystem research from the LTER at the Kellogg Biological Station has been used to inform policy-makers about better management practices, to reduce human impacts on the atmosphere.

I know of no topic that has been more frequently a thorn in the side of LTER principal investigators than criticism of the quality of their data management efforts. Modern analytical and computational power gives us the potential to generate and manage huge datasets, ranging from the genome to the globe. Many authors recognize how geographic information systems have enhanced their research and the organization and interpretation of spatial data. The LTER network has emerged as a leader in the best practices to manage and archive large data sets. In the face of NEON, my hope is that we will not get so caught up in automated data collection that we forget the role of calibration and common sense as governors of what arrives from autonomous networks of sensors.

Similarly, I hope we do not diminish the enthusiasm and creativity of young scientists by focusing the national agenda for science on large projects. As Peterson stated (1993), "the creativity of individual scientists is increased by interaction with other scientists but only up to a point. As the size and complexity of projects increases, a disproportionate level of effort must be devoted to activities that are tangential to science though necessary to the maintenance of the projects."

Has the LTER program accomplished anything that long-term funding of existing institutes and field stations across the country might not have otherwise accomplished? Probably not. I look no further than to my own institute, where Rick Ostfeld has used long-term studies of acorn production, mice populations, and ticks to predict the risk of Lyme disease for humans in the woods (Ostfeld et al. 2006). Long-term field studies of global change phenomenon also stem from Stanford's Jasper Ridge Field Station (e.g., Gutknecht, Field, and Balser 2012), the Rocky Mountain Biological Laboratory (e.g., Harte, Saleska, and Shih 2006), the Hudson River (Strayer et al. 2014), Chesapeake Bay (Drake 2014), and La Selva, Costa Rica (e.g., Clark, Clark, and Oberbauer 2010). What the LTER effort did was to provide some structure and modest standardization to a set of common measurements at a few sites chosen for long-term funding. Colocation of data collection in the five focal areas gives an added value to the program.

The common thread to the most successful LTER sites is leadership by an individual with the passion and commitment to carry the project forward—folks such as David

Foster (Harvard Forest), David Tilman (Cedar Creek), Gus Shaver (Arctic), and Phil Robertson (Kellogg Biological Station). Sites with frequent turnover of leadership have fared less well. The authors of many chapters in this volume lament that current funding limits what might be done and the number of new investigators that can be added to the LTER program. But, experience tells me that whatever your endeavor, there is never enough funding. You muster forward and do the best you can with what you have. With luck, you will discover something important, and have some fun along the way.

ACKNOWLEDGMENTS

The author thanks Peter Groffman and Deb Fargione for helpful comments on the manuscript.

REFERENCES

Baker, L.A., Hope, D., Xu, Y., Edmonds, J., and Lauver, L. (2001). Nitrogen balance for the Central Arizona Phoenix (CAP) Ecosystem. *Ecosystems*, **4**, 582–602.

Brown, J.H., and Zeng, Z.Y. (1989). Comparative population ecology of 11 species of rodents in the Chihuahuan Desert. *Ecology*, **70**, 1507–1525.

Callahan, J.T. (1984). Long-term ecological research. *BioScience*, **34**, 363–367.

Clark, D.B., Clark, D.A., and Oberbauer, S.F. (2010). Annual wood production in a tropical rain forest in NE Costa Rica linked to climatic variation but not to increasing CO_2. *Global Change Biology*, **16**, 747–759.

Drake, B.E. (2014). Rising sea level, temperature, and precipitation impact plant and ecosystem responses to elevated CO_2 on a Chesapeake Bay wetland: Review of a 28-year study. *Global Change Biology*, **20**, 3329–3343.

Edmondson, W.T. (1991). *The Uses of Ecology: Lake Washington and Beyond*. University of Washington Press, Seattle.

Fahey, T.J., and Knapp, A.K., eds. (2005). *Principles and Standards for Measuring Primary Production*. Oxford University Press, New York.

Franklin, J.F., Bledsoe, C.S., and Callahan, J.T. (1990). Contributions of the Long-Term Ecological Research program. *BioScience*, **40**, 509–523.

Goldberg, D.E., and Turner, R.M. (1986). Vegetation change and plant demography in permanent plots in the Sonoran Desert. *Ecology*, **67**, 695–712.

Gutknecht, J.L.M., Field, C.B., and Balser, T.C. (2012). Microbial communities and their responses to simulated global change fluctuate greatly over multiple years. *Global Change Biology*, **18**, 2256–2269.

Harmon, M.E., Silver, W.L., Fasth, B., Chen, H., Burke, I.C., Parton, W.J., Hart, S.C., Currie, W.S., and the LIDET network. (2009). Long-term patterns of mass loss during the decomposition of leaf and fine root litter: An intersite comparison. *Global Change Biology*, **15**, 1320–1338.

Harte, J., Saleska, S., and Shih, T. (2006). Shifts in plant dominance control carbon-cycle responses to experimental warming and widespread drought. *Environmental Research Letters*, **1**, 014001 (doi:10.1088/1748-9326/1/1/014001)

Holmes, R.T. (2007). Understanding population change in migratory songbirds: Long-term and experimental studies of neotropical migrants in breeding and wintering areas. *Ibis* (doi:10.1111/j.1474-919x.2007.00685x)

Huenneke, L.F., Anderson, J.P., Remmenga, M., and Schlesinger, W.H. (2002). Desertification alters patterns of aboveground net primary production in Chihuahuan ecosystems. *Global Change Biology*, **8**, 247–264.

Huxman, T.E., Smith, M.D., Fay, P.A., Knapp, A.K., Shaw, M.R., Loik, M.E., Smith, S.D., Tissue, D.T., Zak, J.C., Weltzin, J.F., Pockman, W.T., Sala, O.E., Haddad, B.M., Harte, J., Koch, G.E., Schwinning, S., Small, E.E., and Williams, D.G. (2004). Convergence across biomes to a common rain-use efficiency. *Nature*, **429**, 651–654.

Knapp, A.K., and Smith, M.D. (2001). Variation among biomes in temporal dynamics of aboveground primary production. *Science*, **291**, 481–484.

Likens, G.E. (2013). *Biogeochemistry of a Forested Ecosystem*. (3rd ed.). Springer, New York.

Likens, G.E., Bormann, F.H., Pierce, R.S., and Reiners, W.A. (1978). Recovery of a deforested ecosystem. *Science*, **199**, 492–496.

McClintock, J., Ducklow, H., and Fraser, W. (2008). Ecological responses to climate change on the Antarctic Peninsula. *American Scientist*, **96**, 302–310.

Melillo, J.M., Butler, S., Johnson, J., Mohan, J., Steudler, P., Lux, H., Burrows, E., Bowles, F., Sith, R., Scott, L., Vario, C., Hill, T., Burton, A., Zhou, Y.M., and Tang, J. (2011). Soil warming, carbon-nitrogen interactions, and forest carbon budgets. *Proceedings of the National Academy of Sciences U.S.A.*, **108**, 9508–9512.

Metson, C.S., Hale, R.L., Iwaniec, D.M., Cook, E.M., Corman, J.R., Galletti, C.S., and Childers, D.L. (2012). Phosphorus in Phoenix: A budget and spatial representation of phosphorus in an urban ecosystem. *Ecological Applications*, **22**, 705–721.

National Research Council (1990). *The U.S. Global Change Research Program: An Assessment of the FY 1991 Plans*. The National Academies Press, Washington, DC.

Nelson, M.P., Vucetich, J.A., Peterson, R.O., and Vucetich, L.M. (2011). The Isle Royale wolf-moose project (1958-present) and the wonder of long-term ecological research. *Endeavour*, **35**, 30–38.

Ostfeld, R.S., Canham, C.D., Oggenfuss, K., Winchcombe, R.J., and Keesing, F. (2006). Climate, deer, rodents and acorns as determinants of variation in Lyme-disease risk. *PLoS Biology*, **4**, e145.

Peterson, B.J. (1993). The costs and benefits of collaborative research. *Estuaries*, **16**, 913–918.

Robertson, G.P., Coleman, D.C., Bledsoe, C.S., and Sollins, P., eds. (1999). *Standard Soil Methods for Long-term Ecological Research*. Oxford University Press, New York.

Schindler, D.W. (1974). Eutrophication and recovery in experimental lakes: Implications for lake management. *Science*, **184**, 897–899.

Strayer, D.L., Cole, J.J., Findlay, S.E.G., Fischer, D.T., Gephart, J.A., Malcom, H.M., Pace, M.L., and Rosi-Marshall, E.J. (2014). Decadal-scale change in a large-river ecosystem. *BioScience*, **64**, 496–510.

Tilman, G.D., Knops, J., Wedin, D., Reich, P., Ritchie, M., and Siemann, E. (1997). The influence of functional diversity and composition on ecosystem processes. *Science*, **277**, 1300–1302.

Woodwell, G.M. (1967). Radiation and the patterns of nature. *Science*, **156**, 461–470.

41 Scholarly Learning in an Ecological Setting

Applying the Knowledge, Attitudes, and Behaviors Framework to Perceived Outcomes from Participation in the Long-Term Ecological Research Program

Mark A. Boyer and Scott W. Brown

IN A NUTSHELL

1. Using the conceptual framework of Knowledge, Attitudes, and Behaviors (KABs), we develop a structured aggregate analysis of the essays in this volume.
2. Building from the KAB analysis, we examine how the Long-Term Ecological Research Program (LTER) program altered career paths (or not), perhaps changed original scholarly directions, and led to more integrative and important research over the course of careers.
3. Our primary finding is that the LTER program has successfully affected ecological research careers, mostly because the involved participants were predisposed to thriving in an interdisciplinary environment.

INTRODUCTION

Every scholar can point to events or experiences in his or her career that had a significant impact on his or her intellectual trajectory. For example, the authors of this chapter began collaborating as a result of a happenstance phone call in 1998 that led to more than 15 years of fruitful scholarly interactions centering around online international studies education in middle school through college environments. About a decade ago, one (Boyer) made a distinct turn in his scholarly agenda away from political economy and toward environmental inquiry. The other (Brown), a psychologist by training, has spent large portions of his career in K–12 and college settings working to promote research-based educational practices. The point is, many scholars develop their careers in ways unimagined early on, some by happenstance, as in our own case.

With this type of "we know intellectual change happens, but how do we understand it?" in front of us, this collection of essays by participants in the LTER program provides a rich body of data on which to develop a macro-level analysis of patterns of intellectual evolution in environmental research. We will begin by laying out the conceptual framework of KABs as a means to develop a structured aggregate analysis of the chapters in this volume. In using this conceptual framework and associated qualitative methodology, we hope to provide insights into whether participation in the LTER program did in fact affect the scholars involved in the enterprise and, if so, in what ways. Using the KAB approach, we will comment on whether involvement in the LTER program altered career paths in subtle or not-so-subtle ways, perhaps changed original scholarly directions, and led to more integrative and important research over the course of a career. As a result, our analyses will be more of a broad overview of the scholarly landscape defined by the LTER program, rather than one focused on the details of any single essay. As a result, we first discuss the KAB evaluation framework and then provide an analytical assessment of the themes of the chapters in this volume.

KNOWLEDGE, ATTITUDES, AND BEHAVIORS

The KAB framework is both an analytic process that can examine learning outcomes across three key interrelated dimensions and an instructional design framework that identifies and targets educational interventions that can result in powerful learning outcomes. The three dimensions have been found to be interrelated through educational research across such domains as problem-based learning, science, social studies, disease prevention, and teacher education (Schrader and Lawless 2004; Schrader and Brown 2008; Brown, Lawless, and Boyer 2013). Simply, attitudes drive behavior (Fishbein 1967), and more recent examinations of self-efficacy (a type of attitude) show that it is a strong predictor of the behaviors "task engagement" and "task persistence" (Bandura 1997). According to Bloom (1976), a behavioral change corresponds to a change in both knowledge and attitudes of the individual, thus implying a general progression from "K" to "A" to "B." Research examining these three dimensions also suggests that the relationships are reciprocal and dynamic (Bandura 1997). Because the KAB model is valuable in assessing behavior, skill, or performance changes, we apply it as a means for understanding scholarly change in the setting of the LTER program.

It is important to note, however, that educational interventions focused on the development of knowledge may not significantly affect attitudes or behaviors (Miller et al. 1990). For example, national educational interventions focused on increasing the wearing of seatbelts will likely be ineffective if the intervention only focuses on increasing the knowledge of operating a seatbelt. A successful intervention must also focus on changing the attitudes of people, because nearly everyone understands how to operate the seatbelt. It is not the operation of the seatbelt that restricts full compliance with wearing it but the attitude of the passengers that results in the poor level of compliance. Therefore, extending this analogy to the LTER program, the increased research opportunities might lead to an increase in a scientist's knowledge base but not necessarily to a change in attitudes about long-term research collaborations or an inclination to develop collaborative work further in one's career.

Knowledge

With respect to Bloom's taxonomy (Bloom 1976), the cognitive domain of learning is concerned with knowledge (comprehension, application, and understanding). Within a field, knowledge embodies all the information and strategies that a person possesses (Alexander, Jetton, and Kulikowich 1995). Importantly, knowledge is field or context specific, and therefore an individual can be highly knowledgeable in one field while possessing little knowledge in an unrelated field (Alexander 1992; Anderson 2000). For example, someone may possess a great deal of knowledge about bats and little or none about invasive plant species. In addition, the amount of knowledge in a field is a strong predictor of the ability to acquire new information from formal or informal instruction.

Attitudes

For the purpose of this application, the primary attitude construct we focused on was self-efficacy. Bandura (1986) defined self-efficacy as "people's judgments of their capabilities to organize and execute courses of action required to attain designated types of performances. It is concerned not with the skills one has, but with judgments of what one can do with whatever skills one possesses." Self-efficacy is thus a fluid construct that changes with a person's experiences (Bandura 1997). A person's self-efficacy guides their behaviors by influencing what is attempted and how much effort is applied to performance. Multiple studies have demonstrated that a person's self-efficacy affects his or her judgments and behaviors (Schunk 1991; Zimmerman and Bandura 1994).

Self-efficacy is a strong predictor of engagement in tasks and behavior, and it is central to a person's level of motivation (Bandura 1986). It also affects performance outcomes such as academic achievement and behavior (Schunk 1991). Compared to people with low self-efficacy, highly efficacious people have greater perseverance, set more challenging goals, and persist at a higher level when faced with difficult barriers and failures (Bandura 1997). In contrast, people with low self-efficacy are more likely to avoid difficult tasks, limit their effort, and give up quickly when difficulties arise. Therefore, an individual's efficacy expectations directly affect willingness to approach a task and the effort used to address it.

Behaviors

Behaviors consist of observable actions, skills, and products such as taking a water sample, writing a laboratory report, assembling a device, or conducting a faculty research project. The behavior may be measured and the qualities of these behaviors assessed.

Although most educational studies investigating educational outcomes rely almost exclusively on gained knowledge, the application of that knowledge may not be demonstrated for a wide variety of reasons. Emmons et al. (1986) found that knowledge regarding AIDS was a factor most strongly associated with the behavioral changes of the variables considered in their study. Participants reverted back to their baseline sets of behaviors within a short period time, after being given information about AIDS. Therefore, although information (knowledge) may have been most important to behavioral changes, information alone does not necessarily result in sustained behavioral changes (Miller et al. 1990). Emmons et al. (1986) concluded that interventions that inform individuals can contribute to behavioral implementation when considering specific tasks and have the potential to positively affect behavior modifications. But more than just information must be provided to sustain changes in behaviors, supporting the contention of the reciprocity of the three dimensions of KABs.

DATA AND ANALYSIS

Applying the KAB framework, the following sections lay out our aggregate analysis. First, using grounded qualitative analyses (Guba and Lincoln 1994) and the NVivo qualitative software (QSR International), the texts of the 36 essays in this book were loaded into a single document. Second, titles, references, tables, figures, and photographs were removed so that repeated words, phrases, and references would not contaminate the analytical process. Third, NVivo provides an option to remove "high frequency" words that are generally excluded for analytical purposes, such as "and," "the," "is," "will," and "but," so that the analytical process can focus on words of greater substantive importance. Fourth, a word frequency analysis was conducted on the combined master chapter file to identify the most frequent words present in the text of the essays.

At the most macrolevel of analysis, we determined the 10 words most frequently used in the combined master chapter file (Table 41.1). The "Stem Word" column includes the stem word and its associates. For example, the most frequent stem word, "Activities," includes the stem word "activities" as well as associates, such as "activity" and "activation." The "Word Count" column includes both the stem word and its associates. The "Weighted Percentage of Total Words" provides the percentage of the total number of words in the combined text that are stem words plus stem–associate words adjusted for the number of words in that cluster. Therefore, when interpreting the weighted percentage of words in the master file of all chapters, if there are a large number of words associated with a stem word, the weighted percentage would be lower than if there are a smaller number of words related with the stem word with the same frequency level.

The fifth step categorized "key words" into NVivo nodes labeled "knowledge," "attitudes," or "behaviors." A review of the most common words in the master file enabled a classification of key words associated with each of these three dimensions. A sample of the key words associated with the KAB dimensions is presented in Table 41.2. With the KAB categories and the top 10 word clusters, we have preliminary evidence of "task-orientation" ("behavior") that emerges across the essays.

The KAB category examples (Table 41.2) were then subjected to the word frequency analyses again but this time focusing on the overall KAB category rather than individual

Table 41.1 Ten Most Frequent Word Clusters[1] and Weighted Percentage of Total Words[2]

Rank	Stem Word	Word Count[1]	Weighted Percentage of Total Words
1	Activities	6,300	1.45%
2	Work	3,909	1.00%
3	Content	5,940	0.95%
4	Think	4,958	0.88%
5	Scientists	675	0.84%
6	Site	1,126	0.83%
7	Act	4,913	0.79%
8	Community	3,253	0.74%
9	Collection	1,662	0.71%
10	Research	1,694	0.68%

[1] World clusters consist of a stem word plus other technically similar words. For example, "Activities" as a cluster includes the following words: action, active, actively, adherence, conduct, and so on. As a cluster, "Activities" has more than 160 associated words, whereas "Site" has 21.

[2] The weighted total is the percentage of the words for a word cluster adjusted for the percentage of the similar words in the total file of chapters. Word clusters that have a high frequency of occurrence but a relatively small number of words in the cluster will have a higher weighted percentage than those with a large number of similar words associated with the stem word and the same total frequency in the master file.

[3] Word count includes the stem word and the associated words in the cluster.

Table 41.2 Knowledge, Attitudes, and Behaviors (KAB) Dimensions and Associated Sample Key Words

Dimension	Sample Key Words
Knowledge	Biology, climate, data, discipline, knowledge, ecological, ecosystem, environment, foundational, ideas, interdisciplinary, learned, natural, resources, science, scientific, systems, technology, think, understanding
Attitudes	Belief, benefits, criticality, culture, diverse, educational, global, good, impact, importance, improvements, interest, involved, like, need, perceive, perspective, successful, values
Behaviors	Acquire, actions, approach, author, career, collaborate, collect, discover, engagement, funding, leadership, management, mentoring, observations, professionals, project, skills, tools, training, work

Table 41.3 Percent of Words Associated with Knowledge, Attitudes, and Behaviors (KAB) Dimensions[1]

Dimension	Percent	Number of References
Knowledge	11.34	8,371
Attitudes	5.29	4,138
Behaviors	11.64	8,444
Total	**28.27**	**20,953**

[1] Note that the total percent is equivalent to the sum percent of words associated with the Knowledge, Attitudes, and Behaviors (KAB) modes. Words not fitting this classification system account for the remaining 71.73%.

word clusters. This analysis enables a different macro-level examination of the patterns of KABs and the evolutionary trends evidenced in the essays. The KAB dimensions accounted for 28.27% of the words in the master text file, as the authors described their experiences, development, challenges, and successes (Table 41.3). Although "knowledge" and "behaviors" were remarkably equal in their presence in these essays, "attitudes" was less than 50% as frequent, suggesting that the scholars focused more on knowledge or behaviors than they did on their discussion of attitudes or values.

We can refine this finding visually by examining the word cloud results (Figure 41.1). These results build from the top 10 word clusters (Table 41.1), expanding the words to the top 20, but nonetheless show consistency with those initial findings. Simply, action-oriented ("behavior") language is highly prominent in this figure, with words such as "activities," "change," "collaborative," "studies," and "research" playing large roles. In addition, also consistent with early findings (Table 41.3), words such as "student," "scientists," and "systems" are knowledge-linked words that show the high loading for "Ks" and "Bs." As we have seen, the high loading of "behavior" language lends credence to the notion that the LTER program provided a very action-oriented set of research opportunities for participants.

Examining each chapter independently for patterns of KAB emphases revealed that the dominant dimension in 24 of the chapters was "behavior" followed by "knowledge" for 9 chapters. We found no chapters where "attitudes" was the dominant dimension, supporting the NVivo analyses presented in Table 41.3 and elsewhere, where "behaviors" ability act **activities** change collaborative collection community concepts **content** create data ecological ecosystem events experiences field ideas information instrumentation investigator involved management **moving** office organization part period place point present program project properties provided quality questions regional **research** science **scientists site** state **students** studies support systems **think** use **work** writing

FIGURE 41.1 Word cloud of the 50 most frequent words appearing in the 36 essays by Long-Term Ecological Research scientists.

and "knowledge" were the dominant themes of the chapters and "attitudes" were approximately half as frequent as "Bs" and "Ks." This is not a surprising finding given the background, experience, and passion that these scientists have for their chosen profession. It does highlight, however, an important factor that these same professionals must address as they share their knowledge, skills, and concerns with the general public. They likely will need to focus more directly on affecting attitudinal change rather than just on affecting the public's knowledge and behaviors. Such attitudinal transformation is essential in sustaining long-term societal change on environmental issues.

To complete our analysis, it is worth examining capsule statements from each author that help provide a richer context to the aggregate statistical data (Table 41.4). We do not argue that the excerpts are a representative or random sample from the essays but rather that these excerpts help to explain what we found above the aggregate level. Simply, given that the KAB framework assumes a progression from "K" to "A" to "B," we need to understand better why "Ks" and "Bs" dominate our analysis. Thus, we felt it necessary to move away from the aggregate and delve into the rich narratives by the authors. In this way, we diverge from the more narrow constructs of the KAB framework to inform why the KAB findings diverge from what is expected theoretically.

When examining the excerpts (Table 41.4) and the corresponding classification notes shown in the second column, we see a great deal of evidence that a new effect of selection bias is occurring when we consider who becomes involved in the LTER program in the first place. Simply, researchers become involved in long-term collaborative settings because they are predisposed to do so. Some of that predisposition can be attributed to their "academic upbringing" (e.g., see Giblin, Shaver, Silver, and Troxler excerpts; Chapters 32, 9, 26, 18, respectively, this book), whereas others may have been drawn to such work because of their particular skills (e.g., see Stafford, Chapter 5) or more widely a desire for scholarly interaction evidenced in many of the essays.

The excerpts (Table 41.4) also show the high value that authors placed on interdisciplinarity and collaboration. We argue that this finding, taken in tandem with the earlier data, indicates that researchers in the LTER program are likely more interested in broader, systemic thinking than perhaps is the case among the population of ecological researchers from outside the LTER program. It would also help to explain why so little emphasis was shown on "attitudes" (see Tables 41.1 and 41.3 as well as Figure 41.1). In essence, the authors did not talk much about attitudinal impact, because their attitudes toward such things as collaboration, innovation, and interdisciplinary were present prior to involvement in the LTER program. This finding complements Flint's assessment of the fundamental interdisciplinary character of the LTER program and also the essayists'

Table 41.4 Long-Term Ecological Research (LTER) Scientists Coded by the Themes Present in Their Chapters[1]

LTER Scientist	Comment Code	Illustrative Excerpt(s)
Bestelmeyer	C, I	The multidisciplinarity inherent in research at LTER sites enables the integrative approaches needed for a landscape perspective. As part of a research team at an LTER site, you are more likely to work with scientists from outside of your core discipline than from within it.
Blair	C, SB	I have been part of the LTER program for most of my career, from graduate student at one LTER site to principal investigator at another … It has contributed to the growing recognition that many of the most important ecological questions, and ecological problems facing society, require collaborative, long-term efforts.
Borer	C	Midway into my career, I see myself as philosophically a "next-generation" scientist from the LTER program—standing on the shoulders of the incredible infrastructure and reenvisioning of ecological science spanning collaboration, data management and sharing, and long-term evaluations of ecosystem structure and functioning put in place by the LTER program.
Childers	C, I	I have always made the point to my students that ecosystem science by its very nature is collaborative. Notably, ecosystem ecologists seldom publish single-authored papers, and when they do they are typically reviews. As my research and worldview has become more interdisciplinary, the diversity, extent, and richness of my collaborations have exploded.
Collins	C, SB	I have long believed in sharing data … Share your data! You will expand your collaborations and might even get involved in working groups pursuing new and exciting synthesis research.
Ducklow	I, P, SB	Joining the Palmer Antarctica site changed my life. (Actually, going to Antarctica for the first time changed my life, but the LTER program gave me the opportunity to go every year.) … I think I was preadapted to work in LTER. As an oceanographer, I have always been fairly collaborative and interdisciplinary. By interdisciplinary, I mean a microbiologist working with physicists, geochemists, and modelers, not just other biologists (but maybe working with penguin and whale scientists counts, too).
Gaiser	C, I	The future of ecosystem science is evolving quickly through persistent integrative, transdisciplinary, and policy-relevant research, so by embracing the concept of community established in the LTER program, young scientists can lead ecologists toward a more effective and brighter future.

(continued)

Table 41.4 Continued

LTER Scientist	Comment Code	Illustrative Excerpt(s)
Giblin	C, SB	I feel as though my graduate student experiences "preadapted" me to become involved in long-term ecological research. I already enjoyed collaborative research and instantly felt comfortable in the LTER system when I first had the opportunities to work in it.
Gosz	I	My teaching of science benefited through recognition of the need for a combination of a deep understanding of each discipline's role in an issue (reductionist approach) and the collaborative need for integrating disciplines to fully understand complexity. No single discipline can answer the complexity in an environmental question.
Gough	C, I, P	My views of the importance of communities and particular species are continually challenged at the Arctic LTER site by my biogeochemist colleagues, in a friendly and constructive way. The biotic interactions that I placed so much weight on as a student, particularly competition, were not considered the drivers of ecosystem carbon and nitrogen cycling in the tundra … This environment helped me to mature my own ideas … My experiences … have had a profound influence on my professional identity.
Gragson	I, SB	I was trained as an ecological anthropologist with an emphasis on behavioral and ecosystem ecology at the University of Montana and the Pennsylvania State University. I have conducted archaeological, behavioral, cultural, and historical research throughout the western and the southeastern United States, several countries in lowland South America, the Dominican Republic, and southern France.
Grimm	I, P	A long-term approach is definitive for my career, which has evolved at a single place over more than 30 years. But the LTER program, and especially its urban research, has broadened my thinking far beyond the boundaries of the ecosystem science tradition in which I was trained … Ultimately, I think that the LTER has changed my professional identity.
Grove	I, P	My experience in the LTER program has affected me profoundly as a scientist. It has been an affirmation of my interdisciplinary training and provided me with the opportunity to pursue interdisciplinary research at a high level.
Hayden	C, I	Collaboration requires a tolerance of the epistemological norms of the participating collaborators … It is not required that collaborators adopt a common epistemological norm. However, respect of the belief system norms of others is critical, as is a welcoming environment … Collaboration between scholars from different disciplines is more likely to suffer epistemological impediments.

Table 41.4 Continued

LTER Scientist	Comment Code	Illustrative Excerpt(s)
Hobbie	SB	My experiences in the LTER program have only increased my belief in the value of collaboration. I have understood for a very long time how this valuable tool can be achieved.
Holbrook	C, I, P	My involvement with two LTER sites has been a transformative experience. Both the Santa Barbara Coastal and Moorea Coral Reef sites provide a platform for collaborative interdisciplinary research on spatial and temporal scales that simply cannot be achieved by individuals or small teams of investigators, and this has affected the type of research questions I can ask as well as the potential to work collaboratively.
Johnson	C, SB	My research at LTER sites has frequently expanded into collaborative projects, both interdisciplinary and cross-site collaborations ... Or alternatively, maybe there is self-selection of researchers, in that those drawn to long-term ecological research might be more pre-disposed to collaborative science.
Knapp	C, I, SB	I have always been interested in basic and fundamental ecological questions. My experience in the LTER program has not changed this and indeed, the breadth of the program has allowed me to stay the course with this interest. My early experiences in LTER gave me a great appreciation for the power of collaboration, large scale and long-term experiments, and cross-disciplinary interactions.
Lodge	C, I	The LTER program has not influenced my basic approach to science ... Although I can easily say that my involvement in research at Luquillo has led me to be more multidisciplinary, collaborative, synthetic, and insightful in my research, I cannot say that my research is more theoretical or comparative because of that involvement ... My involvement in LTER research has not influenced my approach to science or mentoring, but it has greatly expanded my skills, and my opportunities for mentoring and for collaborative research.
Lugo	C, I	The key to ecological understanding of the world is to recognize that both short-term and long-term research is required to advance social and ecological knowledge ... The LTER program is a conduit for research collaboration at all levels of involvement from local to international, within and between institutional colleagues, and for colleagues from different institutions, disciplines, and career types.
Magnuson	I, SB	My students came from a broader mix of subspecialties within limnology, ecology, and environmental sciences ... I have always believed in and worked in interdisciplinary sciences, but the practices were broadened and realized through the LTER program.

(continued)

Table 41.4 Continued

LTER Scientist	Comment Code	Illustrative Excerpt(s)
Moore	I, P	My experiences with the LTER program have greatly enriched my career. My research, education, and outreach activities do not exist in different career silos, but rather are inextricably interrelated.
Morris	I, SB	I was a disciple of long-term interdisciplinary research before my affiliation with the LTER program … The LTER program attracts scientists who are collaborative by nature.
O'Brien	C, P	[O]ur complimentary skill sets and deliberate use of similar computing architecture allows us to efficiently divide work … It is not enough to simply say that my involvement in the LTER program has affected my professional identity over time; I would have to say that it has enabled my development.
Peters	C, I	[T]he LTER network provided me opportunities to interact closely and informally with ecologists from different disciplines.
Pickett	C, I, P	The interdisciplinary LTER project has deepened my understanding of theory, and it has also developed my sensitivity to the role of metaphor across disciplines, with various public, and with professional colleagues in urban and regional policy or management.
Porter	C, P	The LTER program has shaped almost every aspect of my scientific career. It has enabled me to pursue ecoinformatics, a new and growing field, while allowing me to build on my training as an environmental scientist within the context of an intelligent, vibrant, and dedicated team of researchers and collaborators.
Schmitt	SB	As a field experimental ecologist, I have been frustrated with some of the limitations associated with the standard approach to ecology. With a rapidly increasing human footprint on ecological systems globally, most of my colleagues and I recognize the acute need for environmental science in general and ecology in particular to predict the dynamics of linked socioecological systems and the consequences of those dynamics to ecosystem functions of value to society.
Shaver	SB	My experiences in the LTER program have generally confirmed and extended the lessons that I learned as a student about how ecological research is accomplished most effectively.
Shaver	SB	If there is one lesson that I try to impart to aspiring ecologists and students, it is to "be generous—with your data, your time, your ideas, your equipment and supplies, and your friendship." By being generous there is much more to be gained than lost, while selfishness often leads to missed opportunities, limited productivity, and narrow horizons.

Table 41.4 Continued

LTER Scientist	Comment Code	Illustrative Excerpt(s)
Silver	C, I, SB	I grew up, professionally speaking, in the LTER program … The LTER program has helped to foster collaboration by creating a platform (site-based science) for researchers to come together to address broad, complex, interdisciplinary questions.
Smith	C, I	I believe that being involved in LTER from the start of my research career has made me well aware of the benefits and power of collaborative, multidisciplinary research.
Stafford	C, P, SB	The H. J. Andrews Experimental Forest (AND) site has literally defined my research focus and shaped my professional identity … My experience in the LTER program did more than merely influence my ideas about how environmental research should be done—it established the gold standard! At the AND site, collaboration and inclusiveness were canonical principles that I naively thought were applicable to all LTER sites … The early LTER sites did not include the human dimension. It soon became abundantly clear, however, that the integration of social sciences within the LTER network was an excellent idea.
Suding	SB	My involvement at the Niwot LTER site began when I was an undergraduate summer research assistant, and has extended through a postdoctoral fellowship, a tenured professorship, and now a leadership role in the program … I would not like to do science in isolation.
Swanson	C, I, SB	Increasingly, we recognize our science as a cultural enterprise. Our work as scientists and collaborations with those in the arts and humanities can influence public perceptions of our bioregional icons: ancient forests; fast, cold rivers; and volcanoes in the case of the Pacific Northwest … A feature of the LTER program I greatly appreciate has been the incredible room for interdisciplinary work that it encourages.
Tilman	C, I	Our individualistic perspectives gradually broadened as our team interacted and learned from each other.
Troxler	C	Collaboration is not an intuitive undertaking for the critical and independent thinkers we know as scientists. Yet, the LTER program was defined for me, when I became actively involved as a graduate student of the LTER network, as endeavoring to succeed in long-term, collaborative research.

[1]Selection Bias (SB; N = 17), Interdisciplinary (I; N = 23), Collaboration (C; N = 24), and Personal Impact (P; N = 10).

emphasis on "LTER friendships." In this way, LTER is a comfortable setting for those privileged enough to be admitted to the club, even if that admission also requires relevant training, expertise, and creativity. As discussed with the data above and also implied by the conclusions of Flint (Chapter 42), the LTER program had its greatest impact on knowledge and behaviors by the default existence of attitudinal inclination from the start.

Lastly, we noted a highly positive appraisal of the LTER program by the authors, as also remarked upon by Schlesinger (Chapter 40), even while citing a relative lack of cross-LTER site collaboration. The only significant exception to that assessment is the essay by Lodge (Chapter 24) where she states that the LTER "program has not influenced my basic approach to science" among some other similar comments. Otherwise, nearly every other author spoke in more positive terms about the impact of their involvement in the LTER program. In fact, as shown in Table 41.4, excerpts from 10 authors were notable for speaking to the high personal impact of participation in the LTER program on their careers.

CONCLUSION

The value of involvement in the LTER program as a personal and career-impacting phenomenon is abundantly clear throughout this volume. There is no question that the LTER program was and is a seminal experience for nearly all the authors. What is less clear, however, is the identity of the causal drivers in understanding their impact. From the foregoing macroanalysis, we can say with confidence that the LTER program may have been so successful in influencing ecological research careers because the participants were predisposed to thriving in such an environment. As stated above, it was a comfortable setting for scholarly and interpersonal interaction that over the long-term facilitated personal and professional success. What we do not know, and likely never will, is if a similar impact would have been possible with a group of randomly assigned scientists to research environments within and outside of the LTER program. Given the enormity of the human subjects' research concerns in performing such an experimental design, it is safe to say that we will never answer that question definitively. Perhaps the only Achilles heel that we can identify is the degree to which LTER program remains relatively insular, not by design or conscious choice, but rather by the de facto self-reinforcing interdisciplinary collaboration that characterizes the program. The ongoing impact and future success of this "invisible college" may well hinge on the ability of current and past participants to draw in new blood for innovation and growth in a rapidly changing scholarly landscape.

ACKNOWLEDGMENTS

The authors would like to thank Marissa Boyer for research assistance on the data analysis for this chapter. Thanks, as well, to the all those at the Center for Environmental Sciences and Engineering at the University of Connecticut for providing a welcoming and productive intellectual home.

REFERENCES

Alexander, P.A. (1992). Domain knowledge: Evolving themes and emerging concerns. *Educational Psychologist,* **27,** 33–51.

Alexander, P.A., Jetton, T.L., and Kulikowich, J.M. (1995). Interrelationship of knowledge, interest, and recall: Assessing a model of domain learning. *Journal of Educational Psychology,* **87,** 559–575.

Anderson, J.R. (2000). *Cognitive Psychology and its Implications.* Worth, New York.

Bandura, A. (1997). *Self-efficacy: The Exercise of Control.* W.H. Freeman and Co., New York.

Bandura, A. (1986). *Social Foundations of Thought and Action: A Social Cognitive Theory.* Prentice-Hall, Englewood Cliffs, New Jersey.

Bloom, B.S. (1976). *Human Characteristics and School Learning*, McGraw-Hill, New York.

Brown, S.W., Lawless, K.A., and Boyer, M.A. (2013). Promoting positive academic dispositions using a web-based PBL environment: The GlobalEd 2 Project. *Interdisciplinary Journal of Problem-based Learning, 7*, 67–90. Available at: http://dx.doi.org/10.7771/1541-5015.1389

Emmons, C.A., Joseph, J.G., Kessler, R.C., Wortman, C.B., Montgomery, S.B., and Ostrow, D.G. (1986). Psychosocial predictors of reported behavior change in homosexual men at risk for AIDS. *Health Education Quarterly*, **13**, 1-45.

Fishbein, M. (1967). Attitude and the prediction of behavior. In M. Fishbein, ed. *Readings in Attitude Theory and Measurement*. Wiley, New York.

Guba, E.G., and Lincoln, Y.S. (1994). Competing paradigms in qualitative research. *Handbook of Qualitative Research*, **2**, 163–194.

Miller, T.E., Booraem, C., Flowers, J.V., and Iversen, A.E. (1990). Changes in knowledge, attitudes, and behavior as a result of a community-based AIDS prevention program. *AIDS Education and Prevention*, **2**, 12–23.

Schrader, P.G., and Brown, S.W. (2008). Evaluating the first year experience: Students' knowledge, attitudes, and behaviors. *Journal of Advanced Academics*, **19**, 310–344.

Schrader, P.G., and Lawless, K.A. (2004). The knowledge, attitudes, and behaviors approach: How to evaluate performance and learning in complex environments. *Journal of Performance Improvement Quarterly,* **43**, 8–15.

Schunk, D.H. (1991). Self-efficacy and academic motivation. *Educational Psychologist*, **26**, 201–231.

Zimmerman, B.J., and Bandura, A. (1994). Impact of self-regulatory influences on writing course attainment. *American Educational Research Journal, 31*, 845–862.

42 Exploring the Scientific and Beyond-Science Interactions of Long-Term Ecological Research Scientists

Courtney G. Flint

IN A NUTSHELL

1. The essays in this volume are analyzed to assess the degree to which they portray scientific and beyond-science interactions.
2. The Long-Term Ecological Research (LTER) program represents a scientific or intellectual movement based on articulation of the program's highly respected founders, resource allocation for individual and collective pursuits, use of LTER sites for recruitment, and commonly held themes or foci for research.
3. Interdisciplinary scientific interactions within the LTER program have influenced researchers' ideas, networks, and productivity but have also presented challenges, particularly for junior participants.
4. Interactions beyond the scientific community focus on one-dimensional flows of information as well as on collaborative, multidirectional partnerships with a variety of stakeholders.

OVERVIEW

This analytical chapter explores social interactions catalyzed by experiences of scientists associated with the LTER program. I analyze the essays by LTER scientists in this volume using a broad, three-tiered structure: (1) the degree to which insights from the essays suggest that the LTER program represents a scientific or intellectual movement within environmental sciences examining ecological dynamics; (2) the extent of interdisciplinary interactions with scientists across broader fields of study, including associated reactions and challenges; and (3) interactions with others beyond science. Findings are examined across different career stages of respondents. Direct quotations are used to illustrate findings and to provide evidence for conclusions based on the LTER scientists' own words.

INTRODUCTION

The LTER program was initiated 34 years ago (Waide [Chapter 2]; Gholz, Marinelli, and Taylor [Chapter 3]). Given the growth of the LTER program, in terms of the number and geographic distribution of sites, as well as the contributions of engaged scientists and students, there is no doubt of the influence of the LTER program on the science of ecology and general understanding of ecosystems around the world (Robertson et al. 2012). In this chapter, I examine the social interactions of scientists in the LTER program through the lenses provided by their essays in this volume to explore three dimensions—interactions within the environmental sciences focused on ecological dynamics, broader interdisciplinary interactions, and interactions with stakeholders beyond science. The exploration of scientific interactions within environmental sciences focuses on the degree to which the LTER program is a collective, collaborative endeavor or social intellectual movement associated with long-term ecological research. Secondly, I examine the extent to which the essayists discuss interdisciplinary interactions, including associated challenges. Finally, these essays provide an opportunity to look "beyond science" to explore interactions with various stakeholders fostered by experiences in the LTER program.

To address the question about whether the LTER program is a scientific or intellectual movement, my analysis draws on the framework provided by Frickel and Gross (2005), who outlined a number of propositions related to scientific or intellectual movements (SIMs), which they define as "collective efforts to pursue research programs or projects for thought in the face of resistance from others in the scientific or intellectual community." According to these authors, SIMSs challenge a dominant way of thinking and are inherently political in that they require negotiating power and distribution of scarce resources, as well as organization and coordination. Examples of SIMs include the emergence of new fields, disciplines or areas of research such as human ecology, experimental psychology, and sustainability science (Frickel and Gross 2005). The Resilience Alliance is another contemporary example of a SIM (Parker and Hackett 2012).

Four general propositions or influential factors are articulated along with associated components that might elevate a collective activity to be considered as a SIM (quotations from Frickel and Gross 2005). The first proposition claims that SIMs are "more likely to emerge when high-status intellectual actors harbor complaints against what they understand to be the central intellectual tendencies of the day." In addition, a key element of a SIM is the use of references to older, intellectual leaders or mavericks with new ideas challenging the field in the early days as well as new leaders arising from emerging scholars with connections to these early leaders. The second proposition focuses on the importance of providing key resources and structural conditions to support the SIM efforts. These resources focus on individuals in terms of funding, employment, and intellectual prestige, as well as support for collective organizational communication and coordination. The third proposition emphasizes the importance of local sites or "micromobilizational contexts" for recruiting new SIM members. The fourth proposition suggests that SIMs are more successful when their participants frame ideas in ways that "resonate with the concerns of those who inhabit an intellectual field or fields." References to a collective identity and a particular historical narrative describing the founders and development of the movement over time are factors that suggest the development of a SIM.

Interdisciplinary interactions of scientists across fields or disciplines have become a contemporary norm, particularly related to grappling with challenges associated with environmental or earth systems (Reid et al. 2010). Interdisciplinarity is associated with both benefits and challenges. Cummings and Kiesler (2005) highlight empirical findings from interdisciplinary studies funded by the National Science Foundation (NSF), including consequences related to ideas, tools, training, and outreach. Additional elements associated with collaboration across fields or disciplines include influences on productivity (Lee and Bozeman 2005), expanded network connectivity (Goring et al. 2014), and challenges as well as other opportunities (Pooley, Mendellsohn, and Milner-Gulland 2013; Redman, Grove, and Kuby 2004; Romolini et al. 2013). Given that Michener and Waide (2008) posited that the future success of the LTER program would hinge on its ability to foster interdisciplinary collaborations, an assessment of this dimension of interactions is timely and warranted.

Another trend often discussed in relation to ecological and conservation science is the increasing call for interactions with nonscientific stakeholders, particularly to address the gap between science and policy (Kinzig et al. 2013). The essays in this volume provide an opportunity to examine the degree to which the authors identify with interdisciplinary and beyond-science interactions or discuss associated benefits and challenges. Furthermore, the question posed by this analysis is: if there is evidence of substantive interactions beyond science by participants in the LTER program, is it characterized by a unidirectional flow of information or by a multidirectional or partnership model?

All of the essays that inform my analyses are available in this volume for others to draw their own conclusions. This chapter represents my interpretation of quotations from various essays in this book with explicit mention of the framework elements outlined with respect to SIMs, interdisciplinarity, or beyond-science interactions. Other conclusions are quite possible, as is the potential that essay authors who did not explicitly mention something would still agree with such sentiments on further inquiry. In short, my conclusions are not intended to be the only possible interpretation regarding social interactions of scientists in the LTER program. Furthermore, because of the element of subjectivity in my approach, broad qualitative statements are used to convey the proportion of the 36 authors on any given areas of inquiry (e.g., "few" describes sentiments by 9 or fewer authors, "less than half" describes between 10 and 18 authors, "more than half" describes between 19 and 24 authors, and "many" or "more than three fourths" describes more than 24 authors). Assigning career stage to each author was done in consultation with the volume editors and corresponds more or less with academic position, years in the field, and to some extent, years associated with the LTER program (but not always). Based on this categorization, 5 authors were deemed to be "junior" or early career, 7 as mid-career, 16 as mid-late career, and 8 as "senior." This apparent skew arises in part because invitations to write essays were extended to those with more experience in the LTER program.

FINDINGS

The chapters in this volume clearly indicate that involvement in the LTER program had a profound impact on the authors. From professional development to the very lives of these scientists, the chapters are filled with positive reflections on their experiences. Grimm (Chapter 15) writes, "No other single event in my career has had a greater impact on my development as a scientist than the decision to lead the CAP [Central Arizona–Phoenix] project." Also capturing the significance of the experience, Ducklow (Chapter 31) writes, "Joining PAL [Palmer LTER site] changed my life." Yet the chapters raise concerns as well, highlighting challenges, past and present, as well as recommendations for the future. A common refrain was heard of budgetary challenges and a need to plan for the future of the program, as illustrated by Stafford (Chapter 5):

> The LTER program is facing a real conundrum. Optimizing efforts to support the long-term research conducted on the scale of decades or centuries, is a nontrivial challenge, especially if the goal is to support hypothesis-driven inquiry, not just the accumulation of data and information. This requires a commitment to work today to support tomorrow's science, which will require committing resources to future studies that could otherwise be directed toward current investigations. This conundrum is exacerbated in today's research funding climate—with shrinking budgets, lower funding rates, and a general decline in public understanding and support for science.

Overall, it is clear that involvement with the LTER program not only shaped the science and professional development of these authors, but it also influenced their scientific and beyond-science interactions. It is these interactions that are the focus of this chapter.

The LTER Program: A Scientific Intellectual Movement?

As suggested by Collins (Chapter 35), "There is an LTER culture!" There is a clear sense by the essay authors that they have been a part of something with great consequence, not only for themselves, but also for the understanding of ecosystem dynamics around the world. Based on the first proposition of Frickel and Gross (2005), it is quite clear that the role of founding members of the LTER network are held in high regard as "visionary" (Ducklow, Chapter 31) and "forward thinking" (Troxler, Chapter 18) individuals who provided a "solid footing to embark on the next steps and to train the next generation of scientists" (Suding, Chapter 29). More than half of the authors (mostly those in mid-late and senior career stages) mention founders and mentors and links to following generations of scientists. However, only a few explicit references are about these early leaders as mavericks differentiating from prevailing views at the time. Magnuson (Chapter 30), as one of those early founders, highlights the exciting element of discovery and original rationales put forward for the LTER program about "long-term changes and dynamics" and "complicated system behavior." In addition, the general tenor of the chapters indicates that the long-term focus was a considerable "shift" for ecology (Schmitt, Chapter 28) and thus may support the notion of the LTER program as a scientific or intellectual movement regarding the first proposition.

The chapters are also full of insights supporting the second proposition about providing resources for funding, employing, and supporting individual scientists, as well as collective and collaborative efforts. Most of the authors are grateful for the career opportunities provided by funding from the LTER program for themselves and for their students. A few authors note the longer 6-year cycle of funding as valuable—although somewhat in tension with even longer-term research objectives (Peters, Chapter 20). Although only mentioned by a few authors, some clearly offer claims that the LTER program was influential for "standing, locally and in the broader ecological community" (Blair, Chapter 21), "visibility" (Blair, Ducklow; Chapters 21 and 31, respectively), career boost (Pickett, Chapter 11), placement in leadership of emerging field (Grimm, Chapter 15), and research productivity (Suding, Chapter 29. As articulated by Knapp (Chapter 22), such personal prestige benefits from involvement in the LTER program have broader impacts to others, "Scientists within the LTER network are among the most successful and influential in the world, and thus associating with them has many positive professional and personal consequences."

The importance of resources from the LTER program for collective efforts, including communication and coordination, are mentioned by nearly half of the authors, and represent scientists from each career stage. The following quote from Swanson (Chapter 6) highlights the value of LTER All Scientists Meetings as "a model of open culture interactions where participants of all career stages share ideas and enthusiasm in workshops and informal interactions on dozens of topics of shared interest." Support for networking extends to communication and data sharing as well, making it "possible to create more robust and general solutions than would otherwise be possible" (Porter, Chapter 39).

Despite the gratitude and recognition of the value of resources from the LTER program for individuals and collective efforts, strong voices are communicated by many of the same authors about resource challenges for maintaining site and network responsibilities. Resources fund the meetings, but often not the participant's time, as suggested by Suding (Chapter 29), "Although meetings can be productive, the productivity often depends on preparation prior to and completion after the meeting: an unfunded responsibility for a very busy group of people." Furthermore, funds for research are noted as "meager," particularly for cross-site comparisons (Lodge, Chapter 24) and spread "too thinly" for the diverse array of activities related to the LTER program (Giblin, Chapter 32). The following quote by Hayden (Chapter 38) acknowledges the benefit of expanding the LTER network activities but also adds that perhaps "mission creep" is related to resource challenges, "LTER sites cannot be all things to all people. Mission creep, or the growth of expectations, is and has been an impediment to the research planning process." Holbrook (Chapter 27) notes that challenges for data management components of scientific collaboration are not just financial, but also technical and cultural. Shaver (Chapter 9) calls for the LTER program to reach out to more people but claims, "the LTER program is already trying to do too much with too few resources." The resource concerns discussed by the essay authors suggest potential risks for the LTER program as a scientific or intellectual movement.

The third proposition regarding scientific or intellectual movements focuses on the key role of sites for recruiting or "micromobilization contexts." The many comments by authors about their own experiences as students in the LTER network, as well as about those of the students they mentor, support this dimension. The following quotation by Troxler (Chapter 18) articulates a common sensitivity about the LTER program: " … engendering a spirit of collaboration among graduate students as key to building a network of scientists that could address various grand challenge questions." Echoing this statement, other authors feel that the LTER program has supported their interests and efforts in preparing the "next generation" (Silver, Morris; Chapters 26, 33, respectively), "new people" (Ducklow, Chapter 31), and "new blood" (Gough, Knapp, Smith; Chapters 7, 22, 23, respectively).

There is both optimism and concern about the future of new recruits. The following quotation by Gaiser (Chapter 17) reveals the optimist perspective. She states: "The LTER program has a bright future ahead, mainly because of the attention that has been paid to students (at all levels). Students in the LTER program are now better trained in creating theoretically motivated research in a collaborative, persistent context; communicating their work through innovative outlets; and in the skills of collaborative science leadership than before." Some essays highlight the inclusivity and welcoming nature of the LTER program, but for others, there are concerns about a "pipeline problem" (Hayden, Chapter 38) and the perception of the LTER program as "insiders" (Borer, Chapter 12) having a "perceived exclusivity" (Moore, Chapter 37) or being a "closed system" (Pickett, Chapter 11), "closed community" (Smith, Chapter 23), or "club" (Gough, Knapp; Chapters 7, 22, respectively). Suding (Chapter 29) claims that "it remains difficult to come in from the outside" and the following quotation by Johnson (Chapter 4) points out a relatively common concern: "To those not affiliated with a LTER site, it is not obvious how research projects are proposed and approved at a site or how LTER research funding is allocated." Hayden (Chapter 38) states that, "diverse cultures and ethnic groups are underrepresented at most LTER sites and in the LTER network in general." Although not unanimous across the essayists, this concern about the closed nature of the community supported via the LTER program is more prevalent in the essays than expressions of perceived inclusivity.

A few authors pose solutions to this recruiting threat. Holbrook (Chapter 27) suggests an "open policy" is needed for bringing new scientists into leadership positions. Others recommend changes to the funding mechanisms (Collins, Chapter 35) or other creative efforts as indicated by Smith (Chapter 23): "Thus, there is a need for the LTER network to be more proactive and creative in the ways that it attracts new researchers to get involved in the site-based or network-level research. Ultimately, the LTER network will only benefit from increased involvement by new investigators, who also could serve the role of leading the LTER network in the future." To summarize the micromobilization context of the LTER sites and network, there is strong recognition among the essayists of the opportunities created by the LTER program in the past but considerable concern about the future ahead.

The essays in this volume highlight a number of common themes that compose a prominent collective identity for participants in the LTER network, as suggested in the fourth SIM proposition. It comes as no surprise that the focus on "long-term" research is the most common theme. This emphasis is followed closely by a focus on cross-site or multisite, network, synthesis, or meta-analytical research by about half of the authors from all career stages. Despite this cross-site emphasis, place- or site-based research is also clearly articulated as important by slightly fewer than half the authors. Socioecological research is voiced as a theme of the LTER program by fewer than one fourth of the authors, particularly by mid-late or senior authors. Although some acknowledge the inclusion and importance of the social sciences in the mission of the LTER program, Moore (Chapter 37) questions the extent of this integration by suggesting that the social sciences "while appreciated have yet to be fully embraced by the community as a whole as part of the LTER mission." In addition, data and information management appear to be considered a core component of identity in the LTER program by a small cluster of essayists, as indicated by O'Brien (Chapter 34), who states, "When I started in the LTER program, cataloging and publishing data represented a novel concept. Now, data management is no longer an afterthought. The LTER program has fostered that."

In addition to the collective identity around a few core research themes, there is also representation of a historical narrative for just under half of the authors, mostly by those in mid-late or senior career stages. This pattern of responses suggests some threat to the forward movement of the scientific or intellectual movement if junior colleagues are less cognizant or mindful of the historical roots of the LTER program than are senior scientists. Often, this historical narrative is articulated in essays in terms of one's personal involvement in the history, suggesting that an independent or ubiquitous historical narrative may not be commonly held.

Resonating with the framing of such movements by Frickel and Gross (2005), it is clear that the LTER program has, to some degree, challenged normative practices and perspectives within ecology across the nation and the world (Hamlin, Chapter 43). Although the authors in this volume do not appear to be particularly revolutionary, there may be some recognition that leaders of the LTER program, at least in the past, have been drawn from highly respected colleagues from within the broader ecological community. In these ways and others, the essays in this volume suggest substantial collective evidence that the LTER program has elements of a scientific or intellectual movement. However, there are clear concerns by this cadre of scientists that the LTER network faces considerable challenges as it moves forward.

LTER as an Interdisciplinary Experience

In this section, I focus on the degree to which authors claim that interdisciplinary interactions have influenced their research and professional development. Given the request to comment on impacts of experience in the LTER program on collaborations, it comes as no surprise that this is a common theme across the essays. More than three fourths of the authors describe interdisciplinary engagement as part of the culture in the LTER program. Most of the related comments focus on the effect of these cross-disciplinary collaborations on scientific ideas as articulated by Suding (Chapter 29) who states, "The most interesting questions sit on edges between disciplines, and this experience has allowed me to better understand those edges." Silver (Chapter 26) further comments on interdisciplinary interactions, stating, "This ongoing need for new skill sets is one of the primary drivers of collaboration. The LTER program has helped to foster collaboration by creating a platform (site-based science) for researchers to come together to address broad, complex, interdisciplinary questions." The influence of interdisciplinary interactions on scientific ideas, despite accompanying challenges, is echoed by Bestelmeyer (Chapter 19) who states, "As part of a research team at an LTER site, a scientist is more likely to work with scientists from outside his or her core discipline than from within it. Such multidisciplinarity might have the disadvantage of diluting focus on particular topics, but it enables opportunities for developing novel questions and insights."

About half of those commenting about interdisciplinary interactions focus on how these efforts have expanded their networks as suggested by Childers (Chapter 14) who states, "As my research and worldview has become more interdisciplinary, the diversity, extent, and richness of my collaborations have exploded. Today my colleagues are more likely to be social scientists, economists, planners, designers, humanists, and even artists." The notion of incorporating social science into the LTER program is also a key part of the interdisciplinary discussion by a cluster of authors as illustrated by the following quotation by Swanson (Chapter 6) who states, "It is easy to think that science findings guide decisions about management of natural resources, but personal and societal values are at the root of such decisions and social sciences and humanities help people understand those values." In the following quotation, Stafford (Chapter 5) looks back on the decision to integrate social sciences in the LTER program:

> At early IM [information management] meetings, I often mused that we needed a sociologist-in-residence to watch how well, or how poorly, we were doing and how well sites were addressing the human dimensions of the ecological questions under study. The early LTER sites did not include the human dimension. It soon became abundantly clear, however, that the integration of social sciences within the LTER network was an excellent idea.

Whether the inclusion of social sciences is deemed important for the LTER network in the future remains uncertain. The essays express little opposition to the inclusion of human dimensions, but enthusiasm from only a few.

Although a few authors mention the effects of interdisciplinary interaction on their productivity (Gaiser, Silver, Gosz; Chapters 17, 26, 36, respectively), "tool kit" (Gough, Chapter 7), or involvement with policy (Tilman, Childers, Magnuson; Chapters 13, 14, 30, respectively), these are not dominant themes across the full set of essays. The clearest collective observation about interdisciplinary interaction from the authors is the challenge it presents for junior colleagues. Interestingly, these concerns are raised by about one fourth of the authors, but only by one junior author and by no mid-career authors. These concerns focus mostly around balancing interdisciplinary commitments and time with the need to establish an "independent research agenda or line of scholarship" (Moore, Chapter 37). Additional related concerns are raised about how to evaluate collaborative research for promotion and tenure, as well as for allocating authorship credit. Most, however, conclude that the interdisciplinary interactions are worth the effort despite the challenges.

LTER Interactions Beyond Science

Nearly all of the authors mention some involvement with stakeholders beyond the scientific community. The most commonly cited engagement activities are involvement with K–12 students and teachers as facilitated by the Schoolyard LTER Program (along with some involvement with NSF's Research Education for Undergraduates Program). Beyond this educational interaction, quite a few authors mention interactions with a diverse array of stakeholders as indicated by Stafford (Chapter 5), who states, "interactions with various stakeholder groups including state, federal, international, and nongovernmental organizations were commonplace and part of everyday site business." Lugo (Chapter 25) expands this list by stating, "socioecological research uncovers a new cadre of urbanites, nongovernmental organizations, communities, government agencies, and academic disciplines of unprecedented magnitude and scope." Although resource management and government agencies combine to form the next highest group for engagement beyond science, various public, community, business, media, and nongovernmental entities are also mentioned.

About half of the authors claiming interactions beyond science are focused predominantly on a unidirectional flow of information from science to stakeholders in the form of communicating to the public (Hobbie, Chapter 8), "outreach" (Gough, Holbrook, Collins; Chapters 7, 27, 35, respectively), or presenting information (Bestelmeyer, Chapter 19). Site locations are said to constrain such engagement by some as captured in the following quote by Shaver (Chapter 9), who states "This physical isolation makes it difficult to devise an applied research program that is well integrated with our other research activities." But for many, communicating scientific research is a key component of their role in the LTER site or in the LTER network as stated by Tilman (Chapter 13):

> As our research at the CDR [Cedar Creek] LTER site became more relevant to major environmental problems, I grew more interested in communicating our findings with the media, governmental and corporate leaders, and the public ... The LTER Schoolyard Program initiated an interest, now shared by most CDR researchers, in passing on the love of science and concern for the environment to the next generation.

For the other half of the authors involved in beyond-science interactions, the emphasis is on multidirectional partnerships or the coproduction of knowledge with managers and other stakeholders as indicated in the following quotations by LTER scientists.

> The Coweeta Listening Project was established in 2011 in partial response to this situation to engage scientists in the LTER program and the public in southern Appalachia in coproducing usable knowledge. The members participate in diverse boundary-spanning activities to achieve credibility, salience, and legitimacy in diverse settings. (Gragson, Chapter 16)
>
> Watching leadership at the AND [H. J. Andrews Experimental Forest] site cultivate, nurture, and build these enduring partnerships provided invaluable lessons for working effectively across a broad array of stakeholders and external constituents. My office was housed in the Pacific Northwest Research Station, and we worked in a National Research Forest. This provided me with a first-hand view of "researchers" working with "managers." Although some of these interactions initially were contentious, over time I saw both sides come to respect, listen to, and work with each other effectively. (Stafford, Chapter 5)
>
> Perhaps the most significant changes in "outreach" as a result of my experiences at the BES site [Baltimore Ecosystem Study] has been to realize that one-directional information transfer is far less successful than two-way dialogue with those interested in ecological knowledge about cities and urban regions. We quickly shifted from a "delivery model" of science in Baltimore to a model of community engagement through continual dialogue, in which our research community

could learn as well as teach. Hence, abandoning literal outreach was one of my first and most important lessons from the urban LTER site. (Pickett, Chapter 11)

There is no particular pattern to beyond-science engagements by looking across career stages. Multidimensional interactions with stakeholders are not limited to the urban LTER sites, as might be expected. Those more fully engaged in dynamic relationships with stakeholders come from a variety of LTER sites and from scholars with different disciplinary backgrounds. Despite the prevalence of beyond-science engagements, a number of authors suggest a need for more focus and training in this dimension as captured in the following quotation:

> Nonetheless, there is a near absence of LTER network-wide resources, guidelines or training opportunities to build the capacity at sites for increasing their effectiveness in communicating with the public or engaging in service and outreach. These activities at best are seen as necessary evils, and at worst they are ignored completely. Their momentary burden, however, is ultimately critical to spanning the boundaries to achieve credibility, salience, and legitimacy among those who we eventually hope will care about or benefit from our research; or whose children will become the next generation of researchers. (Gragson, Chapter 16)

CONCLUSION

The interpretive analysis of the essays of this volume suggests that experiences in the LTER program have had a profound impact on the authors. Furthermore, their experiences and observations suggest that the LTER program has the hallmarks of a SIM as described and framed by Frickel and Gross (2005). There is a clear sense that highly respected scientific founders pushed the boundaries of ecology to create the LTER program and that their mentorship paved the way for new leaders over time. The LTER network has also provided key resources for individual professional development and collective coordination, as well as site-specific and network-wide recruiting opportunities of students and scientists. Although there are certainly some challenges highlighted by the authors, the program clearly has a collective identity and a prominent place within the broader scientific community.

The degree to which the perception of the LTER network as a closed community poses a risk to the LTER program or movement remains to be appreciated by its members. Frickel and Gross (2005) do not discuss the role of exclusivity in SIMs. But the recognition of this issue raised by a considerable number of essayists in this volume will hopefully foster a healthy dialogue about the future of the LTER network. The authors also paint a clear picture of expanding interactions fostered by the experience in the LTER program, across disciplines and beyond science. These interactions have led not only to new ideas and new collaborators but also to increased productivity and new knowledge from beyond the bounds of science.

There is a strong interdisciplinary current running through the interactions of LTER program scientists. These interactions appear to influence the expansion of scientific ideas and the networks and productivity of participants. Inclusion of social science in the LTER program is noted as important by some, but by no means all, of the essayists. This appears to be a question to be more fully addressed as the LTER program moves forward. Although there is no doubt that the LTER experience has fostered beyond-science interactions, it is my sense that there is work to be done in moving these endeavors from unidirectional, science-dominant flows of information to more collaborative partnerships with nonscience stakeholders to bring more collective knowledge to bear in addressing global to local socioecological challenges.

One element of social interactions emerging clearly from the authors is the influence of the LTER program on their personal relationships. More than three fourths of the authors mentioned friendships emerging from their experiences in the LTER program. Grove (Chapter 10) captures the importance of these interactions by stating, "More importantly, I have developed lifelong friendships because we work on a long-term project, meet with scholars from other LTER sites at annual Science Council Meetings and triennial LTER All Scientists Meetings. Furthermore, these friendships are multigenerational, and I feel as if I am part of an extended family." Ducklow (Chapter 31) extends the discussion of the influence of personal relationships enhanced by the LTER program experience:

> Besides the nature of the ocean itself, working on ships also brings us closer together. We are in a very intimate, close-knit environment and society aboard a research vessel. Think about standing at the rail of a small ship in a heaving sea, throwing up, with your advisor, student, technician, or colleague standing at your side, sick as a dog. One of the things I always loved about oceanography is the way it forces your life and work together, literally in the same boat. That experience fosters collaboration like nothing else.

These strong bonds formed by participating in the LTER program suggest that this scientific movement is important in the lives of participants beyond research. The essayists in this book recognize the LTER program's profound influence and are clear advocates. The endurance and success of the LTER program in the future likely hinges on its ability to expand this opportunity to others in an inclusive way while still maintaining the allocation of resources for individual and collective goals.

REFERENCES

Cummings, J.N., and Kiesler, S. (2005). Collaborative research across disciplinary and organizational boundaries. *Social Studies of Science*, **35**, 703–722.

Frickel, S., and Gross, N. (2005). A general theory of scientific/intellectual movements. *American Sociological Review*, **70**, 204–232.

Goring, S.J., Weathers, K.C., Dodds, W.K., Soranno, P.A., Sweet, L.C., Cheruvelil, K.S., Kominosky, J.S., Rüegg, J., Thorn, A.M., and Utz, R.M. (2014). Improving the culture of interdisciplinary collaboration in ecology by expanding measures of success. *Frontiers in Ecology and the Environment*, **12**, 39–47.

Kinzig, A.P., Ehrlich, P.R., Alston, L.J., Arrow, K., Barrett, S., Buchman, T.G., Daily, G.C., Levin, B., Levin, S., Oppenheimer, M., Ostrom, E., and Saari, D. (2013). Social norms and global environmental challenges: The complex interaction of behaviors, values, and policy. *BioScience*, **63**, 164–175.

Lee, S., and Bozeman, B. (2005). The impact of research collaboration on scientific productivity. *Social Studies of Science*, **35**, 673–702.

Michener, W.K., and Waide, R.B. (2008). The evolution of collaboration in ecology: Lessons from the U.S. Long-Term Ecological Research Program. In G.M. Olson, A. Zimerman, and N. Bos, eds. *Scientific Collaboration on the Internet*, pp. 297–309. MIT Press, Cambridge, Massachusetts.

Parker, J.N., and Hackett, E.J. (2012). Hot spots and hot moments in scientific collaborations and social movements. *American Sociological Review*, **77**, 21–44.

Pooley, S.P., Mendellsohn, J.A., and Milner-Gulland, E.J. (2013). Hunting down the chimera of multiple disciplinarity in conservation science. *Conservation Biology*, **28**, 22–32.

Redman, C.L., Grove, J.M., and Kuby, L.H. (2004). Integrating social science into the Long-Term Ecological Research (LTER) Network: Social dimensions of ecological change and ecological dimensions of social change. *Ecosystems*, **7**, 161–171.

Reid, W.V., Chen, D., Goldfarb, L., Hackmann, H., Lee, Y.T., Mokhele, K., Ostrom, E., Raivio, K., Rockström, J., Schellnhuber, H.J., and Whyte, A. (2010). Earth system science for global sustainability: Grand challenges. *Science*, **330**, 916–917.

Robertson, G.P., Collins, S.L., Foster, D.R., Brokaw, N., Ducklow, H.W., Gragson, T.L., Gries, C., Hamilton, S.K., McGuire, A.D., Moore, J.C., Stanley, E.H., Waide, R.B., and Williams, M.W. (2012). Long-term ecological research in a human-dominated world. *Bioscience*, **62**, 342–353.

Romolini, M., Record, S., Garvoille, R., Marusenko, Y., and Geiger, R.S. (2013). The next generation of scientists: Examining the experiences of graduate students in network-level socio-ecological science. *Ecology & Society*, **18**, 42.

43 Long-Term Ecological Research Over the Long Term
A Historian's Perspective

Christopher Hamlin

IN A NUTSHELL

1. There are many precedents for long-term research in the history of science.
2. Long-Term Ecological Research (LTER) program's current identity reflects significant change—intended and accidental, both consensual and conflictual—from research concerns that were prevalent in the 1980s.
3. LTER program has pioneered modes of research organization and professional norms that are increasingly prominent in many areas of research and that belong to a significant transformation in the social relations of scientific research.

INTRODUCTION

The essays in this volume explore the impact of the LTER program, a generation after its founding, on both the practice of ecological science and the careers of scientists. The authors have applied the agenda of long-term scrutiny to their own careers as LTER researchers. They have recognized the LTER program as distinct, even perhaps unique, both in the ways that it creates knowledge and in the ways that it shapes careers. They have reflected on how they have taught (and were taught) in LTER settings, on how they interact with one another and with the public, and on how research in the LTER program has affected them "as persons." A rationale for this volume is LTER's distinctiveness. In many of the chapters, and in other general treatments of the LTER program, beginning with Callahan (1984), one finds a tone of defensiveness. Sometimes the concerns are explicit: authors (e.g., Stafford, Knapp, Lugo, Morris; Chapters 5, 22, 25, 33, respectively) bemoan colleagues who dismiss LTER as mere monitoring instead of serious science or who resent LTER's independent funding stream. But more broadly, there is concern that various groups, ranging from other bioscientists to the public at large, may not appreciate the importance of long-term, site-specific environmental research.

Accordingly, my hope here is to put LTER into several broader contexts. I do so in three ways. First, to mainstream LTER within the history of science, I show that the LTER program is not a new and odd way of doing science but rather exemplifies research agendas that have been recognized at least since the seventeenth century in the biosciences and beyond. Second, I highlight LTER's contingency and hybrid nature. It is hardly the only or obvious institutional structure for undertaking long-term ecological research, but rather an unstable creature of recent US science policy. Much of what LTER researchers find rewarding as well as what they find frustrating reflects the ongoing messiness of its history. It represents no single scheme or agenda. As it has evolved, so, presumably, it will continue to do so. Third, I consider values and norms. Given the many ways of living a scientific life, what values (e.g., in governance, agenda-setting, and research practices) does LTER represent and reinforce? I shall suggest that regardless of the significance of the research findings, LTER represents a distinct exemplar of a set of emerging scientific norms: its most lasting impact may be as a normative and institutional experiment. In brief, these three sections explore the past, present, and future of the LTER program.

A PEDIGREE FOR LTER: LONG-TERM AND SITE-SPECIFIC RESEARCH IN THE HISTORY OF SCIENCE

Reviewing the initial rationale for the LTER program, Waide (Chapter 2), following Callahan (1984), noted the general aspects of nature it had been established to study (see also Gholz, Marinelli, and Taylor; Chapter 3). Foremost was an expanded temporal scale: both variation and directional change may occur on scales that are long with regard to the convenience of the observer. One may not know which phenomena will exhibit such variability; that discovery itself will require studying a single site for some considerable length. Study of more general changes will require compatible inquiries at multiple sites, and ability to explore the relations of multiple variables. There will be need for both uniformity and continuity of data.

In the history of science, observational studies of significantly varying or relatively slowly changing environmental phenomena have indeed been more rare than one might expect, given their current importance. Sometimes regulatory principles of reasoning, like the geologists' seminal principles of uniformitarianism (or actualism or gradualism), were taken to sanction the adequacy of singular or short-term observation (Kitts 1977). Or faith in equilibria warranted an expectation that with the exception of obvious cycles of the year or a species' lifespan, phenomena were time-independent for most practical purposes: what holds at one time holds for any (Wise 1989; Lodge and Hamlin 2006). Hence, one can readily imagine a scenario in which my ex-boss, Charles David Keeling, would have been satisfied with a single year's data on atmospheric CO_2 levels from Mauna Loa Volcano in Hawaii. Of course, had that been the case, the state of the environmental sciences would be profoundly different than it is: we would not know that fossil fuel emissions were changing the composition of the atmosphere and probably affecting climate.

What follows is a list of antecedents for elements of the LTER program. These elements are organized empirical research responsive to policy concerns; site-specific research, including large-scale mapping of environmental variables and analysis of casual processes from such data; and, finally, development of means for grappling with the stubborn problems of calibrating and coordinating data taken by many persons in many places and over a long time. One must hunt for these antecedents. They do not stand out in most histories of science. Indeed, at some key stages in the history of the environmental biosciences they have been neglected: where we might expect to find them, we do not. The list below is of accreting features, which bring us ever closer to the kind of research institution the LTER represents.

Organization of Empirical Environmental Research Responsive to Policy Concerns

Here the obvious starting point is the English philosopher Francis Bacon (1561–1626). At a time when most philosophers were fixated on metaphysics, Bacon advocated empiricism, the collection of observations on a wide variety of natural features. In his posthumous utopian work, *The New Atlantis* (1627), Bacon not only outlined a general public research institution, "Salomon's House," but recognized many of the component skills that LTER projects typically include. Along with "merchants of light," who traveled abroad to collect data, were "compilers" who reduced and tabulated it, and "dowrymen" responsible for knowledge-transfer: they applied science to public needs. Those known as "lamps" planned the research agenda following "divers meetings and consults of our whole number" (as in LTER All Scientists Meetings). Finally, some, "interpreters of nature" were charged with inferring general laws. Bacon was prescient in recognizing a "big" science that would require coordination of many types of expertise, a theme that arises in many of the chapters of this book.

Site-Specificity

Contemporaneous with Bacon was the golden age of chorography. This now discarded term referred to empirical inquiry, characterized neither in terms of topic nor method but simply place. One sought to chronicle all changes in human and natural phenomena over as long a period as possible (Jankovic 2000). Often dismissed as "antiquarians," chorographers have been marginalized in the history of science. Although their works may be valuable in reconstructing environmental change, generally they saw themselves as committed to the intrinsic value of documenting a unique place, usually a town or region, rather than as contributing to general geographic knowledge.

The main antecedent to chorography was epidemiological. Environmental causation of disease was a mainstay of seventeenth- and eighteenth-century medical theories. Whether pathological effects were attributed to the integrated environment (e.g., climate, landforms, soils, built environment, human activities) or to some single environmental component was not clear, but the presumption was that disease was a function of place and time. Long-term chronicling of prevailing disease, chiefly in conjunction with meteorological factors, might disclose local variability and, possibly, more general laws of the relation of climate to disease. The best known chroniclers were Thomas Sydenham, who identified five distinct states in the prevailing diseases of London in the 1660s and 1670s; John Huxham, who compiled disease incidence and meteorological data in Plymouth, England, from 1724 to 1748; and John Rutty, who studied similar phenomena in Dublin from 1725 to 1765 (Riley 1987). The precedent of long-term, site-specific research on disease incidence would continue into the nineteenth century and become the work of public agencies. In colonial India, statistical officers such as James Bryden and H. W. Bellew correlated multiple meteorological variables with disease incidence, integrating multiple sites in a search for laws of cholera pandemicity (Hamlin 2014).

The documentation of site-specific natural history is exemplified by the clergyman and naturalist Gilbert White of Selbourne, Hampshire, United Kingdom. White's letters on botany and zoology to metropolitan colleagues with more professional involvement in natural history have become famous. White was not focused on environmental change per se, but because he kept diaries of natural phenomena (e.g., blooming of plants, bird migration) in the mid-to-late eighteenth century, he was setting a precedent for collection of data that would demonstrate change (White 1891).

The work of Huxham, Rutty, and White was site-specific because their careers kept them in one place. Through publication they contributed to natural knowledge, but they were part of no coordinated campaign of environmental inquiry analogous to the LTER program or network. By the mid-nineteenth century, some national scientific organizations, notably the new British Association for the Advancement of Science, were sponsoring coordinated multisite inquiries. Usually these studies relied on amateur on-site volunteers responding to a questionnaire or utilizing a common protocol, with results being tabulated by a professional committee. Although some of this was long-term monitoring, often spatial considerations were more prominent. Such approaches were inspired by the explorer–geographer Alexander von Humboldt (1769–1859) and have come to be known as Humboldtian science. On the basis of such collected data, Humboldt pioneered the introduction of isopleth maps (with lines of equal value for a particular parameter). Taking geophysical and botanical data in Central and South America, he was able to show, for example, how elevation was a proxy for latitude with regard to species distribution. Humboldtian approaches would become most prominent in geophysics (e.g., gravitational variability), but they also formed the foundation of modern descriptive climatology, and, in turn, helped to undergird biogeography (James 1972). Such approaches would become important in oceanographic and polar research. Within the biosciences, the coordination of amateur observations has been most conspicuous in ornithology (Allen 1994).

During the Progressive Era (1890–1920) in the United States, state and federal resource management agencies focused on water-, timber-, and fish- and game-related issues. Already, the medical officers at remote army posts had been charged with compiling weather data, and in some places the US Army Corps of Engineers was gauging streams. Many LTER sites and scientists are direct legatees of that heritage, through connections with the US Forest Service and similar agencies. In documenting temporal and spatial variability, such rich data sets could invite explanatory hypotheses. Yet monitoring did not invariably beget research. Data might be taken and never analyzed or if analyzed never examined with regard to dynamic questions.

Calibration and Coordination

In the nineteenth century, as in the LTER program in recent decades, integration of observations made over long periods of time by many observers, with instruments whose calibration and skilled use could rarely be guaranteed, raised issues of data quality. The problems were most systematically addressed in astronomy, where there was need to reconcile many stations and interest in slowly changing phenomena (e.g., variable stars). In some cases, experience was paramount. John Herschel might be able to discern 60 different degrees of brightness in stars, but it was hardly realistic to recruit a corps of observers with such skill (Case 2014). To neutralize observer variability, mathematicians and astronomers (Legendre, Gauss, and LaPlace) developed statistical techniques, notably the least squares approach (1805–1810) (Desrosières 1998).

As well as antecedents, there were also areas of the environmental biosciences where site-specific, long-term research lagged. One is evolutionary biology. In the immediate post-Darwin decades, belief that evolution occurred only on geological time scales discouraged efforts to observe it. Even in ecology, the hallmark of directional change, Cowles' concept (Cowles 1901) of succession was not actually the product of lengthy study. In the dunelands at the foot of Lake Michigan, Cowles hypothesized, in a transect across the landscape, the several temporal stages of succession.

I offer these antecedents of LTER practices and principles as cultural resources. It is true that they have been only peripheral constituents of the better known heritage of

Western science, one that (1) emphasizes the universality rather than the uniqueness of time and space and (2) highlights critical experiments and individual acts of genius rather than collaborative development of long-term data sets. Nevertheless, as I shall suggest below, these elements are becoming more characteristic of contemporary science. The LTER program itself deserves some credit for that change.

EVOLUTIONARY AGENDA OF THE LTER PROGRAM

The LTER program is a child of the National Science Foundation (NSF). Its distinctiveness from other NSF programs is most conspicuous in its 6-year funding cycles. Six years is not "long" for many of the questions scientists want to address, but as Callahan (1984) notes, expectations of continuity in monitoring programs were a key element in LTER site selection.

Issues such 6-year funding cycles do raise the question of how far LTER's evolution reflects contingent and often arbitrary institutional factors rather than an inherent research trajectory. For example, would a more autonomous program have evolved in the way LTER has done? And the LTER program has evolved. Although this is not always clear to some of the younger authors, who represent LTER as an unproblematic and reasonably coherent institutional framework, it is strikingly different in 2014 than it was in the early 1980s. Waide (Chapter 2) and Gholz, Marinelli, and Taylor (Chapter 3) draw attention both to the changes and to the areas of ongoing tension. Sometimes that change came from outside. Notwithstanding its professed concern for continuity, NSF has hardly been stable in its expectations. What Hayden (Chapter 38) labels as "mission creep" has included both explicit policy change as well as more subtle pressures and changes in outlook. The LTER network now includes urban sites and embraces social science, policy, and outreach, but these changes were neither simple nor uncontroversial (Gholz, Marinelli, and Taylor; Chapter 3). In the mid-1980s, many ecologists still aspired to study a nature that existed relatively independently from human activity. Most now acknowledge the phenomena they study as somehow anthropogenic. Essayists have varied views on these changes. Sometimes their assessments of what the LTER program ideally should be appear to reflect the length of their own involvement and the kinds of work that they do.

Still, many do recognize an implicit baseline—how ecology used to be. Two decades before the formation of the LTER program, that baseline was apparent in concerns about the establishment of the International Biological Program (IBP). Not only was this LTER's immediate predecessor, it arose at a signal time in the institutionalization of American ecology. In his very personal history of the IBP, W. Frank Blair notes his and others' initial lack of enthusiasm for making ecology group-oriented and global. Ecologists, Blair suggests, were cantankerous individualists who did research in their own ways (Blair 1977). There were few of them and no clear need for many more. They enjoyed no obvious or stable funding stream, and usually their work was marginal to major issues of public policy. Although institutional cultures varied, a high rate of knowledge production through support of a research group of students and postdoctoral scholars had not yet become a standard criterion of professional success. Hence, ecologists shied from the prospect of the incessant meetings needed to plan big and dispersed projects. One, Tom Park, held that they would be "better off in the years to come … if science evolved in its own way without the seductive pressures of funding …, organization charts, preconceived concepts, and 'glamour'" (quoted in Blair 1977). But to others the IBP mandate seemed not too constraining, and, noted Blair, there were few who "wouldn't object to getting more support for continuation of work they already are doing." The skeptical Blair would himself become chair and chief salesman of the American IBP effort (Blair 1977).

A half century later, one may revisit Park's concern, for analogous tensions pervade these essays: do funding structures enhance or inhibit good research? Pure, context-free science does not exist, but one may examine the effects of the pervasive trade-offs. A number of such tradeoffs recur in these essays:

1. *Site selection.* At the program's inception, as Schlesinger (Chapter 40) notes, institutional considerations of long-term viability were in tension with comprehensive representation of regions or biomes. Only as the network expanded would there be a fuller geographic coverage. But the sites, some suggest, are not always ideal for many cross-site comparisons, one of the LTER program's priorities.
2. *Site-based funding.* Site-based funding and site-specific intellectual identity have enhanced site-based collaborative research at the cost of cross-site collaborations, some suggest. To some, the larger LTER network seems mainly an administrative entity and a necessary evil, while to others it is or should be a nucleus of integration toward which researchers should focus their efforts (Shaver, Peters, Silver; Chapters 9, 20, 26, respectively).
3. *Continuity versus innovation.* NSF has demanded the coexistence of continuity (maintenance of the gathering of long time data) with expectations of novelty (renewal requires fresh ideas to explore and hypotheses to be tested). Here the 6-year cycle may be absurdly brief (Giblin, Chapter 32; see also Gholz, Marinelli, and Taylor; Chapter 3).
4. *Big versus little ecology.* Editors M. R. Willig and L. R. Walker invited respondents to consider the impact of collaboration in their research (Chapter 1). This comes in many forms, including types of interdisciplinarity, and responses range widely. Plainly, a key rationale for funding single sites, and making them parts of a network, was and is to encourage interdisciplinary collaborative projects. But do practices intended to serve ends become ends in themselves? Although some see such "big ecology" as the appropriate evolution of the field (Shaver, Schmitt; Chapters 9, 28, respectively), others are ambivalent. Interdisciplinarity can result in new kinds of creativity (Giblin, Chapter 32), but inclusion of many authors and disciplines is laborious (Pickett, Peters, Knapp, Silver; Chapters 11, 20, 22, 26, respectively). There are worries, too, about the epistemic issues that arise in coordinating the many ways of knowing (Stafford, Morris, Hayden; Chapters 5, 33, 38, respectively). Some (e.g., Swanson; Chapter 6) suggest that collaborative approaches acquire their own inertia. Many suggest that their own research careers would not have been so collaborative had they evolved outside the auspices of the LTER program. But they do not see that as an unambiguous good (Lodge, Chapter 24).
5. *Independence versus conformity.* Both the emphasis on collaboration and on long and large data sets mandate uniformity in methods. But continuity is not an unproblematic virtue. Changing analytical technology, newly important parameters, and individual interests and creativity threaten that stability. An objector to the IBP had already worried about "junk data that will be scrapped when the next 'standard' method comes along" (quoted in Blair 1977). LTER's data scientists (Giblin, O'Brien; Chapters 32, 34, respectively) must somehow reconcile individual initiatives and needs with the demands of continuity and broader data applicability. In fact, notes Borer (Chapter 12), there is less uniformity than one might expect.
6. *Human-dominated landscapes and the social sciences.* Inclusion of urban and other human-dominated sites in the LTER network reflected the broadening of ecology, but it also involved inclusion of social scientists who studied human land use. However warranted, inclusion of the human variable initiated a profound change,

not only in the scope of research, but in understandings of environmental dynamics (Swanson, Grimm, Lugo; Chapters 6, 15, 25, respectively). As Gholz, Marinelli, and Taylor (Chapter 3) note, that extension did not come about all at once, nor does it prevail uniformly across all sites.

7. *Applicability.* Editors M. R. Willig and L. R. Walker (Chapter 1) also ask whether LTER research has been applicable. Traditionally, the justification of long-term research was that one would not know its utility until one had collected the data over the long term. And yet, as with the demand for novelty, LTER must compete in an applicability-driven funding environment. For most respondents, contexts of resource management or environmental quality are already part of their work; that expectation is probably less problematic than it might otherwise be. But the specter of unrealistic expectations and of the tradeoff between basic science and expediency remain. An objector to the IBP had worried that it would lead to "over-optimistic hopes in the mind of the public, who might think we are going to double the world's food supply or something" (quoted in Blair 1977).

8. *Public interface.* In the past, the communication of science was entrusted to museums, journalists, and educators. Growing pressures that publicly funded research be accessible and accountable has meant that researchers must be both willing and skilled, not only in responding to public concern, but in reaching out. The trade-off is more than one of time: issues of autonomy may arise. Where public policy matters are involved, administrators of the LTER program may be in the position of responding as well as communicating (Swanson, Pickett; Chapters 6, 11, respectively).

The challenge of meeting these many missions is a common theme of these essays. The tensions and new missions are not necessarily bad. Yet such "mission creep" does reflect the trials of maintaining (and expanding) a program funded by a single source. Although some (e.g., Blair, Chapter 21) complain that the burden of organizing research infringes on research, others find that they constitute the realities of scientific practice. Some take pride: administrative accomplishment has become scientific accomplishment. And yet, to the question of whether LTER accomplished anything that could not have been achieved by funding existing field stations, Schlesinger answers, "probably not."

LTER: A TEMPLATE FOR THE SCIENCES OF THE FUTURE?

From my peculiar perspective, the LTER program's most profound achievement has been its influence on the practice and norms of science. By its very nature, commitment to long-term research is incompatible with the focus on discrete episodes of individual discovery that have typically been the most valued indices of scientific achievement. The characteristic practices of knowledge-creation in the LTER program challenge conventional academic practices. Several authors (Stafford, Borer, Troxler, Knapp, Magnuson, Hayden; Chapters 5, 12, 18, 22, 30, 38, respectively) note that tenure standards have changed or must change to accommodate group work in which candidates will not be lead author or principal investigator. They challenge the sociological norms of science, too. The LTER program exemplifies the emerging value of open access to data sets—a generation ago, these were much more likely to be carefully guarded private property. And, under "mentoring," respondents note that they are now training students not only to work in interdisciplinary groups, but to think in transdisciplinary and even perhaps antidisciplinary ways. Traditionally, graduate education has involved the acquisition of disciplinary methods to be used on disciplinary problems, with the faint hope that these might somehow be collected to address transdisciplinary problems. Recognition that most practical problems are transdisciplinary has suggested to some the need to give greater priority to the acquisition of these problem-solving skills (for a pioneering view, see Johnson 1984; Hamlin and Shepard 1993).

There are then risks for young researchers in connecting to LTER sites. Those who do are exemplifying in extreme forms some of the famous "Cudos" norms articulated in 1942 by the sociologist Robert Merton as essential to the practice of science. Merton's norms are "communalism" (common access to knowledge), "universalism" (faith that scientific truths are universal across cultures), "disinterestedness," and "organized skepticism." Of these, the first and third have been the most controversial (Merton 1976), and there is deep disagreement within the contemporary biosciences as to the appropriate norms of professional conduct.

Traditionally, communalism was practiced through journal publication and shared offprints. But (at least prior to publication), one's hard-won results were property, not necessarily to be shared. The maturation of the Internet changed that. Not only did data-openness become practical, it became, in the views of many, ethically obligatory (Olson, Zimmerman, and Bos 2008). But in other parts of the biosciences, the pull was in the opposite direction. Disinterestedness had usually meant avoidance of excessive partisanship on behalf of one's own ideas as well as disparagement of patenting of discoveries. Then, in 1984, the Bayh-Dole Act, passed just as the LTER program was gearing up, that made it possible to patent products of publically funded research. This legislation, along with legal decisions creating genetic forms of intellectual property, has resulted in some quarters in the virtual abandonment of the disinterestedness norm, and the accompanying rejection of communalism. A central theme of modern research ethics is data ownership (compare successive editions of the National Academy of Science orientation guide for young scientists, "On Being a Scientist; 1989, 1995, 2009).

At the same time, the LTER program has been developing an ethos that data sharing is not an incidental virtue but a seminal and obligatory tool of research. There is recognition that students must be inculcated into these new habits of sharing—of ideas as well as of data—and cultivation of a network of trust to ensure adequate recognition (Schmitt, Suding; Chapters 28, 29, respectively). These strong commitments to communalism and disinterestedness thus represent one wing in an emerging bifurcation of the norms of scientific practice.

Remarkably, that disinterestedness (Waide, Chapter 2; he terms it "altruism") carries over to attitudes toward disciplinarity. It appears as if many in the LTER program operate on the principle that multiple ways of knowing have intrinsic and not merely instrumental utility, and they hold that value even when they perceive that available resources are insufficient for their own disciplinary work. Given a long-standing legacy of disciplinary rivalry in the history of science, one often exacerbated by resource scarcity in recent decades, this enthusiasm for interdisciplinary inclusivity is remarkable. It represents election of big, messy problems over discrete disciplinary problems. It suggests a sea change in professional identity, a departure from the paradigm-based model of science articulated by Thomas Kuhn (Kuhn 1962). In Kuhn's view, graduate mentors were usually replicators of their own disciplinary research methods. But, if we take some of the senior respondents at face value, they seem to be saying that the future of good science is that it *not* be done by people like me, (or not just by people like me).

What motivates these new norms? Recognizing that researchers in the LTER program are a self-selected population, it is likely that social aspects and governance structures have much to do with creating a climate of normative reinforcement. Within the

LTER program, freedom and friendship are hallmarks. Many comment on bottom-up or grassroots governance—or on the importance of LTER-based friendships. There are even allusions (Swanson, Troxler; Chapters 6, 18, respectively) to the "culture" of LTER. There is a sense of LTER setting its own agenda and of the ongoing democratic balance of program, site, and individual agendas. The contrast between the LTER approach and more top-down approaches to long-term research (e.g., National Ecological Observatory Network) seems clear (Gough, O'Brien, Hayden; Chapters 7, 34, 38, respectively). At least for many, much work at an LTER site takes place away from one's home campus; the romance of field work and the adventure of travel are important elements in that culture.

But the modes of scientific interaction that the LTER program embodies are not exclusive to it. Notwithstanding the self-consciousness and defensiveness that pervades these essays, most respondents work within multiple networks beyond the LTER program, as Hayden notes. How well those other networks reinforce or accommodate those values is less clear, but if the same individuals are involved in those networks, we may assume that they represent at least some of the same values. For these individuals are, after all, the grassroots of LTER, just as they will be in other institutional frameworks.

CONCLUSION

The LTER program represents a distinct mode of reconciling research goals with research structures. Although its elements are not unprecedented, it represents an innovative model for the organization of a kind of environmental research that is otherwise likely to be undervalued, both as a funding priority and in terms of academic stature. The satisfactions researchers find as a consequence of participation in the LTER program, however, involve constant balancing of rival goods—site versus network, individual versus group research. To date, the LTER program has maintained itself as an attractive locus for some researchers (we hear nothing about those who have left). Its success, however, has and will continue to rely on inculcation of norms of network-produced knowledge. At a time when traditional norms of science are being pulled in incompatible directions between individualism and communalism, that success is problematic unless ways continue to be found to recognize the distinct importance of long-term interdisciplinary group work.

REFERENCES

Allen, D.E. (1976). *The Naturalist in Britain: A Social History* (2nd ed.). Princeton University Press, Princeton.
Bacon, F. (1915). *New Atlantis*. Clarendon Press, Oxford.
Blair, W.F. (1977). *Big Biology: The US/IBP*. Dowden, Hutchinson, and Ross, Stroudsburg, Pennsylvania.
Callahan, J.T. (1984). Long-term ecological research. *BioScience*, **34**, 363–367.
Case, S.R. (2014). *Making Stars Physical: John Herschel's Stellar Astronomy, 1816-1871* (PhD dissertation). University of Notre Dame.
Committee on Science, Engineering and Public Policy, NAS, NAE, and IOM (1995). *On Being a Scientist: Responsible Conduct in Research*. 2nd ed. National Academy Press, Washington, DC.
Committee on Science, Engineering and Public Policy, NAS, NAE, and IOM. (2009). *On Being a Scientist: Responsible Conduct in Research*. 3rd ed. National Academy Press, Washington, DC.

Committee on the Conduct of Science, NAS. (1989). *On Being a Scientist*. National Academy Press, Washington, DC.

Cowles, H.C. (1901). The physiographic ecology of Chicago and vicinity: A study of the origin, development and classification of plant societies. *Botanical Gazette*, **31**, 73–108, 145–182.

Desrosières, A. (1998). *The Politics of Large Numbers: A History of Statistical Reasoning*. Harvard University Press, Cambridge.

Hamlin, C. (2014). Surgeon Reginald Orton and the pathology of deadly air: The contest for context in environmental health. In J.R. Fleming and A. Johnson, eds. *Toxic Airs: Body, Place, Planet in Historical Perspective*, pp. 23–49. University of Pittsburgh Press.

Hamlin, C., and Shepard, P.T. (1993). Deep Disagreement in U.S. Agriculture: Making Sense of Policy Conflict. Westview Press. Boulder, Colorado.

James, P.E. (1972). *All Possible Worlds: A History of Geographical Ideas*. Wiley, New York.

Janković, V. (2000). *Reading the Skies: A Cultural History of English Weather, 1650-1820*. University of Chicago Press.

Johnson, G.L. (1984). Academia needs a new Covenant for servicing Agriculture. Mississippi Agriculture and Forestry Experiment Station Special Publication. Mississippi State University.

Kitts, D.B. (1977). *The Structure of Geology*. SMU Press, Dallas.

Kuhn, T.S. (1962). *The Structure of Scientific Revolutions*. University of Chicago Press.

Lodge, D., and Hamlin, C. (2006). *Religion and the New Ecology: Environmental Responsibility in a World of Flux*. University of Notre Dame Press.

Merton, R.K. (1976). The ambivalence of scientists. In R.K. Merton, ed., *Sociological Ambivalence and Other Essays,* pp. 56–64. Free Press, New York.

Olson, G.M., Zimmerman, A., and Bos, N. (2008). *Scientific Collaboration on the Internet*. MIT Press, Cambridge.

Riley, J.C. (1987). *The Eighteenth Century Campaign to Avoid Disease*. MacMillan, London.

White, G. (1891). *Natural History and Antiquities of Selborne* (New ed.). Macmillan, London.

Wise, M.N. (1989). Work and Waste. *History of Science,* **27**, 221–61, 263–301, 391–449.

44 Trade-offs of Participation in the Long-Term Ecological Research Program

Immediate and Long-Term Consequences

Lawrence R. Walker and Michael R. Willig

INTRODUCTION

For those who may have skipped to this chapter and not read the 3 introductory chapters, the 36 essays, or the 4 evaluative chapters of this book, the answer to the burning question "Does participation in the Long-Term Ecological Research (LTER) program change scientists?" is an unequivocal "Yes!" As Boyer and Brown (Chapter 41) point out, however, those changes are mostly in the realms of knowledge acquisition and behavior adoptions in the practice of science. Participation in the program did not appear to have a substantial effect on the development of attitudes. Could such changes have occurred outside of the LTER program? Schlesinger (Chapter 40) thinks so. He suggests that the LTER program provides "some structure and modest standardization to a set of common measurements" but that it has not substantially broadened or deepened the ecological sciences. Yet the effect of the LTER program on science, while a fascinating and often-addressed question, is not the focus of this book (see Willig and Walker, Chapter 1). Of course, to address how scientists change also involves understanding how they approach and conduct science. In addition, personal change occurs in a broad societal context. For example, the LTER program has coincided with and helped promote a transition in ecology from research done by one or a few investigators on a particular organism or process in a particular habitat to investigations involving multidisciplinary teams working together to test models about how ecosystem dynamics unfold across large spatial and temporal scales. However, going to "big programs" and "big data sets" does not mean losing a sense of place or being divorced from the natural history of particular organisms. Even as spatial and temporal scales increase, ecological research is ideally still "place aware" (Bestelmeyer, Chapter 19).

Using the essays of this book as a rich source of information to address fundamental questions about the nature of scientists, we provide some final thoughts on how the LTER program has affected its participants, particularly on how they view time and space, collaboration, and communication. We end with reflections on the future of ecology and society, based on the views expressed in this book and on our own participation in the LTER program.

TIME AND SPACE

The strongest common theme in the essays is the effect of conducting long-term research on the scientists themselves (Flint, Chapter 42). Participants in the LTER program have internalized three lessons. First, it is humbling to realize that conclusions about the properties of an ecosystem based on a short (e.g., 2- to 4-year) study make poor predictors of decadal or longer temporal patterns (Magnuson's Invisible Present; Magnuson 1990). Second, it is similarly humbling to discover that no matter how deeply one examines a particular location, data from a variety of locations are needed for general understanding (Swanson's Invisible Place; Swanson and Sparks 1990). Third, we are humbled when we try to understand what is happening at a particular moment and realize that we need data from many sources and disciplines (Porter's Invisible Moment; Porter, Chapter 39). Once these lessons are internalized, scientists working on long-term research might well become more patient with long-term data collection, explore more comparative aspects of ecosystem dynamics, or become more interdisciplinary in their research. The rewards of such changes include expanded perceptions of interactions across space or through time (Bestelmeyer, Chapter 19); an integrated coupling of short- and long-term phenomena, such as the potential for the reversibility of coral–algal shifts (Schmitt, Chapter 28); and a better understanding of the complex responses to global change, such as grassland responses to altered fire regimes (Blair, Chapter 21).

COLLABORATION

A second strong theme underlying the essays was that participation in the LTER program requires at least some collaboration: among fellow ecologists, other environmental scientists, or social scientists within one's site and across the LTER network. Collaboration is most commonly found in disciplines studying complex systems (e.g., oceanography, ecosystems) and can involve sharing equipment, skills, or ideas. One example that exemplifies all three types of sharing is when information managers in the LTER program collaborate on software development (O'Brien, Chapter 34). Collaboration can make one into an amalgam of unlikely parts or skills, much like a platypus (Grove, Chapter 10), and is best suited to those who prefer to play in an ensemble, rather than to those who prefer duets or solos (Grove, Chapter 10). Advantages to collaboration include the ability to disentangle interactions within complex ecosystems (Tilman, Chapter 13), conduct meta-analyses (Borer, Chapter 12), share syntheses (Gaiser, Chapter 17), and build more comprehensive stories (Silver, Chapter 26). Disadvantages include uneven contributions from members of a research team (Troxler, Chapter 18) and overcoming the challenge of data collection protocols that are so site-specific that they rarely match across sites. This latter hurdle likely arose because the LTER program originated and evolved primarily as a site-based rather than network-based model (Waide [Chapter 2]; Gholz, Marinelli, and Taylor [Chapter 3]), with very few sites in similar ecosystems types (e.g., only one site in a tropical rainforest). The effort to collaborate within, but especially across, disciplines is huge, requiring time, mutual respect, and a willingness to learn about other's areas of expertise. The LTER program does not offer formal training in how to collaborate (Silver, Chapter 26), but the essayists were mostly positive about the benefits of doing so over the course of a career. More organized cross-site visits of some duration (e.g., sabbaticals) were suggested as a programmatic catalyst to promote network connectivity. Encouragement and support for coordinated, long-term research efforts that are designed to enhance cross-site comparison, synthesis, and theory are needed to provide the intellectual stimulus to advance ecological understanding, especially within the core areas of the LTER program. Such coordination can also provide an effective mechanism to engender a multidisciplinary culture of collaboration among scientists with interests in complex spatiotemporal system dynamics, whether or not they are senior personnel on an LTER project.

COMMUNICATION

A third theme addressed how science is communicated, within circles associated with the LTER program, with the broader scientific community, and with the public. A related topic was whether research in the LTER program is perceived as being either basic or applied. For some (Gragson, Silver; Chapters 16, 26, respectively), the LTER program is focused on basic research; for others (Gaiser, Chapter 17), applied research is a substantive part of the core mission of the LTER program. A third viewpoint is that the LTER program helps integrate basic and applied research (Giblin, Chapter 32) into an amalgam that represents actionable and predictive science and that by its nature, informs policy by providing long-term understanding.

Communication within LTER circles is strengthened by the degree of collaboration and a sense of a shared mission. However, this shared bond among scientists in the LTER program can lead to a degree of insularity (Boyer and Brown, Chapter 41) and can be seen as excluding others not in "the club." Entry into such tightly knit research groups with a long history of collaboration may be perceived as intimidating. Gragson (Chapter 16) is an anthropologist and notes how LTER scientists could be considered as a distinctive "ethnic group" within a particular tribal organization! Flint (Chapter 42) observes that many of the essayists are aware of this "clubbiness," and Hamlin (Chapter 43) detects a certain degree of defensiveness about the topic. Aside from entry into the system as a graduate student, suggestions on how to reduce this (perceived or real) insularity include increased efforts to invite researchers to participate and attempts to make databases more accessible. This was somewhat surprising, given the extent to which the LTER program has pioneered the development of metadata standards and the publication of resulting information on publicly accessible electronic media (Stafford, Porter; Chapters 5, 39, respectively). Indeed, each LTER site is evaluated at least twice per funding cycle on the extent to which data and metadata are incorporated into long-term repositories and are available to the scientific community and public via the Web. Nonetheless, the insularity issue will remain to some extent as long as (1) scientists in the LTER program bond among themselves through work and friendship and (2) those outside of the program remain uninformed of its data and metadata management activities.

Communicating science to the public is a much-valued but challenging part of the LTER program. Many LTER projects now have social scientists on their research team and recognition grows among ecologists that all ecosystems are influenced by human activities. The essayists (e.g., Lugo, Collins; Chapters 25, 35, respectively) describe several outreach programs, notably the Schoolyard Project, but also such approaches as webcams linking Antarctica to classrooms around the United States (Ducklow, Chapter 31). The degree to which outreach is integrated into a particular LTER site's activities varies, in part based on the degree of physical isolation between a site and the public. Some sites, such as H. J. Andrews Experimental Forest, have strongly emphasized the socioecological aspects of their site and have invited musicians and poets to interact with scientists at the site. Some LTER scientists receive training (e.g., from programs such as the Leopold Leadership Program and the Google Science Communication Program) on how to communicate to the public with minimal scientific jargon (Silver, Chapter 26). Yet, such integration is not happening fast enough for some (Lugo, Chapter 25). Pickett (Chapter 11) notes the need to abandon outreach and the delivery model of science in "favor of continual dialogue and mutual learning and listening." Flint (Chapter 42) terms this approach the "coproduction of knowledge."

THE FUTURE OF THE LTER PROGRAM, ECOLOGY, AND SOCIETY

Scientists in the LTER program are clearly having fun. They have access to valuable, long-term data sets, facilities and support staff, mentors, and friends, as well as to opportunities to travel and explore other ecosystems and cultures. All of this occurs in a minimally hierarchical organization based on mutual trust and informality. There are, however, a few caveats. Scientists in the LTER program must raise supplemental resources for most of their research. Although budgets for each LTER project are large compared to those in other programs in NSF's Division of Environmental Biology, where one or several investigators collaborate, per-collaborator funding in budgets at LTER sites is limited by the costs of maintaining infrastructure, core measurements, and support staff. Scientists in the LTER program also must figure out how to get the site refunded every 6 years. Long-term research plots must sometimes be sacrificed in proposals in favor of initiating new experiments that address cutting-edge questions. Stable, visionary leadership must be found and new recruits trained to replenish aging scientists within each site. Despite these caveats, the benefits are clearly substantial. But so what? Do the achievements of this contented (and very accomplished) group matter for ecology, science, or society? Will all of the collaboration and long-term focus lead to the development of new fields such as fish acoustics (Magnuson, Chapter 30), support ongoing developments in ecogeosciences (Swanson, Chapter 6) or ecoinformatics (Porter, Chapter 39), elucidate the relationship between ecology and time (Walker and Wardle 2014, Wolkovich et al. 2014), or enable the development of a general (Scheiner and Willig 2011) or efficient (Marquet et al. 2014) theory of ecology?

Lugo (Chapter 25) argues that the LTER program must embrace socioecology and all its inherent messiness to remain relevant and to become more integrated with society. Hamlin (Chapter 43) notes that science is developing down two divergent paths, the traditional individualistic approach and a more recent communal approach. The former, originally characterized by the single-investigator style of studying a particular site or suite of organisms and only sharing data in publications, is currently exemplified by the trend to obtain patents on biological discoveries. The latter is represented by research in the LTER program, where communal approaches dominate, data and metadata are rapidly and widely shared, and interdisciplinary and (temporal and spatial) scalar divisions become obsolete. Hamlin (Chapter 43) finds the fact that scientists in the LTER program embrace "big, messy problems over discrete disciplinary problems" to be remarkable, representing a sea change from "the paradigm-based model of science articulated by Thomas Kuhn." In contrast, Schlesinger (Chapter 40) argues that such changes (e.g., the deepening and broadening of scientific pursuits) are not likely to be affected by the existence of the LTER program. Whatever the larger societal effects, the essayists contributing to this volume are generally enjoying the challenges that "big, messy problems" present!

We undertook this book project because we thought that the effect of participation in the LTER program on LTER scientists had not been evaluated adequately. We do not have a non-LTER control group and know little about conducting sociological research.

Nevertheless, the essayists (and accompanying evaluators who have the requisite historical and social science credentials) provide a rich source of personal reflections and assessments based on substantive participation in the LTER program. Some scientists found it difficult to talk about their attitudes. Others, perhaps embedded in the LTER tribe for their entire professional lives, found it difficult to evaluate the effect of LTER on those attitudes. Nonetheless, most essays demonstrate a remarkably strong imprint of the "LTER experience" on each scientist. We therefore conclude that experience in the LTER program leads to multiple changes in how scientists think about time and space, collaborate, and communicate. Nonetheless, documentation and assessment of the extent to which participation by scientists in the "cultural experiment" embodied in LTER program remains a challenge for the future. We suggest that there are also broader societal implications of the LTER program, as many scientists adopt its multiscalar, multidisciplinary, and networked approach. Critical among these implications is the ability of ecology to adapt to a rapidly changing world. Will the LTER-style collaborative approach, for example, become an effective tool to address our many environmental challenges? As science continually adapts to societal conditions, its relevance and acceptance by society at large may well depend on how well it can integrate the pursuit of intellectual curiosity and the solution of practical concerns. The extent to which the LTER program has and will contribute substantially to such integration remains to be determined. Indeed, such integration represents a challenge for all those who manage and administer programs dedicated to the advancement of fundamental and mission-specific scientific programs.

ACKNOWLEDGMENTS

We thank the many contributors to this book, the outside reviewers, and NSF for funding the LTER program over many years. We gratefully acknowledge support from the NSF LTER program (grants DEB-1239764 and DEB-1313788). In addition, L. R. Walker was supported by a sabbatical from the University of Nevada Las Vegas and M. R. Willig was supported by NSF grant DEB-1354040 to the University of Connecticut.

REFERENCES

Magnuson, J.J. (1990). Long-term ecological research and the invisible present. *BioScience*, **40**, 495–501.

Marquet, P.A., Allen, A.P, Brown, J.H., Dunner, J.A., Enquist, B.J., Gillooly, J.F., Gowaty, P.A., Green, J.L., Harte, J., Hubble, S.P., O'Dwyer, J., Okie, J.G., Ostling, A., Ritchie, M., Storch, D., and West, G.B. (2014). On theory in ecology. *Bioscience*, **64**, 701–710.

Scheiner, S.M., and Willig, M. R., eds. (2011). *Theory of Ecology*. University of Chicago Press.

Swanson, F.J., and Sparks, R.E. (1990). Long-term ecological research and the invisible place. *BioScience*, **40**, 502–508.

Walker, L.R., and Wardle, D.A. (2014). Plant succession as an integrator of contrasting ecological time scales. *Trends in Ecology and Evolution*, **29**, 504–510.

Wolkovich, E.M., Cook, B.I., McLauchlan, K.K., and Davies, T.J. (2014). Temporal ecology in the Anthropocene. *Ecology Letters* (doi:10.1111/ele.12353)

Index

Note: Page numbers in italics denote figures.

acid rain 30, 302
adaptive management 58, 75
Alaska. *See* ARC
All Scientists Meeting 38, 57, 78, 115, 161, 206, 208, 247, 257, 272, 329, 339, 379, 420, 425
alpine. *See* nutrients
AND 9, 55–70, *59*, 73–80
  Ecological Reflections Program 78–79
Antarctica 392. *See also* PAL
anthropology 112, 168–172
Appalachia 14
ARC *10*, 83–105, *84*, *101*
  permafrost 10
Arctic Circle. *See* ARC
Argentina 202
Arizona. *See* CAP
artists 78–79, 181
Asia 202
Australia 202, 273

bacteria 92, 96, 113, 179, 244, 310, 320
Baltimore. *See* BES
beaver 328
benthic research 178–179
BES *11*, 109–126, *114*, *124*
  Baltimore City Department of Recreation and Parks 110
biogeochemistry 46, 85, 87, 112, 156, 216–217, 260–262, 291, 310, 313, 393
biomes 5, 6, 24, 36, 39, 47, 74, 85, 428
birds 346, 393, 425
Botswana 202, 221

C4 grassland 16. *See also* KNZ
cactus 12
CAP *13*, 145–163, *157*

carbon 77, 180, 186, 216–217, 261, 310, 313, 391, 393
carbon dioxide 11, 158, 313
  increases 30, 139
CDR *12*, 129–144
Chesapeake Bay 312. *See also* BES, VCR
China 69, 210, 221, 230
Chihuahuan Desert 23. *See also* JRN, SEV
Colorado. *See* NWT
coral reefs *18*, *271*. *See also* MCR
  children's book 274
CREON 272–273
CWT *13*, 166–174, *171*

data. *See* information management
decomposition 76, 77, 245, 263, 393
deforestation 14, 30, 67, 74
disturbances 102, 279, 327, 392. *See also* LUQ
drought 11–12, 15–17, 23–24, 139, 207, 235, 237, 239, 358, 392. *See also* CDR, KNZ, LUQ, SEV

Ecological Reflections program 78–79
EcoTrends Project 206, 208, 293
England 244
epifluorescence 92
EPSCoR 359, 361
estuary. *See* FCE, PIE, SBC
Europe 77, 110, 113, 124, 170, 328, 383, 385
Everglades 149. *See also* FCE

FCE *14*, 177–193, *179*
fertilization 92–93, 100, 292, 295, 320–321
fire 10, 14, 30, *218*, 327, 392
fish. *See* MCR, NTL

floods 17, 74, 77, 252, 320
Florida. *See* FCE
food webs 56, 58, 85, 93, 179, 329, 366–367, 370
Forest Service, U.S. 29, 43, 67, 74–75, 79, 252–255, 266, 359
forestry, social 110, 115
French Polynesia 17, 279–282. *See also* MCR
French Zone Ateliers 121–122
fungi 93–95, 244–245

geographic information systems 123
geology 78, 94
grasslands. *See* JRN, KNZ, SEV, SGS
grazing 30, 179, 198, 217. *See also* KNZ
Great Barrier Reef 273
guidelines for contributors 5–7

herbivory 24, 130, 139–140, 219, 244, 366
High School Student's Program 151
Hubbard Brook 31, 34, 100, 111, 322, 327, 391–392
Hungary 206, 210
hurricanes 17, 56. *See also* LUQ
Hutchinson's paradox 138

IBP 64, 74, 79, 92, 100, 260, 356, 367, 392
IGERT 159–160, 360
ILTER 169, 172, *179*, 186, 188, 220–221, 273, 279, 282, 357, 360, 383
informatics 77
information management 64–70, 78, 264, 335–342, 382
ITEX 102

Jamaica 244
JGOFS 315
JRN 15, 197–212

KAB 397–402
Kansas. *See* KNZ
karst wetlands 179
kelp 22–23. *See also* MCR, SBC
KNZ *16*, 215–239, *218*
 Tallgrass Prairie National Preserve 220
Kuparuk River 100

lakes. *See* NTL
landscape architecture 120
landslides xii, 17, 244–245
LIDAR 78

LIDET 260, 329
LINX 57–58, 159, 393
limnology. *See* NTL
litter. *See* decomposition
LMER 45–46, 92
logging. *See* deforestation
LTER network xi, 5, 26, 35–37, 47–48, 125, 131, 143, 414
 book syntheses 3, 96
 exclusivity 88, 133–134, 230, 330, 393, 415, 423, 435
 future planning 37–39, 429–431, 436–437
 governance 48
 history 30–37, 424–427
 network office 4, 48
 new sites 45–47, 65
 probation 33–34
 sites 7, 8. *See also* AND, ARC, BES, CAP, CDR, CWT, FCE, JRN, KNZ, LUQ, MCR, NTL, NWT, PAL, PIE, SBC, SEV, SGS, VCR.
 themes 30–32, 35, 44, 49, 415
LTREB 328, 330, 391
LUQ xi–xii, *17*, 30, 243–266

marine. *See* MCR, PAL, SBC
marshes *326*. *See also* FCE, PIE
Maryland. *See* BES, VCR
Massachusetts. *See* PIE
MBL 92, 95–96
MCR *18*, 269–285, *271*
Mediterranean climate 22
Mexico 210
Minnesota. *See* CDR
Mongolia 200
monsoon 23
Mount St. Helens 74, 78
Mozambique 210
mycorrhizae. *See* fungi

Namibia 221
National Science Foundation 43–51, 79, 85, 100, 139, 143
Native Americans 94, 102, 172
natural history 94
NCA 159
NCEAS 130–131, 159, 235
NEON 48–49, 130, 162, 207, 209, 239, 266, 351–352, 378, 393
New Mexico. *See* JRN, SEV
nitrogen 11–12, 77, 95, 139, 216, 261, 292, 392

North Carolina 14, 171, 252
North Slope 102
NPS 187
NTL *20*, 299–306
NutNet 131, 133, 235, 239
nutrients 92, 186, 244, 313
NWT *19*, 289–296, *290*

Oak Ridge xii, 34, 206, 386
oceanography 96, 300–301, 313–315, 387, 420, 434
old-growth forests 67, 77. *See also* AND
Oregon. *See* AND
owls 67

Pacific Northwest. *See* AND
Pacific Ocean. *See* MCR
PAL *21*, 309–316, *311*
paradigm shift 7
patch dynamics 111
PDTNet 85–88
penguins 392
permafrost 93. *See also* ARC
Phoenix *13*. *See also* CAP
phosphorus 180, 244, 391–392
phycology 178–179
PIE *22*, 319–331, *326*
prairies. *See* CDR, KNZ
programming language 340
Puerto Rico xi–xii, 17, 29–30. *See also* LUQ

radiation experiment 253
reefs. *See* coral reefs
remote sensing 47, 85, 115, 199, 202, 301, 304, 360
resilience 18, 113, 124, 270–271, 280–281
restoration 11, 15, 26, 111, 122, 179–181, 189, 216, 220, 255, 273, 293, 328
REU 246, 264, 379
riparian 58, 122, 254, 350
roots
  historical 77, 254, 367, 416
  plants 93, 95, 100

salmon 77
SBC *23*, 335–342, *336*
Schoolyard LTER Program 34, 59, 67, 124, 201, 208, 237, 255, *256*, 282, 328, 350, 359, 367, 369, 379, 394, 417, 435

sea grass. *See* VCR
sea ice 21. *See also* PAL
SEV *24*, *210*, 345–362
SGS *25*, *235*, 365–372
shrimp 254
Silent Spring 322
SIMs 412–420
snow 18–19
sociological research 44–45, 65–66, 74–78, 149, 158, 169, 200, 257, 368, 392, 416–417, 428, 435–436. *See also* BES, CAP
soil 93, 216–217, 246, 366–367
  carbon 293
  chemistry 156, 261
  microbes 95
  organic matter 93
  warming 93
Sonoran Desert 12
South Africa 112, 122, 210, 219, 221, 230, 236, 348
stable isotopes 216
steppe. *See* SGS
streams 158. *See also* AND
  ecology 56–57
  fertilization 93
succession
  ecological xii, 77, 138–139, 245, 426
  leadership 70
sustainability 40, 124, 148–152, 159
systems ecology 149

Taiwan 273, *384*, 386–387
TBS 100
Texas 87, 209
Thailand 273, 282
Toolik Lake. *See* ARC
tundra *290*. *See also* NWT

ULTRA-Ex 120
urban. *See* BES, CAP
  heat island 11–12
Uruguay 221

VCR *26*, 375–388
Virginia. *See* VCR
volcanoes 74–75, 78

watershed 58, *59*
wetlands. *See* FCE, PIE
whales 21. *See also* PAL
Wisconsin. *See* NTL

ACRONYMS USED IN INDEX:

LTER sites:
  AND: H.J. Andrews Experimental Forest
  ARC: Arctic
  BES: Baltimore Ecosystem Study
  CAP: Central-Arizona-Phoenix
  CDR: Cedar Creek Ecosystem Science Reserve
  CWT: Coweeta
  FCE: Florida Coastal Everglades
  JRN: Jornada Basin
  KNZ: Konza Prairie
  LUQ: Luquillo
  MCR: Moorea Coral Reef
  NTL: North Temperate Lakes
  NWT: Niwot Ridge
  PAL: Palmer Antarctica
  PIE: Plum Island Ecosystem
  SBC: Santa Barbara Coastal
  SEV: Sevilleta
  SGS: Shortgrass Steppe
  VCR: Virginia Coastal Reserve
CREON: Coral Reef Environmental Observatory Network
EPSCoR: Experimental Program to Stimulate Competitive Research
IBP: International Biological Program
IGERT: Integrative Graduate Education and Traineeship
ILTER: International Long-Term Ecological Research
ITEX: International Tundra Experiment
JGOFS: Joint Global Ocean Flux Study
KAB: Knowledge, Attitudes, and Behavior
LIDAR: Light Detection and Ranging
LIDET: Long-Term Intersite Decomposition Network
LINX: Lotic Intersite Nitrogen Experiments
LMER: Land-Margin Ecosystem Research
LTREB: Long-Term Research in Environmental Biology
MBL: Marine Biological Laboratory
NCA: National Climate Assessment
NCEAS: National Center for Ecological Analysis and Synthesis
NEON: National Ecological Observatory Network
NPS: National Park Service
NutNet: Nutrient Network
PDTNeT: Productivity-Diversity-Traits Network
REU: Research Experience for Undergraduates
SIMs: Scientific or Intellectual Movement
TBS: Tundra Biome Study
ULTRA-Ex: Urban Long-Term Research Areas–Exploratory